TRANSACTIONS OF THE MOSCOW MATHEMATICAL SOCIETY FOR THE YEAR 1969

ТРУДЫ МОСКОВСКОГО МАТЕМАТИЧЕСКОГО ОБЩЕСТВА

ТОМ 20 (1969)

VOLUME 20

This translation was prepared jointly by the
American Mathematical Society
and the
London Mathematical Society

Published by the American Mathematical Society
Providence, Rhode Island
1971

International Standard Book Number 0-8218-1620-1
Library of Congress Card Number 65-4713

Printed in the United States of America

TRANSACTIONS OF THE MOSCOW MATHEMATICAL SOCIETY

(Труды Московского Математического Общества Том 20, 1969)

TABLE OF CONTENTS

Page

Maĭkov, E. V. τ-smooth functionals. [Майков, Е. В. τ-гладкие функционалы, 3–42] 1

Levin, V. L. Tensor products and functors in categories of Banach spaces defined by KB-lineals. [Левин, В. Л. Тензорные произведения и функторы в категориях банаховых пространств, определяемые KB-линеалами, 43–82] 41

Raševskiĭ, P. K. Tensors with a given invariance group and spherical functions on homogeneous spaces. [Рашевский, П. К. Тензоры с данной группой инвариантности и сферические функции на однородных пространствах, 83–110] 79

Veĭsfeĭler, B. Ju. Some properties of singular semisimple algebraic groups over nonclosed fields. [Вейсфейлер, Б. Ю. Некоторые свойства особых полупростых алгебраических групп над незамкнутыми полями, 111–136] 109

Maz′ja, V. G. On weak solutions of the Dirichlet and Neumann problems. [Мазья, В. Г. О слабых решениях задач Дирихле и Неймана, 137–172] 135

Maljutov, M. B. On the Poincaré boundary value problem. [Малютов, М. Б. О краевой задаче Пуанкаре, 173–204] 173

Dubinskiĭ, Ju. A. Mixed problems for a certain class of partial differential equations. [Дубинский, Ю. А. Смешанные задачи для некоторых классов дифференциальных уравнений с частными производными, 205–240] 205

Volevič, L. R. and Kagan, V. M. Hypoelliptic pseudodifferential operators in the theory of functional spaces. [Волевич, Л. Р. и Каган, В. М. Псевдодифференциальные гипоэллиптические операторы в теории функциональных пространств, 241–275] 243

Trudy Moskov. Mat. Obšč.
Tom 20 (1969)

Trans. Moscow Math. Soc.
Vol. 20 (1969)

τ-SMOOTH FUNCTIONALS

E. V. MAĬKOV

Contents

INTRODUCTION 2

CHAPTER 1. The τ-continuity and τ-differentiability of a functional 4
§11. The space of trajectories 4
§12. τ-continuous functionals 6
§13. τ-differentiability 8
§14. τ-derivatives of higher orders 13

CHAPTER 2. τ-polynomials and τ-analytic functionals 16
§21. Differentiation of a τ-polynomial 16
§22. The general form of a τ-polynomial 20
§23. τ-analyticity 23
§24. An algebra of τ-analytic functionals 24

CHAPTER 3. τ-measures 28
§31. Generalized measures 28
§32. The quasi-measure and its nth variation 29
§33. Integration of a τ-polynomial with respect to a quasi-measure . . . 32
§34. The algebra A as a base space 33

CHAPTER 4. Conclusion 35
§41. On the development of the concepts of τ-smoothness 35
§42. τ-linearity 36
§43. τ-differentiability in the calculus of variations 37
§44. τ-measures. The Feynman measure as a τ-measure 38

REFERENCES 39

The present article is devoted to a study of the properties of τ-smooth functionals $f(\xi)$ over the space Ξ of piecewise continuous functions $\xi = \xi(\tau)$, where τ is a real variable; by a τ-smooth functional we mean one whose increase depends in a smooth (continuous, differentiable, etc.) manner on the τ-perturbations of $\xi(\tau)$, that is, on the perturbations concentrated on a small interval of variation of τ (but the perturbations are not necessarily small in magnitude).

In addition we give an application of τ-smooth functionals to integration in a functional space.

Let us remark that the concept of τ-smoothness arose in our search for a class of functionals, suitable for certain questions in the theory of integration;

later the concept acquired independent interest. We hope that it will be useful in other branches of mathematics.

I take this opportunity to express my gratitude to my teacher, Sergei Vasil′evič Fomin, who in 1958 turned my attention to the problem of the theory of integration in functional spaces and who subsequently directed my investigations.

Introduction

In this Introduction we refer to the whole circle of problems touched upon in the paper, briefly present the contents of each chapter and also make some remarks concerning the character of the presentation, the system of notation and of references.

The paper contains four chapters, each of which consists of four sections. The sections are indexed by two numbers, the first of which indicates the chapter and the second the section in the particular chapter. The numbering of formulas, theorems and other statements in each section is independent of that used in other sections. The system of referencing is clear from the following example: the first theorem in §24 is called Theorem 1 in this section and Theorem 24.1 when referred to in other sections.

The initial concepts are τ-continuity and the τ-derivative of a functional.

Let $f(\xi)$ be a complex-valued functional defined on the set Ξ of piecewise-continuous functions $\xi(\tau)$, $\tau \in [0;1]$. Let $\xi^0 = \xi^0(\tau)$ be some function from Ξ. We fix $t \in [0;1]$ and take the interval with the endpoints t and $t+\eta$. We alter $\xi^0(\tau)$ on this interval so that it coincides with $\xi = \xi(\tau)$ on the interval; we call $\xi = \xi(\tau)$ a *perturbing* function. Let ξ^η denote the *perturbed* function obtained in this manner. If the increment $\Delta f = f(\xi^\eta) - f(\xi^0)$ of $f(\xi)$ is infinitely small together with η for any ξ^0, ξ and t, then $f(\xi)$ is said to be *τ-continuous*; for example, $f(\xi) = \int_0^1 \xi(\tau)\,d\tau$ is τ-continuous (see Example 12.1).

The concept of *τ-differentiability*, the precise definition of which the reader will find in §13, is connected with the existence of the limit

$$\lim_{\eta\to 0} \frac{f(\xi^\eta) - f(\xi^0)}{|\eta|}.$$

These concepts are the subject of attention in the first chapter. Here we present the properties of τ-continuous and τ-differentiable functionals, examine the problem of reconstructing a functional from its τ-derivative and study the properties of functionals that are n times τ-differentiable.

We begin Chapter 2 with a study of τ-polynomials. By analogy with polynomials in a real variable, a functional $f(\xi)$ is called a *τ-polynomial* if it is continuously τ-differentiable any number of times and, starting from some order, its τ-derivatives are identically zero. Theorems 21.1 and 22.2 on the general form of τ-polynomials establish that the set of τ-polynomials is the same as the set of functionals that can be represented in the form

$$g(\xi) = \int_0^1 \cdots \int_0^1 \phi(t_1, \xi(t_1), \cdots, t_n, \xi(t_n))\, dt_1 \cdots dt_n,$$

where $\phi(t_1, x_1, \cdots, t_n, x_n)$ is a continuous function of $2n$ real variables. We study certain properties of τ-polynomials; in particular, we develop a formula that allows us to reconstruct a τ-polynomial from the values of its successive derivatives at a point $\xi^0 \in \Xi$; because of the clear analogy we call it Taylor's formula for a τ-polynomial. This formula is used in §23 to introduce the concept of a τ-analytic functional. After definitions and examples we consider in detail (§24) a class of τ-analytic functionals distinguished by certain bounds on their τ-derivatives (the algebra A).

If aspects of the differential calculus of τ-smooth functionals are studied in the first two chapters, the following chapter refers to the integral calculus of these functionals.

In Chapter 3 we examine an application of τ-analytic functionals to the problem of integration in functional spaces. We are interested in the case when instead of measures we have to have additive set functions that do not have finite or σ-finite variation; this case is interesting, for example, because this kind of "measure" has recently occupied an important position in quantum physics. The usual Lebesgue theory of integration cannot be constructed for these set functions, but we can ask for the integration of a class of functionals that does not include all the bounded measurable functionals. For example, such a question was raised by Gel'fand [**1**]. Problems connected with this question have been considered by Daleckiĭ [**2**], Krylov [**3**] and several other authors.

A theorem on τ-measures, Theorem 34.1, which we regard as the main theorem in Chapter 3, is concerned with this question.

By a *τ-measure* we mean an arbitrary continuous linear functional If, defined on some fundamental space of τ-smooth functionals $f(\xi)$.

The theorem on τ-measures is preceded by the definition of the nth variation of a quasi-measure. In the terminology of [**2**] a *quasi-measure* is an additive function μ of sets in a function space such that all finite-dimensional projections of μ are σ-additive; μ itself may or may not be σ-additive; the upper bound V_n of the variations of the n-dimensional projections of μ is called the nth *variation* of μ. (The total variation V of a quasi-measure is defined by $V = \lim_{n\to\infty} V_n$.)

If a quasi-measure is nonnegative, then it follows from the well-known theorem of Kolmogorov (see [**4**], Russian p. 39) that it can be extended to a σ-additive measure on some σ-algebra of sets in a function space. When the total variation of a quasi-measure is infinite (the nonnegativeness condition is not fulfilled), it cannot be extended to a measure. It turns out that if the nth variation of a quasi-measure μ is such that $V_n \leqq v^n$ $(v < +\infty)$ for all n, then μ can be extended to a τ-measure over the algebra A introduced in §24.

This result is proved in Theorem 34.1 which completes Chapter 3.

In the fourth and final chapter we consider some possible developments of the results obtained in the previous chapters. We indicate ways of applying the concept of τ-smoothness to problems in the calculus of variations (variations of directions compatible with τ-perturbations were used, for example, in [**5**]). In §44 we indicate what fundamental space of τ-smooth functionals must be taken so that the τ-measures over this space include the "Feynman measure", which is well known in physics (see [**6**]).

We remark that some of the results of the present article were briefly published in [**11**]—[**13**].

A few words regarding the character of the presentation. To avoid cumbersome constructions we have separated some sections of proofs into lemmas and remarks. Remarks are frequently located not directly before the theorem in which they are used, but earlier where they most naturally follow from adjacent material. Some statements, analogous either to well-known theorems in analysis or to results proved earlier in the text, are made without proof or only with a brief sketch of a proof. We hope that this will not cause any difficulty to the reader.

We use the standard notation of set theory. $\{x \in X; P(x)\}$ denotes the set of $x \in X$ for which the statement $P(x)$ holds. When X is clear from the text, we write briefly $\{x; P(x)\}$. The union of the sets A and B is $A \cup B$; when A and B are disjoint, their union is denoted by $A + B$. Similarly, $\bigcup_j A_j$ denotes the union over all j of the sets A_j, and $\sum_j A_j$ denotes the union of the disjoint sets A_j. The sets of elements belonging to A but not to B (independently of whether B is contained in A) is denoted by $A - B$.

CHAPTER 1

THE τ-CONTINUITY AND τ-DIFFERENTIABILITY OF A FUNCTIONAL

§11. The space of trajectories

We introduce our notation.

Let D be the interval $[0; 1]$. We denote points of this interval by τ, t, and sometimes by θ (with or without a subscript).

R denotes the set of real numbers.

$\xi(\tau)$ is a function on D with values in R, continuous everywhere except possibly at a finite number of interior points of D, where it has a discontinuity of the first order, with $\xi(\tau + 0) = \xi(\tau)$. We call such functions *trajectories*.

Ξ denotes the set of all trajectories.

We denote a trajectory by the letter ξ, with or without a superscript; ξ with a subscript always denotes the value of $\xi(\tau)$ for a fixed $t \in D$ with the same subscript. For example, $\xi_j = \xi(t_j)$, $\xi_j^\alpha = \xi^\alpha(t_j)$.

We denote the set of points of continuity of $\xi(\tau)$ by C_ξ.

It is obvious that each trajectory is bounded on D; we denote the upper bound of $|\xi(\tau)|$ by M_ξ.

Subintervals of D. The subintervals of D to be discussed are either left-closed semi-intervals whose right-hand endpoint is not 1, or closed intervals whose right-hand endpoint is 1. Thus the form of an interval is uniquely determined by its endpoints t_1 and t_2. It is convenient to have the same notation for these intervals, independent of whether they are closed or half-open. We write them as $[t_1; t_2)$; in more detail, if $t_1 < t_2$ then

$$[t_1; t_2) = \begin{cases} \{\tau; t_1 \leqq \tau < t_2\} & \text{for } t_2 \neq 1, \\ \{\tau; t_1 \leqq \tau \leqq t_2\} & \text{for } t_2 = 1. \end{cases}$$

If $t_1 > t_2$, we set $[t_1; t_2) = [t_2; t_1)$. If $t_1 = t_2$, by definition the set $[t_1; t_2)$ is empty (including the case when $t_1, t_2 = 1$). This notation is the same as the usual one if $t_1 \leqq t_2 < 1$ or $t_1 = t_2 = 1$.

We take $\xi^0, \xi^1, \cdots, \xi^n \in \Xi$, $0 = \tau_0 \leqq \tau_1 \leqq \cdots \leqq \tau_n \leqq \tau_{n+1} = 1$, and put

$$(1) \qquad (\xi^0; \tau_1; \xi^1; \cdots; \tau_n; \xi^n) = \xi(\tau) = \begin{cases} \xi^j(\tau) \text{ for } \tau \in [\tau_j; \tau_{j+1}) \\ (j = 0, 1, \cdots, n). \end{cases}$$

It is easy to check that the function $\xi(\tau)$ formed in this way belongs to Ξ. We frequently use the notation on the left-hand side of (1).

A τ-segment. We take $\xi^\alpha, \xi^\beta \in \Xi$. For $0 \leqq \theta \leqq 1$ we write

$$(2) \qquad \xi^\theta = \xi^\theta(\tau) = (\xi^\beta; \theta; \xi^\alpha).$$

We call the set of all ξ^θ the *τ-segment* that joins ξ^α to ξ^β. We denote a τ-segment by $[\xi^\alpha; \xi^\beta]$.

τ-perturbations. We fix a trajectory $\xi^0 \in \Xi$ and call it the *initial trajectory*. Next we take another trajectory $\xi(\tau)$ (the *perturbing trajectory*) and fix some point $t \in D$ (the *perturbing point*). We put

$$(3) \qquad \xi^\eta = \xi^\eta(\tau) = \begin{cases} \xi(\tau) & \text{for } \tau \in [t; t+\eta), \\ \xi^0(\tau) & \text{for the remaining } \tau \in D. \end{cases}$$

The trajectory ξ^η is said to be *perturbed* relative to ξ^0. In (3), η is variable (the perturbation parameter); we assume that $t + \eta \in D$. We may take η to be arbitrarily close to 0, since in our arguments only the limit as $\eta \to 0$ is essential.

By analogy with the above, by an *n-multiply perturbed* trajectory relative to ξ^0 we mean the trajectory

$$(4) \qquad \xi^{\bar\eta} = \xi^{\bar\eta}(\tau) = \begin{cases} \xi(\tau) & \text{for } \tau \in [t_j; t_j + \eta_j), \quad j = 1, \cdots, n, \\ \xi^0(\tau) & \text{for the remaining } \tau \in D, \end{cases}$$

where $t_1, \cdots, t_n$ are n perturbing points, $t_i \neq t_j$, and $\bar\eta = (\eta_1, \cdots, \eta_n)$ is a set of n numbers such that $t_j + \eta_j \in D$.

The concepts of a τ-segment and of τ-perturbations are our main tools in this and the following chapters.

§12. τ-continuous functionals

Let $f(\xi)$ be a complex-valued functional defined over Ξ.

We take $\xi^0 \in \Xi$, fix the perturbing point $t \in D$, and form the perturbed trajectory ξ^η. If the equation

$$\lim_{\eta \to +0} f(\xi^\eta) = f(\xi^0), \tag{1}$$

holds independently of the choice of the perturbing trajectory $\xi(\tau)$, then $f(\xi)$ is said to be *τ-continuous from the right* on the trajectory ξ^0 *with respect to the perturbing point* $t \in D$. Similarly, if for any $\xi(\tau)$

$$\lim_{\eta \to -0} f(\xi^\eta) = f(\xi^0), \tag{2}$$

then $f(\xi)$ is *τ-continuous from the left* on the trajectory ξ^0 *with respect to the perturbing point* $t \in D$.

If, on the trajectory ξ^0, $f(\xi)$ is τ-continuous from the right and from the left with respect to all $t \in D$, it is said to be *τ-continuous at* ξ^0.

A functional that is τ-continuous on all trajectories $\xi^0 \in \Xi$ is said to be a *τ-continuous functional*. We denote the set of all τ-continuous functionals by T_0.

Note that if $t = 1$, the function ξ^η in (1) is not defined. With respect to this t we can only speak of τ-continuity from the left. An analogous remark can be made about $t = 0$. In what follows such obvious reservations are not stated.

Let us illustrate the concept of τ-continuity by examples.

EXAMPLE 1. The functional $f(\xi) = \int_0^1 \xi(\tau)\, d\tau$ is τ-continuous. In fact,

$$|f(\xi^\eta) - f(\xi^0)| = \left| \int_t^{t+\eta} (\xi(\tau) - \xi^0(\tau))\, d\tau \right| \leqq (M_\xi + M_{\xi^0}) \cdot |\eta|,$$

whence (1) and (2) follow for all $\xi^0, \xi \in \Xi$.

EXAMPLE 2. Let $\xi, \xi^* \in \Xi$. Let $\rho(\xi, \xi^*)$ denote the Lebesgue measure l of the set of points belonging to D on which these trajectories do not coincide:

$$\rho(\xi, \xi^*) = l\{\tau; \xi(\tau) \neq \xi^*(\tau)\}. \tag{3}$$

The functional $f(\xi) = \rho(\xi, \xi^*)$, where ξ^* is fixed, is τ-continuous, because obviously $|f(\xi^\eta) - f(\xi^0)| \leqq |\eta|$.

EXAMPLE 3. The functional $f(\xi) = \xi(t_0)$, where $t_0 \in D$ is fixed, is not τ-continuous on any trajectory $\xi^0 \in \Xi$. For if the perturbing point t coincides with t_0 and $\xi(t_0) \neq \xi^0(t_0)$, then $f(\xi^\eta) - f(\xi^0) = \xi(t_0) - \xi^0(t_0)$ for all $\eta > 0$; that is, (1) is not fulfilled. Note that with respect to $t \neq t_0$ this functional is τ-continuous, because if we take $|\eta| < |t - t_0|$, then $\xi^\eta(t_0) = \xi^0(t_0)$; that is, $f(\xi^\eta) = f(\xi^0)$.

We state some properties of τ-continuous functionals.

PROPERTY 1. *A functional $f(\xi)$ is τ-continuous if and only if it is continuous* *on any τ-segment; that is, if and only if for any $\xi^\alpha, \xi^\beta \in \Xi$, the following is a continuous function of θ:*

$$F_{\xi^\alpha,\xi^\beta}(\theta) = f(\xi^\theta), \tag{4}$$

where $\xi^\theta = (\xi^\beta; \theta; \xi^\alpha)$ (the same as in (11.2)). (The subscripts in $F_{\xi^\alpha,\xi^\beta}(\theta)$ are omitted when this does not lead to misunderstanding.)

PROOF. Suppose that $f(\xi)$ is τ-continuous. We fix arbitrary $\xi^\alpha, \xi^\beta \in \Xi$ and show that $F(\theta)$ is continuous at $\theta = \theta_0$ ($\theta_0 \in D$ is arbitrary). For let ξ^{θ_0} be an initial trajectory, θ_0 be a perturbing point, and ξ^β a perturbing function. Then $f(\xi^0) = F(\theta_0)$ and when $\eta > 0$, $f(\xi^\eta) = F(\theta_0 + \eta)$, and from (1) we see that

$$\lim_{\eta \to +0} F(\theta_0 + \eta) = \lim_{\eta \to +0} f(\xi^\eta) = f(\xi^0) = F(\theta_0),$$

that is, $F(\theta)$ is continuous from the right at each $\theta_0 \in D$. The continuity of $F(\theta)$ from the left is proved in exactly the same way.

Conversely, suppose that $F(\theta)$ is continuous on any τ-segment. We first prove that $f(\xi)$ is τ-continuous from the right with respect to t on ξ^0. We put $\xi^\alpha = (\xi^0; t; \xi)$, $\xi^\beta = \xi$ and form $F(\theta)$ for $[\xi^\alpha; \xi^\beta]$. Since $F(t) = f(\xi^0)$ and $F(t + \eta) = f(\xi^\eta)$ when $\eta > 0$, it follows from the continuity of $F(\theta)$ at $\theta = t$ that

$$\lim_{\eta \to +0} f(\xi^\eta) = \lim_{\eta \to +0} F(t + \eta) = F(t) = f(\xi^0);$$

that is, $f(\xi)$ is τ-continuous from the right; the τ-continuity from the left is proved similarly.

PROPERTY 2. *If $f_1, f_2 \in T_0$, then $\lambda_1 f_1 + \lambda_2 f_2 \in T_0$ (λ_1 and λ_2 are complex constants), $f_1 \cdot f_2 \in T_0$, and when $f_2 \neq 0$, $f_1 : f_2 \in T_0$.*

PROPERTY 3. *If $f(\xi) \in T_0$ and $w(z)$ is a continuous function of the complex variable z, then $w(f(\xi)) \in T_0$.*

These properties follow trivially from Property 1 and from standard theorems on continuous functions.

PROPERTY 4. *For any complex numbers λ_j and distinct trajectories $\xi^j \in \Xi$ ($j = 1, \cdots, n$) there exists a τ-continuous functional $f(\xi)$ such that $f(\xi^j) = \lambda_j$.*

By Property 2 it is sufficient to find a functional such that $f(\xi^j) = \delta_{jk}$ (for $k = 1, \cdots, n$); it is given by $f(\xi) = \Phi(\rho(\xi, \xi^k))$, where ρ is as in (12.3) and $\Phi(\rho)$ is a continuous function of a real variable such that $\Phi(0) = 1$, and $\Phi(\rho) = 0$ when $\rho \geqq \min_{j \neq k} \rho(\xi^j, \xi^k)$ (the τ-continuity of $f(\xi)$ follows from Example 2 and from Property 3).

The connection between τ-continuity and other definitions of continuity of functionals. Let C_0 denote the set of functionals continuous in the sense of the uniform metric on Ξ. Examples 1—3 show that none of the sets $C_0 \cap T_0$, $\overline{C}_0 \cap T_0$, $C_0 \cap \overline{T}_0$ is empty (the bar over a set denotes the complement). It is obvious that $\overline{C}_0 \cap \overline{T}_0$ is also not empty.

If we endow Ξ with the topology of L_p and assume that $f(\xi)$ is continuous in this topology, then it is easy to prove (1) and (2) (obviously

$$\left| \int_0^1 (\xi^\eta(\tau) - \xi^0(\tau))^p \, d\tau \right| \to 0$$

as $\eta \to 0$). In other words, the continuity of $f(\xi)$ in L_p implies its τ-continuity. The converse is not true (Example 2).

§13. τ-differentiability

As before, let $f(\xi)$ be a complex-valued functional over Ξ.

Suppose next that $\xi^0 \in \Xi$ is a fixed initial trajectory, t is a perturbing point, $x \in R$ is fixed, and $\xi(\tau)$ is a perturbing function such that $\xi(t) = x$. If the limit

(1) $$\lim_{\eta \to +0} \frac{f(\xi^\eta) - f(\xi^0)}{\eta} = f'_+(\xi^0, t, s)$$

exists, is independent of the choice of $\xi(t)$, and is finite, then $f(\xi)$ is said to be *τ-differentiable from the right* on the trajectory ξ^0 *with respect to the perturbing pair* (t, x), and the limit (1) is called the *right τ-derivative of* $f(\xi)$ at $\xi = \xi^0$ *with respect to* (t, x). Analogously, if the following limit exists, is independent of the choice of $\xi(\tau)$ under the condition that $\xi(t-0) = x$, and is finite:

(2) $$\lim_{\eta \to -0} \frac{f(\xi^\eta) - f(\xi^0)}{-\eta} = f'_-(\xi^0, t, x),$$

then it is said to be the *left τ-derivative* of $f(\xi)$ at $\xi = \xi^0$ *with respect to the pair* (t, x).

Let us define the τ-differentiability of $f(\xi)$ at $\xi = \xi^0$ *with respect to the perturbing point* t. If $t \in C_{\xi^0}$, this means that the limits (1) and (2) exist for all $x \in R$ and that $f'_+(\xi^0, t, x) = f'_-(\xi^0, t, x)$ (for $t = 0$ or $t = 1$ simply that (1) or (2) exists for all x). If $t \notin C_{\xi^0}$, this means that the limits (1) and (2) exist for all $x \in R$ and that the following *discontinuity condition* holds:

(3) $$f'_+(\xi^0, t, \xi^0(t-0)) + f'_-(\xi^0, t, \xi^0(t)) = 0.$$

If $f(\xi)$ is τ-differentiable at $\xi = \xi^0$ with respect to t, the function of x that is equal to $f'_+(\xi^0, t, x)$ for $t \neq 1$ and to $f'_-(\xi_0, t, x)$ for $t = 1$ is called the *τ-derivative of* $f(\xi)$ *on the trajectory* ξ^0 *with respect to* t; we denote it by $f'(\xi^0, t, x)$.

(The preference for the right τ-derivative in the last definition does not have any significance; firstly, all the statements concerning f'_+ made below can be easily restated for f'_-, and secondly, f'_+ and f'_- coincide everywhere except possibly at a finite number of points $t \in D$.)

If $f(\xi)$ is τ-differentiable at $\xi = \xi^0$ with respect to all $t \in D$, we say that $f(\xi)$ is *τ-differentiable on the trajectory* ξ^0.

If $f(\xi)$ is τ-differentiable for all $\xi^0 \in \Xi$, we call it a *τ-differentiable functional*, and call the function $f'(\xi, t, x)$ defined on $\Xi \times D \times R$ its *τ-derivative*. Sometimes we use the abbreviated notation $f'(\xi)$ (or simply f') for the τ-derivative of the functional $f(\xi)$.

We denote the set of all τ-differentiable functionals by T_1.

EXAMPLE 1. Let us find the τ-derivative of

$$f(\xi) = \int_0^1 \xi(\tau)\,d\tau \tag{4}$$

(we consider the same functional as in Example 12.1). We easily find that

$$f'_+(\xi^0, t, x) = \lim_{\eta\to+0} \frac{1}{\eta} \int_t^{t+\eta} (\xi(\tau) - \xi^0(\tau))\,d\tau = x - \xi^0(t),$$

$$f'_-(\xi^0, t, x) = \lim_{\eta\to+0} \frac{1}{-\eta} \int_{t+\eta}^{t} (\xi(\tau) - \xi^0(\tau))\,d\tau = x - \xi^0(t-0).$$

It is clear that for $t \in C_{\xi^0}$ the τ-derivatives of (4) from the right and from the left coincide. The discontinuity condition (3) holds, because

$$(\xi^0(t-0) - \xi^0(t)) + (\xi^0(t) - \xi^0(t-0)) = 0.$$

Thus, $f(\xi) \in T_1$ and the following holds everywhere on $\Xi \times D \times R$:

$$f'(\xi, t, x) = x - \xi(t). \tag{5}$$

EXAMPLE 2. We consider the functional

$$f(\xi) = \int_0^1 \phi(t, \xi(t))\,dt, \tag{6}$$

where the function $\phi(t, x)$ is continuous on $D \times R$. Just as in the previous example it can easily be proved that this functional is τ-differentiable and that its τ-derivative is given by

$$f'(\xi, t, x) = \phi(t, x) - \phi(t, \xi(t)). \tag{7}$$

EXAMPLE 3. The functional $f(\xi) = \xi(t_0)$, where $t_0 \in D$ is fixed, is τ-differentiable for $t \neq t_0$; its τ-derivative is zero, because for $|\eta| < |t - t_0|$ we have $\xi^\eta(t_0) = \xi^0(t_0)$; that is, $f(\xi^\eta) = f(\xi^0)$ and the numerators in (1) and (2) are zero. As regards the τ-derivative from the right of $f(\xi)$ at $t = t_0$, $x \neq \xi(t_0)$, the numerator in (1) is a nonzero constant $x - \xi(t_0)$ and so no finite limit in (1) exists.

EXAMPLE 4. The functional $f(\xi) = \Phi(\xi_1, \cdots, \xi_n)$, depending only on the values of the trajectory ξ at a finite number of points $t_1, \cdots, t_n$ (we call these functionals *finite-dimensional*), is τ-differentiable with respect to $t \neq t_1, \cdots, t_n$ and its τ-derivative is zero. This result is established in the same way as in Example 3.

We list some properties of τ-differentiable functionals.

PROPERTY 1. *Each τ-differentiable functional is τ-continuous (that is, $T_1 \subset T_0$).*

This is proved by directly comparing (1) and (2) with (12.1) and (12.2), in exactly the same way as the corresponding theorem on functions of a real variable.

Property 2. *We can calculate the τ-derivative $f'(\xi^0, t, x)$ of the τ-differentiable functional $f(\xi)$ by the ordinary differentiation of a certain function of a real variable connected with f.*

We do this as follows. Let ξ^0, t, ξ^η be as in the definition of the τ-derivative. We assume that $\eta \geqq 0$ for $t \neq 1$ and that $\eta \leqq 0$ for $t = 1$. By writing $F(\eta) = f(\xi^\eta)$ for $t \neq 1$ and $F(\eta) = -f(\xi^\eta)$ for $t = 1$ we immediately obtain from (1) or (2) that

$$f'(\xi^0, t, x) = dF/d\eta\,|_{\eta=0}. \tag{8}$$

Property 3. *If $f(\xi) \in T_1$, then on any τ-segment $[\xi^\alpha; \xi^\beta]$ the function $F(\theta) = f(\xi^\theta)$ is differentiable for $\theta \in C_{\xi^\alpha} \cap C_{\xi^\beta}$ and for the remaining $\theta \in D$ it has the right and left derivatives F'_+ and F'_-. Here*

$$F'_+(\theta) = f'_+(\xi^\theta, \theta, \xi^\beta(\theta)), \tag{9}$$

$$F'_-(\theta) = -f'_-(\xi^\theta, \theta, \xi^\alpha(\theta - 0)). \tag{10}$$

Proof. We first use the definition of the right τ-derivative at ξ^θ, putting $t = \theta$ and taking ξ^β as the perturbing function. We obtain

$$F'_+(\theta) = \lim_{\eta \to +0} \frac{F(\theta + \eta) - F(\theta)}{\eta} = \lim_{\eta \to +0} \frac{f(\xi^{\theta+\eta}) - f(\xi^\theta)}{\eta} = f'_+(\xi^\theta, \theta, \xi^\beta(\theta)).$$

So we have established (9); (10) is proved in a similar manner. It follows from the discontinuity condition (3) that the expressions (9) and (10) are equal when $\theta \in C_{\xi^\alpha} \cap C_{\xi^\beta}$.

Property 4. *If $f_1(\xi)$ and $f_2(\xi)$ are τ-differentiable then so are*: (i) *their linear combinations with complex coefficients*; (ii) *their product*; (iii) *the quotient $f_1 : f_2$ when $f_2 \neq 0$. The usual formulas hold*:

$$(\lambda_1 f_1 + \lambda_2 f_2)' = \lambda_1 f_1' + \lambda_2 f_2', \quad (f_1 \cdot f_2)' = f_1' \cdot f_2 + f_1 \cdot f_2',$$
$$\left(\frac{f_1}{f_2}\right)' = \frac{f_1' \cdot f_2 - f_1 \cdot f_2'}{f_2^2}. \tag{11}$$

Property 5. *If $f(\xi) \in \mathrm{T}_1$ and the function $w(z)$ is differentiable, then $f_1(\xi) = w(f(\xi)) \in \mathrm{T}_1$ and*

$$f_1'(\xi, t, x) = w'(f(\xi)) \cdot f'(\xi, t, x). \tag{12}$$

Properties 4 and 5 are proved in the same way as the corresponding properties of functions of a real variable. We do not dwell on this detail. The verification of specific details, such as the discontinuity condition, is left to the reader as a straightforward exercise.

Property 6. *The following equality holds identically for any τ-differentiable functional*:

$$f'(\xi^0, t, \xi^0(t)) = 0. \tag{13}$$

For when calculating the τ-derivative in (13) we can take $\xi^0(t)$ as the perturbing function. Then the numerator in (1) turns out to be zero.

An analog. We take a functional $f(\xi)$ over Ξ and a function $F(\zeta) = F(\zeta_1, \cdots, \zeta_n)$ defined in the n-dimensional space R^n, $\zeta = (\zeta_1, \cdots, \zeta_n) \in R^n$. It is well known that the variational derivative $\delta f/\delta \xi$ of $f(\xi)$ is a generalization to the infinite-dimensional case of the partial derivative $\partial F/\partial \zeta_j$ of $F(\zeta)$ defined in a finite-dimensional space. In other words, $\partial F/\partial \zeta_j$ is the finite-dimensional analog of the variational derivative. We ask: what is the finite-dimensional analog of the τ-derivative? Equations (1) and (2) show that we can take this analog to be the partial difference

$$\Delta F = F^1(\zeta, j, x) = F(\zeta_1, \cdots, \zeta_{j-1}, x, \zeta_{j+1}, \cdots, \zeta_n) - F(\zeta_1, \cdots, \zeta_n) \tag{14}$$

(note that (14) does not involve a limiting process and that the limit sign occurs only in the infinite-dimensional case).

Under certain restrictions on $f(\xi)$ we can prove that if we fix ξ^0 and t in its τ-derivative and then differentiate it with respect to x for $x = \xi^0(t)$, the resulting number turns out to be the variational derivative of $f(\xi)$ at ξ^0 with respect to $\tau = t$. (For an example compare the τ-derivative $f'(\xi^0, t, x) = \phi(t, x) - \phi(t, \xi^0(t))$ of the functional (6) with its variational derivative, which is equal to $\phi'_x(t, \xi^0(t))$.)

Continuity of the τ-derivative. As we have already remarked, the τ-derivative of a functional is a function on $\Xi \times D \times R$. It would be natural to say that a function $F(\xi, t, x)$ on this set is continuous at (ξ^0, t_0, x_0) if, for any perturbing function ξ and any perturbing point t_1 of the initial trajectory ξ^0, as $(\eta, t, x) \to (0, t_0, x_0)$ we have

$$\lim F(\xi^\eta, t, x) = F(\xi^0, t_0, x_0). \tag{15}$$

However, by defining the continuity of a τ-derivative we somewhat weaken (15). We wish the functional $f = \int_0^1 \xi(t)\,dt$ to go into a number of continuously τ-differentiable functionals, whereas its τ-derivative $f' = x - \xi(t)$ does not satisfy (15), if only because $\xi(t_0)$ is not τ-continuous (see Example 12.3). We next impose a restriction on the method of convergence of (η, t, x) to its limit (however, the restriction only concerns η and t); when this restriction is observed, (15) holds for $F(\xi, t, x) = x - \xi(t)$.

For given ξ^0, t_1 and ξ we say that a passage to the limit $(\eta, t, x) \to (0, t_0, x_0)$ is *admissible* if

$$\lim \xi^\eta(t) = \xi^0(t_0). \tag{16}$$

EXAMPLE 5. If $t_0 \in C_{\xi^0}$ and $t_1 \neq t_0$, then any limit passage is admissible; for when $|\eta| < |t_1 - t_0|$ we have $\xi^\eta(t) = \xi^0(t)$ and

$$\lim \xi^\eta(t) = \lim \xi^0(t) = \xi^0(t_0).$$

It is not difficult to give an explicit expression for the restrictions on η and t that characterize admissible limiting passages. We do not need this expression and so do not give it.

So we make the following definition: the τ-derivative $f'(\xi,t,x)$ of the τ-differentiable functional $f(\xi)$ is said to be *continuous* if for any $\xi^0, t_0, x_0, t_1, \xi$ the following equation holds for all admissible limiting passages of (η,t,x) to $(0,t_0,x_0)$:

$$\lim f'(\xi^\eta,t,x) = f'(\xi^0,t_0,x_0). \tag{17}$$

EXAMPLE 6. The τ-derivative of the functional $f(\xi) = \int_0^1 \phi(t,\xi(t))\,dt$ in Example 2 is continuous. For $f'(\xi^\eta,t,x) = \phi(t,x) - \phi(t,\xi^\eta(t))$ tends to $\phi(t_0,x_0) - \phi(t_0,\xi^0(t_0)) = f'(\xi^0,t_0,x_0)$ for each admissible limit passage; this result follows from the continuity of $\phi(t,x)$ and from (16).

The indefinite τ-integral. Can we reconstruct the functional f from a given τ-derivative f'? The following theorem gives an answer to this question.

THEOREM 1. *If $f(\xi)$ has the continuous τ-derivative $f'(\xi,t,x)$, then for any $\xi^0,\xi \in \Xi$*

$$f(\xi) = f(\xi^0) + \int_0^1 f'(\xi^\theta,\theta,\xi(\theta))\,d\theta, \tag{18}$$

where $\xi^\theta = (\xi;\theta;\xi^0)$.

PROOF. We set $F(\theta) = f(\xi^\theta)$. It is obvious that

$$F(0) = f(\xi^0), \quad F(1) = f(\xi). \tag{19}$$

Property 3 asserts that if $\theta \in C_{\xi^0} \cap C_\xi$, then $F(\theta)$ is differentiable and

$$F'(\theta) = f'(\xi^\theta,\theta,\xi(\theta)). \tag{20}$$

Let us prove that F' is continuous for these θ. We fix $\theta \in C_{\xi^0} \cap C_\xi$ and show that

$$\lim_{\eta\to 0} f'(\xi^{\theta+\eta},\theta+\eta,\xi(\theta+\eta)) = f'(\xi^\theta,\theta,\xi(\theta)). \tag{21}$$

The passage to the limit in (21) is of the form (17). The passage of the variables on the left-hand side of (21) to their limits is relative to a number of admissible methods, since $\xi^{\theta+\eta}(\theta+\eta) = \xi^0(\theta+\eta) \to \xi^0(\theta) = \xi^\theta(\theta)$ as $\eta \to 0$. Hence our assumption that f' is continuous implies that (21) is valid and so

$$\lim_{\eta\to 0} F'(\theta+\eta) = F'(\theta).$$

We have shown that $F'(\theta)$ is continuous everywhere except, possibly, for a finite number of points belonging to D. Together with the continuity of $F(\theta)$ (see Property 12.1) this allows us to write $\int_0^1 F'(\theta)\,d\theta = F(1) - F(0)$, and this after substituting from (19) and (20) yields (18). The theorem is proved.

§14. τ-derivatives of higher orders

For fixed (t_1, x_1) the first τ-derivative $f'(\xi, t_1, x_1)$ of the τ-differentiable functional $f(\xi)$ is a functional over Ξ. If this functional is τ-differentiable for each $\xi \in \Xi$ with respect to all $t_2 \in D$ $(t_2 \neq t_1)$, then $f(\xi)$ is said to be *twice τ-differentiable*.

More generally, if for fixed $(t_1, x_1, \cdots, t_{n-1}, x_{n-1})$ the $(n-1)$th τ-derivative $f^{n-1}(\xi, t_1, x_1, \cdots, t_{n-1}, x_{n-1})$ of $f(\xi)$ is τ-differentiable for all $\xi \in \Xi$ with respect to any $t_n \in D$ (t_n is different from $t_1, \cdots, t_{n-1}$), then $f(\xi)$ is said to be *n times τ-differentiable*; the τ-derivative of f^{n-1} is said to be the *nth τ-derivative* of $f(\xi)$ and is denoted by $f^n(\xi, t_1, x_1, \cdots, t_n, x_n)$, or $f^n(\xi)$ or f^n. We sometimes denote the variables in the nth τ-derivative by $\xi, \bar{t}, \bar{x}$ by introducing the abbreviations $\bar{t} = (t_1, \cdots, t_n)$, $\bar{x} = (x_1, \cdots, x_n)$.

We put

$$D^{(n)} = \{ (t_1, \cdots, t_n) \in D^n;\ t_j \neq t_k \text{ for } j \neq k \}. \tag{1}$$

It follows from what has been stated above that the nth τ-derivative of an n times τ-differentiable functional is a function on $\Xi \times D^{(n)} \times R^n$.

EXAMPLE 1. Let us prove that the functional $f(\xi) = \int_0^1 \phi(t, \xi(t))\, dt$ in Example 13.2 has τ-derivatives of all orders and find them.

We have already established that the first τ-derivative of f is given by $f'(\xi, t_1, x_1) = \phi(t_1, x_1) - \phi(t_1, \xi)$. We find the second derivative by differentiating f' with respect to t_2 $(t_2 \neq t_1)$. Since for fixed (t_1, x_1) f' depends only on the value of $\xi(\tau)$ at $\tau = t_1$, by using Example 13.4 we see immediately that f^2 exists and is zero everywhere. Hence it follows that $f(\xi)$ is τ-differentiable any number of times and that all further τ-derivatives are zero. (We may say that $f(\xi)$ is analogous to a linear function $F(\zeta) = a\zeta + b$, $\zeta \in R$, whose second and higher order derivatives are also identically zero.)

PROPERTY 1. *The task of finding the nth τ-derivative of $f(\xi)$ reduces to the calculation of an nth mixed derivative of some function of n real variables.*

Let ξ^0 be an arbitrary initial trajectory, $t_1, \cdots, t_n$ be n perturbing points, and $\xi(\tau)$ a perturbing function such that $\xi(t_j) = x_j$ $(j = 1, \cdots, n)$. We form the n-fold perturbed trajectory $\xi^{\bar{\eta}}$ by assuming that the η_j corresponding to $t_j < 1$ are nonnegative and that if some $t_j = 1$, then the corresponding $\eta_j \leqq 0$. Let $F(\eta_1, \cdots, \eta_n) = f(\xi^{\bar{\eta}})$ if all $t_j < 1$ and $F(\eta_1, \cdots, \eta_n) = -f(\xi^{\bar{\eta}})$ if one of the $t_j = 1$. Then

$$f^n(\xi^0, t_1, x_1, \cdots, t_n, x_n) = \left. \frac{\partial^n F(\eta_1, \cdots, \eta_n)}{\partial \eta_1 \cdots \partial \eta_n} \right|_{\eta_1, \cdots, \eta_n = 0}. \tag{2}$$

This is proved by the n-fold use of Property 13.2.

DEFINITION. The function $\psi(t_1, x_1, \cdots, t_n, x_n)$ defined on $D^n \times R^n$ or on $D^{(n)} \times R^n$ is said to be *simple relative to the trajectory* ξ^0 (for brevity, ξ^0-simple) if for each $j = 1, \cdots, n$ it vanishes for $x_j = \xi^0_j$. A function simple relative to

$\xi^0(\tau) \equiv 0$ is said to be *simple* (without specifying the trajectory).

EXAMPLE 2. Let $v(x)$ be any function on R. Then $\psi = \prod_{j=1}^n (v(x_j) - v(\xi_j^0))$ is ξ^0-simple.

PROPERTY 2. *For a fixed $\xi^0 \in \Xi$ the nth τ-derivative of an n times differentiable functional is an ξ^0-simple function on $D^{(n)} \times R^n$.*

We prove this statement, which generalizes (13.13), with the help of Property 1. We take a perturbing function $\xi(\tau)$ that coincides with $\xi^0(\tau)$ in some neighborhood of t_j. Then $x_j = \xi_j^0$ in (2). Since for this ξ $F(\bar{\eta})$ does not depend on η_j when η_j is small, the right-hand side of (2) is zero, thus proving our assertion.

The continuity of the *n*th τ-derivative. We fix $\xi^0 \in \Xi$, $\bar{t}_0 = (t_{01}, \cdots, t_{0n}) \in \bar{D}$, $\bar{x}_0 = (x_{01}, \cdots, x_{0n}) \in R^n$ and let ξ be a perturbing function. We take n perturbing points $t_{11}, \cdots, t_{1n}$ and form the n-fold perturbation $\xi^{\bar{\eta}}$ of ξ^0. A limiting passage $(\bar{\eta}, \bar{t}, \bar{x}) \to (0, \bar{t}_0, \bar{x}_0)$ is said to be *admissible* if for any $j = 1, \cdots, n$

$$\lim \xi^{\bar{\eta}}(t_j) = \xi^0(t_{0j}). \tag{3}$$

EXAMPLE 3. If only the x_j are varied in a passage to the limit and $\bar{\eta} = 0$, $\bar{t} = \bar{t}_0$, then it is admissible in view of the trivial equation $\xi^{\bar{\eta}}(t_j) = \xi^0(t_{0j})$.

EXAMPLE 4. The limit passage $t_j \to t_{0j} + 0$, which leaves $\bar{\eta} = 0$ and all the other variables unchanged, is also related to an admissible passage: $\lim \xi^0(t_j) = \xi^0(t_{0j} + 0) = \xi^0(t_{0j})$. The same is true for the limit passage $t_j \to 1 - 0$.

If for any admissible limit passage the following equation holds for all $\xi^0 \in \Xi$, $\bar{t}_0 \in D^{(n)}$ and $\bar{x}_0 \in R^n$:

$$\lim f^n(\xi^{\bar{\eta}}, \bar{t}, \bar{x}) = f^n(\xi^0, \bar{t}_0, \bar{x}_0) \tag{4}$$

and, moreover, the left-hand side of (4) exists for $\bar{t}_0 \in D^n - D^{(n)}$, then the nth τ derivative f^n of the n times differentiable functional $f(\xi)$ is said to be *continuous*. We assume everywhere in what follows that the continuous τ-derivative f^n is defined for $\bar{t}_0 \in D^n - D^{(n)}$ by (4) and thus is a function on $\Xi \times D^n \times R^n$.

PROPERTY 3. *If $f(\xi)$ has a continuous nth τ-derivative, then for each fixed $\xi^0 \in \Xi$ it is a function on $D^n \times R^n$, continuous for $t_1, \cdots, t_n \in C_{\xi^0}$; that is, if $t_{01}, \cdots, t_{0n} \in C_{\xi^0}$, then*

$$\lim_{(\bar{t}, \bar{x}) \to (\bar{t}_0, \bar{x}_0)} f^n(\xi^0, \bar{t}, \bar{x}) = f^n(\xi^0, \bar{t}_0, \bar{x}_0). \tag{5}$$

This follows from the admissibility of (5) (since $\lim_{\bar{t} \to t_0} \xi^0(t_j) = \xi^0(t_{0j})$) and from the definition of continuity of the nth τ-derivative.

PROPERTY 4. *If $f(\xi)$ has a continuous $(n+1)$th τ-derivative f^{n+1}, then for any ξ^0, $\xi \in \Xi$ for which $\xi_1^0 = \xi_1, \cdots, \xi_n^0 = \xi_n$,*

$$f^n(\xi) - f^n(\xi^0) = \int_0^1 f^{n+1}(\xi^\theta, t_1, x_1, \cdots, t_n, x_n, \theta, \xi(\theta))\, d\theta, \tag{6}$$

where $\xi^\theta = (\xi, \theta, \xi^0)$.

This property is the analog of Theorem 13.1 for the first τ-derivative and is proved by the same method; we omit the proof.

REMARK 1. If $\xi^0(\tau)$ and $\xi(\tau)$ in (6) are different only when $\tau \in [\tau_1; \tau_2)$, $\tau_1 \leqq \tau_2$, then we can replace $\int_0^1$ in (6) by $\int_{\tau_1}^{\tau_2}$ because by Property 2 the integrand is zero outside $[\tau_1; \tau_2)$.

Let us give an example of the use of the continuity of the nth τ-derivative.

EXAMPLE 5. Let f^n be continuous. We take $\xi^0 \in \Xi$, a perturbing function ξ and n perturbing points $t_1, \cdots, t_n$. We put $x_j = \xi(t_j)$ and suppose, as usual, that $\xi^{\bar{\eta}}$ is an n-fold perturbation of ξ^0. Then

$$\lim_{\bar{\eta}\to+0} f^n(\xi^{\bar{\eta}}, t_1 + \eta_1, \xi(t_1+\eta_1), \cdots, t_n+\eta_n, \xi(t_n+\eta_n)) = f^n(\xi^0, t_1, x_1, \cdots, t_n, x_n). \tag{7}$$

For the variables $(t_j + \eta_j, \xi(t_j+\eta_j))$ tend to (t_j, x_j), and the passage to the limit is admissible because

$$\lim_{\bar{\eta}\to+0} \xi^{\bar{\eta}}(t_j+\eta_j) = \lim_{\bar{\eta}\to+0} \xi^0(t_j+\eta_j) = \xi^0(t_j).$$

We use (7) to prove the following

PROPERTY 5. *The continuous nth τ-derivative f^n of f is a symmetric function of the pairs (t_j, x_j) of its arguments; that is, for any permutation $(\alpha_1, \cdots, \alpha_n)$ of $(1, \cdots, n)$ we have*

$$f^n(\xi, t_1, x_1, \cdots, t_n, x_n) = f^n(\xi, t_{\alpha_1}, x_{\alpha_1}, \cdots, t_{\alpha_n}, x_{\alpha_n}). \tag{8}$$

We conduct the proof by means of Property 1 (by expressing the τ-derivatives in terms of partial derivatives) and in the same notation. First, let all the t_j be distinct and less than 1. Then by (2) equation (8) becomes

$$\frac{\partial^n F}{\partial\eta_1 \cdots \partial\eta_n} = \frac{\partial^n F}{\partial\eta_{\alpha_1} \cdots \partial\eta_{\alpha_n}}. \tag{9}$$

Let us write

$$F^{(n)}_{1,\cdots,n}(\bar{\eta}) = F^{(n)}_{1,\cdots,n}(\eta_1, \cdots, \eta_n) = \frac{\partial^n F}{\partial\eta_1 \cdots \partial\eta_n}.$$

It follows from Property 1 that for $\eta \geqq 0$

$$F^{(n)}_{1,\cdots,n}(\bar{\eta}) = f^n(\xi^{\bar{\eta}}, t_1+\eta_1, \xi(t_1+\eta_1), \cdots, t_n+\eta_n, \xi(t_n+\eta_n)). \tag{10}$$

As we showed in (7), as $\bar{\eta}\to 0$, the right-hand side of (10) (and together with it the left-hand side) tends to $f^n(\xi^0, t_1, x_1, \cdots, t_n, x_n) = F^{(n)}_{1,\cdots,n}(0, \cdots, 0)$. This means that $F^{(n)}_{1,\cdots,n}(\bar{\eta})$ is continuous at $\bar{\eta} = 0$. By the same arguments $F^{(n)}_{\alpha_1,\cdots,\alpha_n}(\bar{\eta})$ is continuous at $\bar{\eta} = 0$ and

$$F^{(n)}_{\alpha_1,\cdots,\alpha_n}(\bar{\eta}) = \frac{\partial^n F}{\partial\eta_{\alpha_1} \cdots \partial\eta_{\alpha_n}}.$$

But, by a well-known theorem of analysis, this is sufficient for (9) (and so

(8)) to hold when the t_j are less than 1 and are distinct. For t_j not satisfying this condition, after proving (8) we extend it to the whole of $\Xi \times D^n \times R^n$ by means of the admissible limit passages $t_j' \to t_j + 0$ and $t_j' \to 1 - 0$ (see Example 4).

REMARK 2. Let $f^n(\xi, t_1, x_1, \cdots, t_n, x_n)$ be continuous. Then the function

$$\psi(t_1, x_1, \cdots, t_n, x_n) = f^n(0, t_1, x_1, \cdots, t_n, x_n),$$

defined on $D^n \times R^n$, is continuous, is simple, and is a symmetric function of the pairs (t_j, x_j). These properties follow from Properties 3, 2 and 5, respectively.

CHAPTER 2

τ-POLYNOMIALS AND τ-ANALYTIC FUNCTIONALS

§21. Differentiation of a τ-polynomial

In the previous chapter we treated as an example the functional $f(\xi) = \int_0^1 \phi(t, \xi(t))\, dt$, which as we noted is analogous to a linear function of a real variable. We now turn to the study of functionals of a somewhat more general form—the τ-polynomials.

A functional $g(\xi)$ is said to be a *τ-polynomial* if it is continuously τ-differentiable any number of times and, starting from some order, its τ-derivatives are identically zero. When $g^n \not\equiv 0$ but $g^{n+1} \equiv 0$, the number n is said to be the *degree* of the τ-polynomial $g(\xi)$.

Theorem 22.2 in the next section asserts that each τ-polynomial can be represented as

$$g(\xi) = \int_0^1 \cdots \int_0^1 \phi(t_1, \xi_1, \cdots, t_n, \xi_n)\, dt_1 \cdots dt_n, \tag{1}$$

where the function $\phi(t_1, x_1, \cdots, t_n, x_n)$ is continuous on $D^n \times R^n$ ($t_j \in D$, $x_j \in R$) and where in (1) we have written $\xi_j = \xi(t_j)$ instead of x_j. The function ϕ is called the *density* of $g(\xi)$. (It will follow from theorems proved below that the degree of the functional (1) is less than or equal to n.)

In this section we study some properties of functionals of the form (1); in particular, we derive formulas for their differentiation and show that any such functional is a τ-polynomial. Thus we establish that the set of τ-polynomials and the set of functionals of the form (1) are the same.

We proceed to study the functionals (1). For the right-hand side of (1) we often use the abbreviated form: $g(\xi) = \int \phi(\xi)\, dt^n$ or $g(\xi) = \int \phi\, dt^n$, omitting the arguments of ϕ and assuming that dt^n is the differential of the Lebesgue measure on D^n; by definition $\int \phi\, dt^0$, when ϕ is a constant function, is equal to ϕ.

PROPERTY 1. *The functionals representable in the form* (1) *form an algebra over the field of complex numbers.*

For if $g_1 = \int \phi_1 dt^m$ and $g_2 = \int \phi_2 dt^n$ $(m \leq n)$, then

$$\lambda_1 g_1 + \lambda_2 g_2 = \int_{D^n} (\lambda_1 \phi_1(t_1, \xi_1, \cdots, t_m, \xi_m) + \lambda_2 \phi_2(t_1, \xi_1, \cdots, t_n, \xi_n)\, dt_1 \cdots dt_n,$$

$$g_1 \cdot g_2 = \int_{D^{m+n}} \phi_1(t_1, \xi_1, \cdots, t_m, \xi_m) \cdot \phi_2(t_{m+1}, \xi_{m+1}, \cdots, t_{m+n}, \xi_{m+n})\, dt_1 \cdots dt_{m+n}.$$

REMARK 1. If we arbitrarily permute the pairs of arguments of ϕ; that is, if instead of ϕ we take $\tilde{\phi} = \phi(t_{\alpha_1}, x_{\alpha_1}, \cdots, t_{\alpha_n}, x_{\alpha_n})$, where $(\alpha_1, \cdots, \alpha_n) = p_n$ is any permutation of $(1, \cdots, n)$, then $g(\xi)$ is unchanged; that is, $\int \phi\, dt^n = \int \tilde{\phi}\, dt^n$. To verify this it is sufficient to make the substitution $t_j = t'_{\alpha_j}$ in the right-hand side of (1).

A function ϕ on $D^n \times R^n$ satisfying the following equation for any permutation

(2) $$\phi(t_1, x_1, \cdots, t_n, x_n) = \phi(t_{\alpha_1}, x_{\alpha_1}, \cdots, t_{\alpha_n}, x_{\alpha_n}),$$

is said to be *symmetric*. Henceforth we always assume that the density ϕ of a functional of the form (1) is symmetric. This does not restrict the generality; if ϕ in (1) is not symmetric, then the corresponding functional can be represented as $g(\xi) = \int S_n\phi\, dt^n$, where $S_n\phi$ is symmetric:

$$S_n\phi = \frac{1}{n!} \sum_{p_n} \phi(t_{\alpha_1}, x_{\alpha_1}, \cdots, t_{\alpha_n}, x_{\alpha_n}).$$

REMARK 2. The representation of $g(\xi)$ in the form (1) is not unique; different densities ϕ can produce the same functional, as follows from Remark 1. The nonuniqueness remains even if we consider only symmetric densities. For example, we may add to the density of $g(\xi)$ a continuous symmetric function $\phi_0(t_1, \cdots, t_n)$, independent of x_j and such that $\int \phi_0 dt^n = 0$, without changing $g(\xi)$. In exactly the same way, if we take $\tilde{\phi} = S_{n+1}\phi$, then $g(\xi) = \int \phi\, dt^n = \int \tilde{\phi}\, dt^{n+1}$.

In the next section we derive a canonical representation of a τ-polynomial which in contrast to (1) is uniquely defined by its values on Ξ.

It is convenient to introduce an abbreviation for writing formulas for the τ-differentiation of the functionals (1). Let $\phi(t_1, x_1, \cdots, t_n, x_n)$ be a complex-valued function on $D^n \times R^n$. The expression

(3) $$\phi[x_{11} \pm x_{12} \pm \cdots] \cdots [x_{n1} \pm x_{n2} \pm \cdots]$$

($\pm$ denotes either the plus or the minus sign) is to be understood as follows: first, regarding (3) as the product of ϕ by the brackets placed after it, we multiply out these brackets by the rule for multiplying algebraic expressions, preserving in each term the order of the sequence of factors. Then, to obtain the final result, we write $\pm\, \phi(t_1, x_{1\beta_1}, \cdots, t_n, x_{n\beta_n})$ instead of each of the derived terms of the form $\pm\, \phi x_{1\beta_1} \cdots x_{n\beta_n}$. For example, $\phi[x_1] \cdots [x_n] = \phi(t_1, x_1, \cdots, t_n, x_n)$ and the derivative (13.7) of $g(\xi) = \int_0^1 \phi(t_1, \xi(t_1))\, dt_1$, which is $g' = \phi(t_1, x_1) - \phi(t_1, \xi(t_1))$, is written in the new notation as $g' = \phi[x_1 - \xi_1]$.

Next we find the τ-derivative of the functional (1). We fix ξ^0, t_1, x_1 and the perturbing function $\xi(\tau)$ (such that $\xi(t_1)=x_1$) and form the perturbed trajectory $\xi^\eta(\tau)$. We study the difference

$$\Delta g = g(\xi^\eta) - g(\xi^0) = \int_{D^n} (\phi(\xi^\eta) - \phi(\xi^0))\, dt^n. \tag{4}$$

Let E_j denote the set of points $(\tau_1, \cdots, \tau_n) \in D^n$ in the strip $\tau_j \in [t_1; t_1+\eta)$. Let $E = \bigcup_j E_j$, let F be the set of points of E that belong to only one of the E_j, and let $G = E - F$. Next let $F_j = E_j \cap F$ and $G_j = E_j - F_j$.

Since the integrand in (4) is nonzero only in $E = F + G$, instead of (4) we can write

$$\Delta g = \int_F (\phi(\xi^\eta) - \phi(\xi^0))\, dt^n + \int_G (\phi(\xi^\eta) - \phi(\xi^0))\, dt^n. \tag{5}$$

LEMMA 1. *The difference* Δg *can be represented as*

$$\Delta g = L(\eta) + o(\eta), \tag{6}$$

where

$$L(\eta) = n \int_{E_1} \phi[\xi_1 - \xi_1^0][\xi_2^0] \cdots [\xi_n^0]\, dt^n. \tag{7}$$

PROOF. First we estimate the second integral on the right-hand side of (5). For any values of η the variables of ϕ do not fall outside the boundary of the compact set $\{|x_j| \leqq \max(M_{\xi^0}, M_\xi),\ j = 1, \cdots, n\}$; hence $|\phi|$ is bounded by a finite constant K. We can write $G = \bigcup_{j\neq k} E_j E_k$, and so its Lebesgue measure is not greater than $n(n-1)\eta^2/2$. Hence the absolute magnitude of the second term in (5) does not exceed $Kn(n-1)\eta^2$, and we can write

$$\Delta g = \int_F (\phi(\xi^\eta) - \phi(\xi^0))\, dt^n + o(\eta). \tag{8}$$

Next we turn to $L(\eta)$. It is obvious that $E_1 = F_1 + G_1$. Since G_1 lies in the interior of G, clearly its measure, as well as the measure of the whole of G, is $o(\eta)$ as $\eta \to 0$. Hence we can write $L(\eta)$ as

$$L(\eta) = n \int_{F_1} \phi[\xi_1 - \xi_1^0][\xi_2^0] \cdots [\xi_n^0]\, dt^n + o(\eta). \tag{9}$$

Next we show that the first terms on the right-hand sides of (8) and (9) are equal. Consider the first term in (8). By using the fact that $\xi^\eta(\tau)$ coincides with ξ on $[t_1; t_1+\eta)$ and with ξ^0 for the remaining values of τ, and by taking the equation $F = F_1 + \cdots + F_n$ into account, we find that

$$\int_F (\phi(\xi^\eta) - \phi(\xi^0))\, dt^n = \sum_{j=1}^n \int_{F_j} \phi[\xi_1^0] \cdots [\xi_{j-1}^0][\xi_j - \xi_j^0][\xi_{j+1}^0] \cdots [\xi_n^0]\, dt^n. \tag{10}$$

We renumber the variables (see Remark 1) in the jth term on the right-hand side of (10) so that the jth pair (t_j, x_j) becomes the first: (t_1, x_1). Then all the n terms turn out to be equal to one another and their sum equal to the first term in (9). The proof of the lemma is completed by comparing (8) and (9).

THEOREM 1. *The functional* (1) *is a* τ*-polynomial.*

We first prove the differentiability of the functional $g(\xi)$ defined by (1). It follows from (6) that

$$g'_+(\xi^0, t_1, x_1) = \lim_{\eta \to +0} \frac{L(\eta)}{\eta} = L'_+(0).$$

For $\eta > 0$ we write $L(\eta)$ in the form

$$L(\eta) = n \int_{t_1}^{t_1+\eta} dt_1 \int_{D^{n-1}} \phi[\xi_1 - \xi_1^0][\xi_2^0] \cdots [\xi_n^0] dt_2 \cdots dt_n. \tag{11}$$

Because, in the expression (11) for $L(\eta)$, η only occurs in the upper limit of integration of the outer integral, by a well-known theorem on the derivative of a definite integral with respect to a variable upper limit we obtain

$$g'_+(\xi^0, t_1, x_1) = n \int_{D^{n-1}} \phi[x_1 - \xi_1^0][\xi_2^0] \cdots [\xi_n^0] dt_2 \cdots dt_n.$$

In exactly the same way we find the left τ-derivative of $g(\xi)$; it is

$$g'_-(\xi\ , t_1, x_1) = n \int_{D^{n-1}} \phi[x_1 - \xi^0(t_1 - 0)][\xi_2^0] \cdots [\xi_n^0] dt_2 \cdots dt_n.$$

It is obvious that g'_+ and g'_- coincide for $t_1 \in C_{\xi^0}$; the discontinuity condition is easily verified (in the same way as in Example 13.1). So we have proved the existence of the first τ-derivative of $g(\xi)$.

$g^2, \cdots, g^n$ are found by a similar calculation. We obtain the following differentiation formulas:

$$g^j(\xi, t_1, x_1, \cdots, t_j, x_j)$$
$$= \frac{n!}{(n-j)!} \int_{D^{n-j}} \phi[x_1 - \xi_1] \cdots [x_j - \xi_j][\xi_{j+1}] \cdots [\xi_n] dt_{j+1} \cdots dt_n \tag{12}$$
$$(j = 1, \cdots, n-1),$$

$$g^n(\xi, t_1, x_1, \cdots, t_n, x_n) = n!\, \phi[x_1 - \xi_1] \cdots [x_n - \xi_n]. \tag{13}$$

Since $g^n(\xi)$ for fixed $t_1, x_1, \cdots, t_n, x_n$, depends only on the values of ξ at the points $t_1, \cdots, t_n$ and the $(n+1)$th τ-derivative is calculated with respect to $t_{n+1} \neq t_1, \cdots, t_n$, we see that g^{n+1} is identically zero (see Examples 13.3 and 13.4).

Next we prove that $g^j(\xi)$ $(j = 1, \cdots, n)$ is continuous. We fix $(\xi_0, \bar{t}_0, \bar{x}_0) \in \Xi \times D^j \times R^j$ and let $\xi^{\bar{\eta}}$ be the j-fold perturbed trajectory. The expression for $g^j(\xi^{\bar{\eta}}, \bar{t}, \bar{x})$ contains under the integral sign (for $j = n$ without the integral sign) a function ϕ in which t_k, x_k, $\xi_k^{\bar{\eta}}$ occupy the places of the arguments. By carrying out the limit passage $(\bar{\eta},\bar{t},\bar{x}) \to (0,\bar{t}_0,\bar{x}_0)$, we bring it first under the integral sign and then under the sign of the continuous function ϕ. By bearing in mind that in an admissible limit passage the above-mentioned variables of ϕ tend respectively to t_{0k}, x_{0k} and ξ^0_{0k}, we find that g^j is continuous.

Thus $g(\xi)$ in (1) is a τ-polynomial (of degree not greater than n because $g^{n+1} \equiv 0$). The proof of our theorem is complete.

We conclude this section by considering τ-polynomials of a particular form that occupy, amongst τ-polynomials, the same place as that occupied by the power functions $a_n(\zeta - \zeta^0)^n$ amongst ordinary polynomials.

A τ-polynomial $h(\xi)$ is said to be *ξ^0-simple* if it can be written as $h(\xi) = \int \psi\, dt^n$, where ψ is a ξ^0-simple function on $D^n \times R^n$ (for the definition of a ξ^0-simple function see §14). In view of Remark 2, no ξ^0-simple density apart from ψ can correspond to an ξ^0-simple τ-polynomial. Let us agree that if h is a ξ^0-simple τ-polynomial, then the equation $h = \int \psi\, dt^n$ always implies that ψ is ξ^0-simple. By definition the functional $h(\xi) = c$, $c \neq 0$, is a ξ^0-simple τ-polynomial of zero degree.

When $\xi^0 \equiv 0$, a ξ^0-simple τ-polynomial $h(\xi)$ is said to be *simple* (without specifying the trajectory).

The following theorem gives a fundamental property of ξ^0-simple τ-polynomials.

THEOREM 2. *If $h(\xi) = \int \psi\, dt^n$ is a ξ^0-simple τ-polynomial, then its τ-derivatives satisfy the equations*

$$h^j(\xi^0, t_1, x_1, \cdots, t_j, x_j) \equiv 0 \quad \text{for } j = 0, 1, \cdots, n-1, \tag{14}$$

$$h^n(\xi^0, t_1, x_1, \cdots, t_n, x_n) = n!\,\psi(t_1, x_1, \cdots, t_n, x_n). \tag{15}$$

To prove this it is sufficient to use the general formulas (12) and (13) for the differentiation of τ-polynomials and to recall the definition of a ξ^0-simple function.

It is clear from (15) that for $\psi \not\equiv 0$ the degree of $h(\xi)$ is equal to n (hence there are simple τ-polynomials of all degrees $n = 1, 2, \cdots$).

§22. The general form of a τ-polynomial

First we prove several auxiliary assertions.

REMARK 1. If $g(\xi)$ is a τ-polynomial of degree n, then for fixed $t_1, x_1, \cdots, t_n, x_n$ its nth τ-derivative $g^n(\xi, t_1, x_1, \cdots, t_n, x_n)$ takes the same values on those trajectories for which $\xi^\alpha_j = \xi^\beta_j$ $(j = 1, \cdots, n)$.

The assertion follows from (14.6) if we remember that $g^{n+1} \equiv 0$ for a τ-polynomial of degree n.

Remark 1 implies that there exists a function $\Psi(t_1, x_1, y_1, \cdots, t_n, x_n, y_n)$ on $D^n \times R^n \times R^n$ such that

$$g^n(\xi, t_1, x_1, \cdots, t_n, x_n) = \Psi(t_1, x_1, \xi_1, \cdots, t_n, x_n, \xi_n). \tag{1}$$

REMARK 2. When $t_1, x_1, \cdots, t_{n-1}, x_{n-1}$ are fixed, the $(n-1)$th τ-derivative of a τ-polynomial of degree n has the following properties: if the trajectories $\xi^\alpha(\tau)$ and $\xi^\beta(\tau)$ differ only on an interval $[t; t+\eta) \in D$ that does not contain $t_1, \cdots, t_{n-1}$, then $\Delta g^{n-1} = g^{n-1}(\xi^\beta) - g^{n-1}(\xi^\alpha)$ is not altered by the same change of ξ^α and ξ^β outside $[t; t+\eta)$ with preservation of their values at $t_1, \cdots, t_{n-1}$.

For by using (14.6) and by taking (1) into account we can write (to be specific we take $\eta > 0$)

$$\Delta g^{n-1} = \int_t^{t+\eta} \Psi(t_1, x_1, \xi_1^\alpha, \cdots, t_{n-1}, x_{n-1}, \xi_{n-1}^\alpha, \theta, \xi^\beta(\theta), \xi^\alpha(\theta))\, d\theta. \tag{2}$$

It remains to note that for fixed $t_1, x_1, \cdots, t_{n-1}, x_{n-1}$ the right-hand side of (2) depends only on the values of ξ^α and ξ^β on the interval $[t; t+\eta)$ and at the points $t_1, \cdots, t_{n-1}$, and our Remark is proved.

LEMMA 1. *We can represent the nth τ-derivative of the τ-polynomial $g(\xi)$ of degree n in the form*

$$g^n(\xi) = \psi[x_1 - \xi_1] \cdots [x_n - \xi_n], \tag{3}$$

where $\psi(t_1, x_1, \cdots, t_n, x_n)$ is a simple symmetric function that is continuous on $D^n \times R^n$.

We also show that for such a function we can take

$$\psi(t_1, x_1, \cdots, t_n, x_n) = \Psi(t_1, x_1, 0, \cdots, t_n, x_n, 0). \tag{4}$$

As a preliminary we note that the function on the right-hand side of (4) is, in fact, continuous, symmetric and simple, because it is equal to $g^n(0, t_1, x_1, \cdots, t_n, x_n)$ in view of (1) (see Remark 14.2).

We turn to the proof of the lemma. Let $[\tau_1; \tau_2)$ be a fixed interval containing t_n, but not containing the other t_j, and let η be such that $t_n + \eta$ lies in $[\tau_1; \tau_2)$ and $\eta \geqq 0$ for $t_n < 1$ and $\eta \leqq 0$ for $t_n = 1$. We take a trajectory ξ^0 that vanishes on $[\tau_1; \tau_2)$, such that $\xi_1^0 = y_1, \cdots, \xi_{n-1}^0 = y_{n-1}$. Let ξ with two numerical superscripts denote the trajectory that is equal to the first of these numbers on $[t_n; t_n + \eta)$, to the second number at the remaining points of $[\tau_1; \tau_2)$, and coincides with ξ^0 outside $[\tau_1; \tau_2)$.

We perform transformations, having fixed $t_1, x_1, \cdots, t_{n-1}, x_{n-1}$ and set $x_n = x$, $y_n = y$:

$$\begin{aligned} g^n(\xi^{yy}, \cdots, t_n, x_n) &= \lim_{\eta \to 0} \frac{g^{n-1}(\xi^{xy}) - g^{n-1}(\xi^{yy})}{|\eta|} \\ &= \lim_{\eta \to 0} \left(\frac{g^{n-1}(\xi^{x0}) - g^{n-1}(\xi^0)}{|\eta|} - \frac{g^{n-1}(\xi^{y0}) - g^{n-1}(\xi^0)}{|\eta|} \right) \\ &= g^n(\xi^0, \cdots, t_n, x_n) - g^n(\xi^0, \cdots, t_n, y_n). \end{aligned} \tag{5}$$

In going from the second to the third part of this equation, using Remark 2 we have replaced ξ^{xy} (ξ^{yy}) in the numerator by ξ^{x0} (ξ^{y0}) and then have added and subtracted $g^{n-1}(\xi^0)$.

By taking account of (1) we obtain from (5) that

$$\Psi(\cdots, t_n, x_n, y_n) = \Psi(\cdots, t_n, x_n, 0) - \Psi(\cdots, t_n, y_n, 0).$$

By applying the result of this calculation n times (for each of the variables t_j, x_j, y_j of Ψ) we find that

$$\Psi(t_1, x_1, y_1, \cdots, t_n, x_n, y_n) = \psi[x_1 - y_1] \cdots [x_n - y_n],$$

where ψ is given by (4), and (3) follows directly from this result. Thus the lemma is proved.

It is of interest to note that instead of taking ψ as in (4) we could have taken $\psi = \Psi(t_1, x_1, \xi_1^0, \cdots, t_n, x_n, \xi_n^0)$, where $\xi^0 \in \Xi$ is any fixed trajectory; here (3) would remain true. We do not dwell on this.

Let us turn to the main theorem of this section.

THEOREM 1 (THE CANONICAL EXPANSION OF A τ-POLYNOMIAL). *Each τ-polynomial $g(\xi)$ can be uniquely represented as a finite sum of simple τ-polynomials of different degrees. This representation takes the form*

$$g(\xi) = h_n(\xi) + h_{n-1}(\xi) + \cdots + h_1(\xi) + h_0, \tag{6}$$

where h_n is a simple τ-polynomial of degree n and each of the remaining h_j either is a simple τ-polynomial of degree j or is zero.

We prove the existence of the expansion by induction on n. The assertion is obviously true when $n = 0$. Suppose that the expansion exists for any τ-polynomial of degree less than n. We prove that then it exists for the τ-polynomial $g(\xi)$ of degree n.

By Lemma 1 the nth derivative of $g(\xi)$ has the form $g^n = \psi[x_1 - \xi_1] \cdots [x_n - \xi_n]$. In view of (21.13) the following simple τ-polynomial of degree n has such an nth derivative:

$$h_n(\xi) = \frac{1}{n!} \int \psi(t_1, \xi_1, \cdots, t_n, \xi_n)\, dt^n.$$

We set $g(\xi) - h_n(\xi) = g_1(\xi)$. Since $g_1^n = g^n - h_n^n \equiv 0$, we see that $g_1(\xi)$ is a τ-polynomial of degree less than n and, by our inductive hypothesis, splits into a sum $g_1 = h_{n-1} + \cdots + h_1 + h_0$, where h_j is either a simple τ-polynomial of degree j or is zero. It follows at once from here that we can represent $g = h_n + g_1$ in the form (6).

Uniqueness of the expansion. We write (6) as

$$g(\xi) = h_0 + \sum_{j=1}^{n} \int \psi^j(t_1, \xi_1, \cdots, t_j, \xi_j)\, dt^j, \tag{7}$$

where ψ^j denotes the density of h_j. It follows from the main property of simple τ-polynomials (Theorem 21.2) that

$$g(0) = h_0,$$

(8) $$g^j(0, t_1 x_1, \cdots, t_j, x_j) = h_j^j(0, t_1 x_1, \cdots, t_j, x_j) = j!\,\psi^j(t_1, x_1, \cdots, t_j, x_j) \qquad (j = 1, \cdots, n).$$

But this means that the terms in (6) are uniquely determined by the values of the τ-derivatives of the appropriate order of $g(\xi)$ when $\xi \equiv 0$. So we have proved the uniqueness of the representation (6), and with it Theorem 1.

The expansion (6) (or (7)) is called the *canonical expansion* of the τ-polynomial $g(\xi)$.

THEOREM 2 (THE GENERAL FORM OF A τ-POLYNOMIAL). *Any τ-polynomial $g(\xi)$ can be represented as*

(9) $$g(\xi) = \int \phi(t_1, \xi_1, \cdots, t_n, \xi_n)\, dt^n,$$

where $\phi(t_1, x_1, \cdots, t_n, x_n)$ is continuous on $D^n \times R^n$.

For by definition each simple τ-polynomial $h_j(\xi)$ is of the form (9). Hence by Property 21.1 $g(\xi) = \sum_j h_j(\xi)$ can also be represented in this form.

It is appropriate to recall here that the representation of a τ-polynomial in the form (9) is not unique (see Remark 21.2). The canonical expansion (7) has the advantage that it is uniquely defined by the τ-polynomial. For example, if $g(\xi)$ is initially given in the form (9), then the expansion (7) is independent of the choice of its representation in this form.

Taylor's formula for a τ-polynomial. By using (8) we can rewrite the canonical expansion of a τ-polynomial as

(10) $$g(\xi) = g(0) + \sum_{j=1}^{n} \frac{1}{j!} \int g^j(0, t_1, \xi_1, \cdots, t_j, \xi_j)\, dt^j.$$

This equation, which expresses the values of the τ-polynomial $g(\xi)$ at any point $\xi \in \Xi$ in terms of the values of $g(\xi)$ and its τ-derivatives at $\xi^0 \equiv 0$, is called the *Taylor expansion* of $g(\xi)$ about the trajectory $\xi^0 \equiv 0$.

We can derive the Taylor expansion of a τ-polynomial about an arbitrary trajectory $\xi^0 \in \Xi$ by the same method as that used to obtain (10). It has the form

(11) $$g(\xi) = g(\xi^0) + \sum_{j=1}^{n} \frac{1}{j!} \int g^j(\xi^0, t_1, \xi_1, \cdots, t_j, \xi_j)\, dt^j.$$

§23. τ-analyticity

We arrive at the concept of τ-analyticity by generalizing (22.11). A complex-valued functional $f(\xi)$ is said to be *τ-analytic* if it is infinitely τ-differentiable and, for any $\xi^0 \in \Xi$, the following equality holds:

(1) $$f(\xi) = f(\xi^0) + \sum_{j=1}^{\infty} \frac{1}{j!} \int f^j(\xi^0, t_1, \xi_1, \cdots, t_j, \xi_j)\, dt^j.$$

We say that (1) is the *Taylor series expansion* of a τ-analytic functional.

It is obvious that any τ-polynomial is a τ-analytic functional (with a finite Taylor series).

We give an example of a τ-analytic functional that is not a τ-polynomial.

EXAMPLE 1. We consider the functional

(2) $$f(\xi) = \exp\left(\int_0^1 v(\xi(\tau))\,d\tau\right),$$

where $v(x)$ is continuous on R. By using the rules for τ-differentiation (see (13.11), (13.12) and Examples 13.2 and 13.4) we obtain

$$f^j(\xi^0, t_1, x_1, \cdots, t_j, x_j) = f(\xi^0) \prod_{k=1}^{j} (v(x_k) - v(\xi_k^0)) \quad (j = 1, 2, \cdots),$$

that is, the functional (2) is infinitely τ-differentiable. Let us check that (1) holds for this functional. We start from the right-hand side of this equation; for the functional (2) it takes the form

$$f(\xi^0) \cdot \left(1 + \sum_{j=1}^{\infty} \frac{1}{j!} \int_{D^j} \prod_{k=1}^{j} (v(\xi_k) - v(\xi_k^0))\,dt^j\right).$$

If we note that the j-fold integral in the jth term of this sum can be written as the jth power of the integral $g(\xi) = \int_0^1 (v(\xi_1) - v(\xi_1^0))\,dt_1$, then by using the ordinary Taylor expansion of e^ζ we obtain

$$f(\xi^0) \cdot \left(1 + \sum_{j=1}^{\infty} \frac{(g(\xi))^j}{j!}\right) = f(\xi^0)\, e^{g(\xi)}$$

$$= \exp\left(\int_0^1 (v(\xi_1^0) + v(\xi_1) - v(\xi_1^0))\,dt_1\right) = f(\xi).$$

The right-hand side of (1) turns out to be equal to the left-hand side; hence the functional (2) is τ-analytic. It is not a τ-polynomial when $v(x)$ is not constant, because in this case none of its τ-derivatives vanishes identically.

§24. An algebra of τ-analytic functionals

We take a sequence $h_0(\xi), h_1(\xi), \cdots, h_j(\xi), \cdots$, where $h_j(\xi) = \int \psi^j dt^j$ is a simple τ-polynomial of degree j (or is zero), whose density ψ^j satisfies the inequality

$$|\psi^j(t_1, x_1, \cdots, t_j, x_j)| \leq c^{j+1}$$

for some constant $c \in [1; +\infty)$ (c is the same for all j). It is obvious that under these conditions $h_j(\xi)$ also satisfies the condition $|h_j(\xi)| \leq c^{j+1}$ for all $\xi \in \Xi$.

We consider the functional defined by the series

(1) $$f(\xi) = \sum_{j=0}^{\infty} \frac{h_j(\xi)}{j!}.$$

This series converges (absolutely and uniformly) on Ξ, because it is dominated by the numerical series $\sum_{j=1}^{\infty} c^{j+1}/j!$ which converges to ce^c.

We denote the set of all functionals of the form (1) by $A(c)$. Obviously, if $c_1 < c_2$, then $A(c_1) \subset A(c_2)$. Let us put

$$A = \bigcup_c A(c).$$

EXAMPLES. Any τ-polynomial whose density is bounded is contained in A; the functional (23.2) belongs to A if $\sup|v(x)| < +\infty$.

This section is concerned with a study of the properties of A.

First we prove that any functional belonging to A is τ-differentiable and τ-analytic.

In proving the τ-differentiability we use the method and results of §21, in which we considered a similar problem for τ-polynomials.

We fix $\xi^0 \in \Xi$, the perturbing pair (t_1, x_1), and the perturbing function $\xi(\tau)$, and form the perturbed trajectory ξ^η.

LEMMA 1. *For* $f(\xi) \in A$, $\Delta f = f(\xi^\eta) - f(\xi^0)$ *can be represented as*

$$\Delta f = L(\eta) + o(\eta), \tag{2}$$

where

$$L(\eta) = \int_{[t_1; t_1+\eta)} dt_1 \sum_{j=1}^{\infty} \frac{1}{(j-1)!} \int_{D^{j-1}} \psi^j[\xi_1 - \xi_1^0][\xi_2^0] \cdots [\xi_j^0] dt_2 \cdots dt_j. \tag{3}$$

PROOF. Suppose, to be specific, that $\eta \geqq 0$. It is obvious that $\Delta f = \sum_{j=1}^{\infty} \Delta h_j / j!$, where $\Delta h_j = h_j(\xi^\eta) - h_j(\xi^0)$. By using the proof of Lemma 21.1 we find that $\Delta h_j = L_j(\eta) + R_j(\eta)$, where

$$L_j(\eta) = j \int_{t_1}^{t_1+\eta} dt_1 \int_{D^{j-1}} \psi^j[\xi_1 - \xi_1^0][\xi_2^0] \cdots [\xi_j^0] dt_2 \cdots dt_j,$$

and $|R_j(\eta)| \leqq c^{j+1} j(j-1)\eta^2$, since $f(\xi)$ belongs to some $A(c)$. Hence

$$\Delta f = \sum_{j=1}^{\infty} \frac{1}{j!} L_j(\eta) + \sum_{j=1}^{\infty} \frac{1}{j!} R_j(\eta). \tag{4}$$

It is obvious that the first summand on the right-hand side of (4) is equal to $L(\eta)$ given in (3) (the interchange of the order of integration and summation is valid, because the series (3) is uniformly convergent). If we denote the second summand in (4) by $R(\eta)$, then the estimate

$$|R(\eta)| \leqq \sum_{j=1}^{\infty} \frac{1}{j!} |R_j(\eta)| \leqq \sum_{j=1}^{\infty} \frac{c^{j+1} j(j-1)\eta^2}{j!} = c^3 e^c \eta^2 = o(\eta)$$

converts (4) into (2). This completes the proof of the lemma.

THEOREM 1. *The functional* (1) *is* τ-*differentiable any number of times.*

PROOF. We first calculate $f'_+(\xi^0, t_1, x_1)$. As in Theorem 21.1 we obtain

$$f'_{+}(\xi^0, t_1, x_1) = L'_{+}(0) = \sum_{j=1}^{\infty} \frac{1}{(j-1)!} \int_{D^{j-1}} \psi^j[\xi_1 - \xi_1^0][\xi_2^0] \cdots [\xi_j^0] dt_2 \cdots dt_j,$$

whence, by taking account of the formula for the differentiation of the τ-polynomials $h_j(\xi)$, we find that

$$f'_{+}(\xi^0, t_1, x_1) = \sum_{j=1}^{\infty} \frac{h'_{j+}(\xi^0, t_1, x_1)}{j!}.$$

Similarly

$$f'_{-}(\xi^0, t_1, x_1) = \sum_{j=1}^{\infty} \frac{h'_{j-}(\xi^0, t_1, x_1)}{j!}.$$

The facts that f'_{+} and f'_{-} coincide for $t_1 \in C_{\xi^0}$ and that the discontinuity condition is fulfilled follow from the validity of the corresponding statements for h'_{j+} and h'_{j-}. Besides, it is easy to verify them directly.

In complete analogy with this result we can prove the existence of the τ-derivatives of higher orders of (1). We arrive at

$$(5) \qquad f^k = \sum_{j=1}^{\infty} \frac{h_j^k}{j!} \quad (k = 1, 2, \cdots),$$

which mean that the series (1) can be τ-differentiated term-by-term any number of times. The theorem is proved.

It follows from (5) that

$$\psi^j(t_1, x_1, \cdots, t_j, x_j) = f^j(0, t_1, x_1, \cdots, t_j, x_j),$$

that is, that the representation of $f(\xi) \in A$ in the form (1) is unique.

THEOREM 2. *Any functional belonging to A is τ-analytic.*

After Theorem 1 it only remains to prove that (23.1) holds for the functional (1). By starting from the right-hand side of this equation we obtain in succession

$$\sum_{j=0}^{\infty} \frac{1}{j!} \int f^j(\xi^0, t_1, \xi_1, \cdots, t_j, \xi_j)\, dt^j$$

$$= \sum_{j=0}^{\infty} \frac{1}{j!} \sum_{k=j}^{\infty} \frac{1}{k!} \int h_k^j(\xi^0, t_1, \xi_1, \cdots, t_j, \xi_j)\, dt^j$$

$$= \sum_{k=0}^{\infty} \frac{1}{k!} \sum_{j=0}^{k} \frac{1}{j!} \int h_k^j(\xi^0, t_1, \xi_1, \cdots, t_j, \xi_j)\, dt^j = \sum_{k=0}^{\infty} \frac{1}{k!} h_k(\xi) = f(\xi).$$

In going from the first part of this equation to the second we have used (5) and we have passed from the second to the third parts by interchanging the order of summation in the double sum; this is valid, because the double series is absolutely convergent—it is dominated by the series

$$\sum_{j=0}^{\infty}\sum_{k=j}^{\infty}\frac{2^j c^{k+1}}{j!(k-j)!}=ce^{3c}.$$

We have then used the Taylor formula (22.11) for a τ-polynomial and, finally, the definition of the functional (1). We have shown that the sum of the Taylor series of the functional (1) is equal to the functional itself, and the theorem is proved.

Below we consider some properties of A.

PROPERTY 1. *A is an algebra over the field of complex numbers.*

PROOF. If $f_1, f_2 \in A$, then we can find a c such that $f_1, f_2 \in A(c)$. Let $f_1 = \sum_{j=0}^{\infty} h_{1j}/j!$ and $f_2 = \sum_{j=0}^{\infty} h_{2j}/j!$. Then $f = \lambda_1 f_1 + \lambda_2 f_2 = \sum_{j=0}^{\infty} h_j/j!$, where $h_j = \lambda_1 h_{1j} + \lambda_2 h_{2j}$ is a simple τ-polynomial of degree j (or is zero) whose density satisfies $|\psi^j| \leqq (|\lambda_1| + |\lambda_2|)\cdot c^{j+1}$; that is, $f \in A(c')$, where $c' = c\cdot \max(1, |\lambda_1| + |\lambda_2|)$.

Analogously, $f = f_1 \cdot f_2 = \sum_{j=0}^{\infty} h_j/j!$, where

$$h_j(\xi) = \sum_{k=0}^{j}\binom{j}{k} h_{1k}\cdot h_{2,j-k}.$$

It is easy to show that $h_j(\xi)$ is a simple τ-polynomial of degree j (or is zero) whose density satisfies

$$|\psi^j| \leqq \sum_{k=0}^{j}\binom{j}{k} c^{k+1}c^{j-k+1} = c^2(2c)^j,$$

whence $f = f_1 \cdot f_2 \in A(2c^2)$.

We define a countable sequence of *norms* in A by

(6) $$\|f\|_m = \max_{0\leqq j\leqq m}\sup_{\xi, t_k, x_k}|f^j(\xi, t_1, x_1, \cdots, t_j, x_j)|, \quad m = 1, 2, \cdots.$$

REMARK 1. It is obvious that for any f we have $\|f\|_1 \leqq \cdots \leqq \|f\|_m \leqq \cdots$.

REMARK 2. If $h = \int \psi\, dt^j$, where $|\psi| \leqq C$, then for all m

(7) $$\|h\|_m \leqq j!\,2^m\cdot C.$$

For $j \geqq m$ the following more precise bound holds:

(8) $$\|h\|_m \leqq \frac{j!}{(j-m)!}2^m\cdot C.$$

These bounds follow from (21.12) and (21.13).

We say that the sequence $f_1, \cdots, f_n, \cdots$ *converges to $f \in A$ in the space A* if there exists a c such that all $f_n \in A(c)$ and for any m

(9) $$\lim_{n\to\infty}\|f_n - f\|_m = 0.$$

A sequence $f_1, \cdots, f_n, \cdots$ is said to be *fundamental in the space A* if it is a fundamental sequence with respect to each norm and if all the f_n belong to some $A(c)$.

REMARK 3. If $f_n = \sum_{j=0}^{\infty} h_{nj}/j!$, where $h_{nj} = \int \psi^{nj} dt^j$, $|\psi^{nj}| \leqq c^{j+1}$, is fundamental in A, then, because of $|\psi^{mj} - \psi^{nj}| \leqq \|f_m - f_n\|_j$ for each j the sequence $\psi^{1j}, \cdots, \psi^{nj}, \cdots$ is fundamental in the sense of the uniform metric on $D^j \times R^j$ and so converges uniformly to some continuous function ψ^j which together with ψ^{nj} satisfies the inequality $|\psi^j| \leqq c^{j+1}$. If we write $h_j = \int \psi^j dt^j$, then for any m

$$\lim_{n \to \infty} \|h_{nj} - h_j\|_m = 0, \tag{10}$$

because by (7) $\|h_{nj} - h_j\|_m \leqq j!\, 2^m \sup|\psi^{nj} - \psi^j| \to 0$ as $n \to \infty$.

PROPERTY 2. *The ring A is complete; that is, every fundamental sequence $f_1, \cdots, f_n, \cdots$ in A converges to some functional $f \in A$.*

We prove this with the help of Remark 3 and in the same notation.

We put $f = \sum_{j=0}^{\infty} h_j/j!$ and show that the sequence f_n converges to f in A. Since all the f_n belong to some $A(c)$, it is only necessary to verify (9).

We fix $\epsilon > 0$ and the integer m. We take $M > m$ so that the remainder $\sum_{j=M}^{\infty} 2^m c^{j+1}/(j-m)!$ of the convergent series is less than $\epsilon/3$. Then we take N so that $\|h_{nj} - h_j\|_m \leqq \epsilon/3M$ for $j = 0, 1, \cdots, M-1$, and $n > N$ (this is possible by (10)). Then for $n > N$

$$\|f_n - f\|_m \leqq \sum_{j=0}^{M-1} \frac{\|h_{nj} - h_j\|_m}{j!} + \sum_{j=M}^{\infty} \frac{\|h_{nj}\|_m + \|h_j\|_m}{j!} \leqq M \cdot \frac{\epsilon}{3M} + 2 \cdot \frac{\epsilon}{3} = \epsilon,$$

which proves our statement.

PROPERTY 3. *The set of τ-polynomials is everywhere dense in A.*

Let $f = \sum_{j=0}^{\infty} h_j/j! \in A(c)$ be arbitrary.

We consider the sequence $g_n = \sum_{j=0}^{n} h_j/j!$ of partial sums of the Taylor series for f. By taking (8) into account, for $n \geqq m$ we obtain

$$\|g_n - f\|_m \leqq \sum_{j=n+1}^{\infty} \frac{\|h_j\|_m}{j!} \leqq \sum_{j=n+1}^{\infty} \frac{2^m \cdot c^{j+1}}{(j-m)!}. \tag{11}$$

The right-hand side of (11) is the remainder of a convergent series and so tends to zero as $n \to \infty$. Hence $\lim_{n\to\infty} \|g_n - f\|_m = 0$. Together with the obvious membership $g_n \in A(c)$ this means that $g_1, \cdots, g_n, \cdots$ converges to f in A. The proof of Property 3 is completed by recalling that all the g_n are τ-polynomials.

CHAPTER 3

τ-MEASURES

§31. Generalized measures

One of the methods of constructing a theory of integration on an abstract set Ω is that of Daniell (see [7]), which is based on axiomatizing the concept of the integral and then introducing a measure through the integral. A similar idea is to regard a measure as a method of integrating some class of functions on Ω, and we use this approach in the present chapter. However, the axioms that Daniell placed on his integral necessarily lead to the result that the corresponding measure is countably additive. This is not applicable when we are just interested in set functions on Ω that are not countably additive; hence our concept of the integral differs from the one described in [**7**] and leads to generalized measures instead of measures.

Let Ω be a set of points and let F be a complete linear space of functions on Ω. By a *generalized measure over* F in Ω we mean an arbitrary complex-valued continuous linear functional over F; that is, any element of the space conjugate to F. We also call this functional an *integral*. In agreement with the terminology of the theory of generalized functions F is called the *base space*.

When Ω is a set of functions of a real variable and F consists of the τ-smooth functionals on this set, the corresponding generalized measure is said to be a *τ-measure*. In other words, a τ-measure is a method of integration for a class of τ-smooth functionals.

In the present chapter we consider the algebra A introduced in §24 as an example of a base space; another example is given in §44.

Note that the term "generalized measure" used by us has a different significance than, for example, in [**8**], where a generalized measure is a countably additive set function without the condition of being nonnegative. From its properties this generalized measure is very close to the ordinary measures (it can be written as the difference of two nonnegative measures). It seems to us convenient to call this simply a measure, and to retain the term "generalized measure" for objects that differ significantly from measures. When we need to distinguish classical measures, we call them nonnegative measures.

§32. The quasi-measure and its nth variation

Let Ω denote the set of all real functions $\omega(\tau)$ on D; following [**9**] we call the following set an *n-dimensional block* in Ω:

$$(1)\qquad b(t_1, x_{01}, x_{11}, \cdots, t_n, x_{0n}, x_{1n}) = \{\omega(\tau);\ x_{0j} \leqq \omega(t_j) < x_{1j},\ j = 1, \cdots, n\},$$

here $(t_1, \cdots, t_n) \in D^{(n)}$ and $-\infty \leqq x_{0j} < x_{1j} \leqq +\infty$ are fixed. Let $B_{t_1,\cdots,t_n}$ be the Borel closure of the set of all blocks (1) for fixed $t_1, \cdots, t_n$, and B the minimal σ-algebra that contains all the $B_{t_1,\cdots,t_n}$.

By a *quasi-measure* we mean a finite additive function $\mu(b)$, defined on all blocks, that can be extended to a measure on each $B_{t_1,\cdots,t_n}$ (we also call this extension a quasi-measure μ). It is obvious that any measure on B is a quasi-measure; the converse is not true—a quasi-measure need not be extendable to a measure on B.

The function

$$(2)\qquad q_n(t_1, x_1, \cdots, t_n, x_n) = \mu(b(t_1, -\infty, x_1, \cdots, t_n, -\infty, x_n))$$

is said to be the *n-dimensional distribution* of μ.

If for each n there is a function $p_n = p_n(t_1, x_1, \cdots, t_n, x_n)$ such that for any n-dimensional block (1) we have

$$\mu(b) = \int_{x_{01}}^{x_{11}} \cdots \int_{x_{0n}}^{x_{1n}} p_n(t_1, x_1, \cdots, t_n, x_n)\, dx_1 \cdots dx_n, \tag{3}$$

then the set $\{p_n;\ n = 1, 2, \cdots\}$ is said to be a *system of finite-dimensional densities* of μ.

The nth variation of a quasi-measure. Let Ω be split into a finite number of disjoint n-dimensional (n is fixed) blocks: $\Omega = \sum_j b_j$; the nth *variation* $V_n(\mu)$ of a quasi-measure μ is defined by

$$V_n(\mu) = \sup \sum_j |\mu(b_j)|,$$

where the upper bound is taken over all the above subdivisions of Ω. It is obvious that the nth variation of a quasi-measure is a nondecreasing function of n. Hence the limit (finite or infinite) $\lim_{n\to\infty} V_n(\mu) = V(\mu)$ exists; we call $V(\mu)$ the *total variation* of μ.

We make three obvious remarks.

REMARK 1. For any nonnegative quasi-measure $V_n(\mu) = V(\mu) = \mu(\Omega)$.

REMARK 2. A quasi-measure whose total variation is infinite cannot be extended to a measure on B.

REMARK 3. If a quasi-measure has a system $\{p_n\}$ of finite-dimensional densities, then its nth variation is equal to

$$V_n(\mu) = \sup_{t_1, \cdots, t_n} \int_{R^n} |p_n|\, dx_1 \cdots dx_n. \tag{4}$$

We give some examples of quasi-measures and calculate their nth and total variations. In these examples the quasi-measures are defined by means of their densities.

EXAMPLE 1. We take a complex-valued function $p(x)$, $x \in R$, such that $\int_R p(x)\, dx = 1$ and $\int_R |p(x)|\, dx = v < +\infty$. Let us write

$$p_n(t_1, x_1, \cdots, t_n, x_n) = p(x_1) \cdots p(x_n). \tag{5}$$

By (3) the family $\{p_n\}$ in (5) defines a quasi-measure μ. We immediately obtain from Remark 3 that its nth variation is given by $V_n(\mu) = v^n$. If $v = 1$, then both the nth variation and the total variation are 1 (in this case μ is nonnegative). When $v > 1$, $V(\mu) = +\infty$.

FURTHER EXAMPLES. Let

$$\partial u / \partial t = Lu \tag{6}$$

be a linear partial differential equation for $u(x, t)$ of the first order in t. Suppose that (6) has the fundamental solution $G(t_0, x_0, t, x)$ which for any t_0, x_0, t satisfies

(7) $$\int_R G(t_0, x_0, t, x)\,dx = 1, \quad \int_R |G(t_0, x_0, t, x)|\,dx < +\infty.$$

We take $(t_1, \cdots, t_n) \in D^{(n)}$. Let $(\alpha_1, \cdots, \alpha_n)$ be a permutation for which $t_{\alpha_1} < \cdots < t_{\alpha_n}$. We write

(8) $$p_n(t_1, x_1, \cdots, t_n, x_n) = \prod_{j=1}^{n} G(t_{\alpha_{j-1}}, x_{\alpha_{j-1}}, t_{\alpha_j}, x_{\alpha_j}),$$

where $t_{\alpha_0} < 0$ and $x_{\alpha_0} \in R$ are fixed. (We could have taken $t_{\alpha_0} = 0$, but then in (8) we must put $G(0, x_{\alpha_0}, 0, x_{\alpha_1}) = \delta(x_{\alpha_1} - x_{\alpha_0})$.) It is not difficult to show that the set of functions (8) defines (according to (3)) a quasi-measure μ. We say that this quasi-measure corresponds to the equation (6). The first of the conditions (7) implies that for any such quasi-measure $\mu(\Omega) = 1$.

We give some partial differential equations of the form (6) that have quasi-measures.

EXAMPLE 2. The diffusion equation

(9) $$\frac{\partial u}{\partial t} = \frac{1}{4}\frac{\partial^2 u}{\partial x^2},$$

for which the fundamental solution has the form

(10) $$G(t_0, x_0, t, x) = \frac{1}{\sqrt{\pi(t - t_0)}}\exp\left(-\frac{(x - x_0)^2}{t - t_0}\right)$$

is well known to satisfy (7). Since $G(t_0, x_0, t, x)$ in (10) is nonnegative, the corresponding quasi-measure μ is also nonnegative, and so $V_n(\mu) = V(\mu) = \mu(\Omega) = 1$. (This quasi-measure can be extended to a Wiener measure on B.)

EXAMPLE 3. The equation

(11) $$\frac{\partial u}{\partial t} = (-1)^{k+1}\frac{\partial^{2k} u}{\partial x^{2k}} \quad (k = 2, 3, \cdots)$$

with the fundamental solution

(12) $$G(t_0, x_0, t, x) = \frac{1}{(t - t_0)^{1/2k}} g_k\left(\frac{x - x_0}{(t - t_0)^{1/2k}}\right),$$

where

$$g_k(y) = \frac{1}{2\pi}\int_{-\infty}^{\infty} \exp(-\xi^{2k} + i\xi y)\,d\xi$$

and $g_k(y)$ does not have constant sign. It is known that as $x \to \infty$ the fundamental solution (12) decreases more rapidly than any power of x. Hence $\int_R |G|\,dx < \infty$ and, as follows from the form of (12), this integral is independent of t_0, x_0 and t and is equal to $v_k = \int_R |g_k(y)|\,dy$ $(1 < v_k < +\infty)$. This yields the following equation for the densities of the quasi-measure of (11):

$$\int_{R^n} |p_n|\, dx_1 \cdots dx_n = \left(\int_R |g_k(y)|\, dy \right)^n = v_k^n.$$

Hence the nth variation of the quasi-measure μ of (11) is given by $V_n(\mu) = v_k^n$, and the total variation by $V(\mu) = +\infty$.

EXAMPLE 4. The equation

$$\frac{\partial u}{\partial t} = \frac{e^{i\alpha}}{4} \cdot \frac{\partial^2 u}{\partial x^2} \quad \left(0 < \alpha < \frac{\pi}{2}\right) \tag{13}$$

has the fundamental solution

$$G(t_0, x_0, t, x) = \frac{\gamma}{\sqrt{\pi(t - t_0)}} \exp\left(-\gamma^2 \frac{(x - x_0)^2}{t - t_0}\right), \tag{14}$$

where $\gamma = e^{-i\alpha/2}$. The quasi-measure μ of (13) is similar to the previous one. For it we have $\int_R |G|\, dx = \sec^{1/2}\alpha > 1$; hence $V_n(\mu) = \sec^{n/2}\alpha$ and $V(\mu) = +\infty$.

In §34 we prove a theorem concerning the extension to τ-measures of a certain class of quasi-measures with an infinite variation; it turns out that all the quasi-measures in Examples 1—4 belong to this class.

§33. Integration of a τ-polynomial with respect to a quasi-measure

It is well known that the integral of the finite-dimensional functional

$$f(\omega) = \Phi(\omega(t_1), \cdots, \omega(t_n)) \tag{1}$$

with respect to a measure μ on B reduces to a Lebesgue-Stieltjes integral on R^n defined by the n-dimensional distribution $q_n(t_1, x_1, \cdots, t_n, x_n)$ of μ ($t_1, \cdots, t_n$ are the same as in (1)):

$$\int f d\mu = \int_{R^n} \Phi(x_1, \cdots, x_n)\, dq_n. \tag{2}$$

We take the τ-polynomial

$$g(\xi) = \int \phi\, dt^n, \tag{3}$$

where $\phi(t_1, x_1, \cdots, t_n, x_n)$ is bounded, and let μ be a measure on B such that $\mu(\Omega - \Xi) = 0$, so that $g(\xi)$ is defined in Ω on a set of full measure.

We introduce in the space $X = D^n \times \Omega$ the measure μ_X, which is equal to the product of the Lebesgue measure in D^n and the measure μ, and calculate the integral over X of $\phi(t_1, \xi_1, \cdots, t_n, \xi_n)$ with respect to μ_X. Using Fubini's Theorem, we obtain

$$\int_X \phi\, d\mu_X = \int_\Omega d\mu \int_{D^n} \phi\, dt^n = \int_{D^n} dt^n \int_\Omega \phi\, d\mu. \tag{4}$$

By using (3) (or (2)) in the second (or third) parts of (4) we obtain

$$\int g\, d\mu = \int_{D^n} dt^n \int_{R^n} \phi(t_1, x_1, \cdots, t_n, x_n)\, dq_n. \tag{5}$$

Thus the integration of a τ-polynomial with respect to the measure μ is reduced to the evaluation of a $2n$-dimensional integral over $D^n \times R^n$.

If now μ is a quasi-measure that cannot be extended to a measure on B, then we define the left-hand side of (5) by the right-hand side of this equation, taking q_n to be the n-dimensional distribution of the quasi-measure.

More precisely, we proceed as follows: if $h(\xi) = \int \psi dt^n$ is a simple τ-polynomial, then by definition for any quasi-measure μ we put

$$\int h d\mu = \int dt^n \int \psi(t_1, x_1, \cdots, t_n, x_n)\, dq_n, \tag{6}$$

and say that h *is integrable with respect to the quasi-measure* μ (μ-*integrable*) if (6) exists and is finite. If $g(\xi)$ is an arbitrary τ-polynomial and $g = \sum_j h_j$ is its canonical expansion, we say that g is μ-*integrable* if each h_j is μ-integrable, and by definition we put

$$\int g d\mu = \sum_j \int h_j d\mu. \tag{7}$$

It would not be convenient to take (5) at once as the definition of the integral of any τ-polynomial $g(\xi)$ with respect to the quasi-measure μ, for it would leave the question of the uniqueness of the definition unsettled, because the representation (3) for a τ-polynomial is not unique. However, we can still use (5) to find $\int g d\mu$; the following theorem, which is stated without proof, refers to this.

THEOREM 1. *Equation* (5) *holds for any* τ-*polynomial* $g = \int \phi dt^n$ *that is integrable with respect to the quasi-measure* μ, *if we take* q_n *to be the* n-*dimensional distribution of* μ.

§34. The algebra A as a base space

We consider the problem of extending the concept of integral we have just introduced to τ-analytic functionals in the algebra A.

We assume that the quasi-measure μ is such that all $V_n(\mu)$ are finite and that each τ-polynomial belonging to A is μ-integrable (for this it is sufficient that $\int \psi(t_1, x_1, \cdots, t_n, x_n)\, dq_n$ should be measurable on D^n for any continuous bounded ψ).

The functional Ig is defined on τ-polynomials in A by

$$Ig = \int g d\mu. \tag{1}$$

REMARK 1. If the density of the τ-polynomial $h_m = \int \psi^m dt^m$ satisfies $|\psi^m| \leqq C$, then

$$|Ih_m| \leqq C V_m(\mu). \tag{2}$$

DEFINITION. The functional If, finite on all $f \in A$, is said to be *continuous* if the convergence of the sequence $\{f_n\}$ to f in A implies that $\lim_{n\to\infty} If_n = If$.

If the functional Ig given by (1) can be extended to a continuous linear functional on the whole of A, we say that the quasi-measure μ can be *extended* to a τ-measure on A.

REMARK 2. This extension is in any case unique, because the τ-polynomials form an everywhere dense set in A (see Property 24.3).

THEOREM 1. *If there exists a finite constant v such that $V_m(\mu) \leqq v^m$ $(m = 1, 2, \cdots)$, then the quasi-measure μ can be extended to a τ-measure on A.*

We show that for each $f = \sum_{m=0}^{\infty} h_m/m! \in A(c)$ this extension is given by

$$If = \sum_{m=0}^{\infty} \frac{Ih_m}{m!}. \tag{3}$$

Firstly, for any τ-polynomial from A the functional (3) is the same as (1) (see (33.7)).

Secondly, it is finite on all $f \in A$, because (2) implies that

$$|If| \leqq \sum_{m=0}^{\infty} \frac{c^{m+1}v^m}{m!} = ce^{cv}.$$

Its linearity is obvious.

Finally, we prove that the functional (3) is continuous in A.

Let $f_n = \sum_{m=0}^{\infty} (1/m!) \int \psi^{nm} dt^m$ $(n = 1, 2, \cdots)$ be a sequence of functionals that converge in A to $f = \sum_{m=0}^{\infty} (1/m!) \int \psi^m dt^m$. We estimate the difference

$$\Delta I = If_n - If = \sum_{m=0}^{\infty} \frac{1}{m!} \int dt^m \int (\psi^m - \psi^{nm})\, dq_m.$$

We fix an arbitrary $\epsilon > 0$ and take M so that

$$\sum_{m=M}^{\infty} \frac{c(cv)^m}{m!} < \frac{\epsilon}{3}.$$

Then

$$|\Delta I| \leqq \left| \sum_{m=0}^{M-1} \frac{1}{m!} \int dt^m \int (\psi^{nm} - \psi^m)\, dq_m \right| + 2\frac{\epsilon}{3}.$$

By using Remark 24.3 concerning the uniform convergence of ψ^{nm} to ψ^m, we take N so that for $n > N$ we have $|\psi^{nm} - \psi^m| < \epsilon/3Mv^M$ when $m = 0, 1, \cdots, M-1$. Then for $n > N$

$$|\Delta I| < \sum_{m=0}^{M-1} \frac{1}{m!} \frac{\epsilon}{3Mv^M} v^M + \frac{2}{3}\epsilon < \epsilon,$$

which means that If is continuous in A. This completes the proof of the theorem.

COROLLARY 1. *Since for each of the quasi-measures μ in the Examples 32.1—32.4 there exists a constant v such that $V_m(\mu) \leqq m^m$, and because all the*

τ-polynomials in A are obviously μ-integrable, all these quasi-measures can be extended to τ-measures on A.

CHAPTER 4

CONCLUSION

§41. On the development of the concepts of τ-smoothness

In this section we wish to indicate, without a detailed discussion, the possible modifications of the definitions of τ-smoothness that were stated in the first two chapters, and to outline certain ways of developing these concepts.

On τ-continuity. The following modifications of the concept of the τ-continuity of a functional may turn out to be useful in certain problems.

We consider the function $\rho(\xi^\alpha, \xi^\beta)$ introduced in (12.3), which is equal to the Lebesgue measure of the set of points of D where these trajectories do not coincide. It is easy to check that $\rho(\xi^\alpha, \xi^\beta)$ is a metric on Ξ. Functionals that are continuous in the sense of this metric on Ξ (or on its completion with respect to this metric) are said to be *absolutely τ-continuous*. It is easy to show that absolute τ-continuity is stronger than τ-continuity. (For example, if a τ-polynomial has a bounded density, then it is absolutely τ-continuous; if its density is not bounded, then it need not be absolutely τ-continuous.)

If we convert Ξ into a topological space by introducing neighborhoods $O(\xi^0, r, M) = \{\xi \in \Xi;\ \rho(\xi^0, \xi) < r,\ |\xi - \xi^0| < M\}$, then continuity in the sense of this topology is weaker than absolute τ-continuity and stronger than τ-continuity.

On τ-differentiability. In the calculus of variations one studies a functional of the form

$$g(\xi) = \int_0^1 \phi(\tau, \xi(\tau), \xi'(\tau))\, d\tau, \tag{1}$$

where $\xi'(\tau)$ is the derivative with respect to τ of $\xi(\tau)$ (we consider $g(\xi)$ on the set of piecewise continuously differentiable trajectories from Ξ). This functional is τ-continuous but not τ-differentiable since (referring, to be specific, to the τ-derivative from the right) the limit

$$\lim_{\eta \to +0} \frac{g(\xi^\eta - g(\xi^0)}{\eta} \tag{2}$$

depends, in general, not only on the pair $t, x = \xi(t)$, but on the derivative from the right $\xi'_+(t)$ of the perturbing function ξ at t.

If we wish to include (1) in the set of τ-differentiable functionals, it is natural to weaken the concept of τ-differentiability by fixing, in the perturbation, not only t, x but also $x' = \xi'_+(t)$ when calculating (2). Then the τ-derivative is a function of the initial trajectory and of the perturbing triad

(t, x, x'). For the functional (1) it is equal to

$$g'_+(\xi^0, t, x, x') = \phi(t, x, x') - \phi(t, \xi^0(t), \xi^{0\prime}_+(t)).$$

We call this derivative a τ-derivative of the first rank.

Similarly we could introduce the concept of τ-differentiability of the mth rank (m is the number of derivatives of the perturbing trajectory $\xi(\tau)$ that are fixed in calculating (2)), in order to include the following functionals in the set of τ-differentiable functionals

$$g(\xi) = \int_0^1 \phi(t, \xi(t), \cdots, \xi^{(m)}(t))\,dt \tag{3}$$

and, more generally, the "τ-polynomials of the nth degree of the mth rank"

$$g(\xi) = \int_{D^n} \phi(t_1, \xi_1, \cdots, \xi_1^{(m)}, \cdots, t_n, \xi_n, \cdots, \xi_n^{(m)})\,dt^n, \tag{4}$$

where $\xi_k^{(j)}$ is the jth derivative of $\xi(\tau)$ at $\tau = t_k$.

§42. τ-linearity

In the second chapter we defined a τ-polynomial to be a functional with certain differentiability properties. We now show how it can be defined by simple properties of an algebraic character. For this the concept of τ-linearity is fundamental.

The functional $f(\xi)$ is said to be *τ-linear* if for all $\xi^1, \xi^2 \in \Xi$ such that $\xi^1(\tau) \cdot \xi^2(\tau) \equiv 0$ we have

$$f(\xi^1 + \xi^2) = f(\xi^1) + f(\xi^2)$$

(in [**10**] (Russian p. 341) functionals satisfying this condition are called *local*).

It is easy to check that any simple τ-polynomial of the first degree (that is, $g(\xi) = \int_0^1 \psi(t, \xi(t))\,dt$, where ψ is continuous and $\psi(t, 0) \equiv 0$) is τ-linear. The following converse is of greater interest.

THEOREM 1. *If $f(\xi)$ has a continuous τ-derivative and is τ-linear, then it is a simple τ-polynomial of the first degree.*

PROOF. First note that for a τ-linear functional $f(0) = 0$. Next, by Theorem 13.1 on the reconstruction of a functional from its τ-derivative, when we put $\xi^0 \equiv 0$ in (13.18) we have

$$f(\xi) = \int_0^1 f'(\xi^\theta, \theta, \xi(\theta))\,d\theta. \tag{1}$$

We transform the integrand in (1) by using the τ-linearity of $f(\xi)$:

$$f'(\xi^\theta, \theta, \xi(\theta)) = \lim_{\eta \to +0} \frac{f(\xi^{\theta+\eta}) - f(\xi^\theta)}{\eta} = \lim_{\eta \to +0} \frac{f(\xi^\theta) + f(\xi^\eta) - f(\xi^\theta) - f(0)}{\eta} =$$

$$= \lim_{\eta \to +0} \frac{f(\xi^\eta) - f(0)}{\eta} = f'(0, \theta, \xi(\theta)).$$

Here $\xi^\eta = (0; \theta; \xi; \theta + \eta; 0)$. Next, by writing $f'(0, t, x) = \psi(t, x)$, we obtain from (1) that

$$f(\xi) = \int_0^1 \psi(t, \xi(t))\, dt,$$

where $\psi(t, x)$ is continuous and simple (see Remark 14.2); that is, $f(\xi)$ is a simple τ-polynomial of the first degree.

If we were to introduce the concept of τ-bilinearity (τ-n-linearity) we would come to τ-polynomials of the second (nth) degree.

So far in the present section we have referred to functionals that have a τ-derivative of zero rank (in the terminology of §41). If we were to consider similarly the functionals that have a τ-derivative of the mth rank, we would obtain the functionals (41.3) and (41.4).

§43. τ-differentiability in the calculus of variations

We have already referred in §13 to the connection between the variational derivative and the τ-derivative. We now wish to show, by a single very elementary example, the relation between the method of determining extrema using Euler's equation, and a method connected with τ-differentiation.

THEOREM 1 (A NECESSARY CONDITION FOR A MINIMUM). *If $f(\xi)$ is τ-differentiable and attains its minimum at $\xi = \xi^0$, then for all t and x we have*

$$f'(\xi^0, t, x) \geqq 0.$$

To prove this theorem it is sufficient to consider the limits (13.1) and (13.2) and to apply the arguments employed in proving the usual theorem of Fermat. We do not dwell on this.

We consider the functional $f(\xi) = \int_0^1 \phi(t, \xi(t))\, dt$, for which Euler's equation takes the form

$$\phi'_x(t, \xi(t)) = 0.$$

Let us now derive this equation by means of τ-differentiation. According to (13.7) the τ-derivative of $f(\xi)$ is given by $f' = \phi(t, x) - \phi(t, \xi(t))$; by Theorem 1, at a minimum point $\xi = \xi^0$ we have

$$\phi(t, x) - \phi(t, \xi^0(t)) \geqq 0;$$

that is, $\phi(t, x) \geqq \phi(t, \xi^0(t))$. This inequality means that, for each fixed t, $\phi(t, x)$ must have a minimum at $x = \xi^0(t)$. Thus we are faced with the problem of finding the minimum of the function $\Phi(x) = \phi(t, x)$ of x; this can be solved by various methods. In particular, if it turns out that $\Phi(x)$ is differentiable at the point where it attains its minimum, then by Fermat's theorem

$\Phi'(\xi^0(t)) = 0$; that is, $\phi'_x(t, \xi^0(t)) = 0$, and we arrive at the above form of Euler's equation.

Note that by using a result analogous to Theorem 1 we can generalize the maximum principle given in [5] to arbitrary τ-differentiable functionals. The basic inequality for this turns out to be a generalization of (1). It is curious that in this approach it would be more natural to call the maximum principle the "minimum principle for a τ-derivative" (this can be seen by comparing (1) and (13.13)).

§44. τ-measures. The Feynman measure as a τ-measure

In Chapter 3 we took as the base space for τ-measures the algebra A of τ-analytic functionals

(1) $$f = \sum_{m=0}^{\infty} \frac{1}{m!} \int \psi^m dt^m,$$

which is distinguished by certain bounds on ψ^m. Note that in the construction of the integral If (see §34) the bounds on ψ^m on the one hand, and on the variation V_m of the m-dimensional distributions q_m on the other hand, were necessary, roughly speaking, only for the integrals $I_m = (1/m!) \int \psi^m dq_m dt^m$ to be finite and for $\sum I_m$ to be convergent. If we wish to select a base space A' of τ-analytic functionals to extend to a τ-measure a system $\{q_m\}$ that does not satisfy the conditions of Theorem 34.1, then we must impose on functionals in A' restrictions that ensure that $I_m < \infty$ and $\sum I_m$ is convergent.

Let us illustrate this by the example of the so-called Feynman measure which corresponds to the Schrödinger equation

$$\frac{\partial u}{\partial t} = \frac{i}{4} \frac{\partial^2 u}{\partial x^2}$$

with the fundamental solution

$$G(t_0, x_0, t, x) = \frac{1-i}{\sqrt{2\pi(t-t_0)}} \exp\left(i \frac{(x-x_0)^2}{t-t_0}\right).$$

If we form a system of m-dimensional densities p_m in terms of this fundamental solution in the same way as in (32.8) (to be specific we take $t_0 = 0$), then starting from the first variation the corresponding mth variations are all infinite because

$$\int |G|\, dx = \int \frac{dx}{\sqrt{\pi(t-t_0)}} = +\infty.$$

Nevertheless,

$$I_m = \frac{1}{m!} \int \psi^m dq_m dt^m = \frac{1}{m!} \int \psi^m p_m dx^m dt^m$$

can be defined in this case as an improper integral, and it is finite if $\int |\psi^m| \, dx^m \leqq C < +\infty$. The latter follows from the estimate

$$\frac{1}{m!} \int |\psi^m p_m| \, dx^m dt^m \leqq \int_{\Delta^m} dt^m \int \frac{|\psi^m| \, dx^m}{\prod_{j=1}^m \sqrt{\pi (t_j - t_{j-1})}}$$

$$\leqq \frac{C}{\pi^{m/2}} \int_{\Delta^m} \frac{dt^m}{\prod_{j=1}^M \sqrt{t_j - t_{j-1}}} = \frac{C}{\Gamma(m/2+1)},$$

where

$$\Delta^m = \{ (t_1, \cdots, t_m) \in D^m; \ t_1 \leqq \cdots \leqq t_m \}.$$

Let A' denote the set of functionals of the form (1) each of which satisfies the inequality $\int |\psi^m| \, dx^m \leqq c^{m+1}$ for some $c < \infty$. Then $\sum I_m$ is convergent because $\sum |I_m| \leqq \sum c^{m+1}/\Gamma(m/2+1) < +\infty$, and $If = \sum I_m$ turns out to be a linear functional defined on the whole of A'. With an appropriate definition of convergence in A' (the analog of the convergence in A given in §24) it is continuous.

We conjecture that the ease of application of τ-smooth functionals to the extension of systems of finite-dimensional distributions to generalized measures has a general explanation; we can regard the τ-smooth functionals that are hardly changed under τ-perturbations to be close to finite-dimensional functionals, which are completely unchanged under almost all τ-perturbations. Because the finite-dimensional functionals are trivially integrable with respect to finite-dimensional distributions (see (33.2)), the proximity to them of τ-smooth functionals must ensure the simplicity of integration of the latter. In this connection it seems to us to be of interest to consider τ-measures over the spaces of all τ-differentiable and of all τ-continuous functionals.

Received 15 JUNE 1967

References

[1] I. M. Gel′fand and A. M. Jaglom, *Integration in functional spaces and its applications to quantum physics,* Uspehi Mat. Nauk **11** (1956), no. 1 (67), 77-114; English transl., J. Mathematical Phys. **1** (1960), 48-69. MR **17**, 1261.

[2] Ju. L. Daleckiĭ, *Functional integrals associated with operator evolution equations,* Uspehi Mat. Nauk **17** (1962), no. 5 (107), 3-115 = Russian Math. Surveys **17** (1962), no. 5, 1-107. MR **28** #4389.

[3] V. Ju. Krylov, *Some properties of the distribution corresponding to the equation* $\partial u/\partial t = (-1)^{q+1} \partial^{2q} u/\partial x^{2q}$, Dokl. Akad. Nauk SSSR **132** (1960), 1254-1257 = Soviet Math. Dokl. **1** (1960), 760-763. MR **22** #9722.

[4] A. N. Kolmogorov, *Grundbegriffe der Wahrscheinlichkeitsrechnung,* Springer, Berlin, 1933; Russian transl., ONTI, Moscow, 1936; English transl., Chelsea, New York 1948; 2nd ed., 1956. MR **11**, 374; **18**, 155.

[5] V. G. Boltjanskiĭ, R. V. Gamkrelidze and L. S. Pontrjagin, *The theory of optimal processes. I: The maximum principle,* Izv. Akad. Nauk SSSR Ser. Mat. **24** (1960), 3-42; English transl., Amer. Math. Soc. Transl. (2) **18** (1961), 341-382. MR **22** #11191; **23** #A1908.

[6] R. P. Feynman, *Space-time approach to non-relativistic quantum mechanics,* Rev. Modern Physics **20** (1948), 367-387. MR **10**, 224.

[7] P. J. Daniell, *A general form of integral,* Ann. of Math. (2) **19** (1917/18), 279-294.
[8] P. Halmos, *Measure theory,* Van Nostrand, Princeton, N. J., 1950; Russian transl., IL, Moscow, 1953. MR **11**, 504; **16**, 22.
[9] G. E. Šilov, *Integration in infinite-dimensional spaces and the Wiener integral,* Uspehi Mat. Nauk **18** (1963), no. 2 (110), 99-120 = Russian Math. Surveys **18** (1963), 97-118. MR **27** # 6161.
[10] I. M. Gel′fand and N. Ja. Vilenkin, *Generalized functions.* Vol. 4: *Applications of harmonic analysis,* Fizmatgiz, Moscow, 1961; English transl., Academic Press, New York, 1964. MR **26** # 4173; **30** # 4152.
[11] E. V. Maĭkov, *τ-smooth functionals and integration in functional spaces,* Uspehi Mat. Nauk **18** (1963), no. 3 (111), 243-244. (Russian)
[12] ———, *τ-continuity and τ-differentiability of a functional,* Dokl. Akad. Nauk SSSR **155** (1964), 266-269 = Soviet Math. Dokl. **5** (1964), 388-393. MR **28** # 3317.
[13] ———, *τ-polynomials and τ-analytic functionals,* Dokl. Akad. Nauk SSSR **156** (1964), 1025-1028 = Soviet Math. Dokl. **5** (1964), 767-772. MR **29** # 2352.

Translated by:
L. W. Longdon

Trudy Moskov. Mat. Obšč.
Tom 20 (1969)

Trans. Moscow Math. Soc.
Vol. 20 (1969)

TENSOR PRODUCTS AND FUNCTORS IN CATEGORIES OF BANACH SPACES DEFINED BY *KB*-LINEALS*

V. L. LEVIN

Contents

INTRODUCTION 41
§1. Introductory information on tensor products and functors in categories of Banach spaces 43
§2. Properties of the tensor product $E \mathbin{\widetilde{\otimes}} X$ 51
§3. Properties of the functor Φ_E 62
BIBLIOGRAPHY 77

Introduction

Let M be a locally compact space with a Radon measure, X a Banach space. Then, as is well known (see [**2**]), the projective tensor product $L^1_M \mathbin{\hat{\otimes}} X$ is naturally isometric to the space $L^1_M(X)$ of all summable functions taking values in X. This result is also true for the space C_Q of functions continuous on a compactum Q, but with the weak tensor product instead of the projective tensor product; in other words, there is a natural isometry between the spaces $C_Q \mathbin{\hat{\hat{\otimes}}} X$ and $C_Q(X)$, where $C_Q(X)$ is the space of functions continuous on Q and taking values in X (see [**2**]). However, for other function spaces, in particular for L^p_M $(1 < p < \infty)$, this is no longer true for either of the two tensor products. In this paper[1)] we introduce a special norm n_E on the algebraic tensor product $E \otimes X$ of any KB-lineal E and Banach space X, which is a cross-norm in the sense of [**1**], and we study the completion $E \mathbin{\widetilde{\otimes}} X$ of $E \otimes X$ by means of n_E. The tensor product $E \mathbin{\widetilde{\otimes}} X$ is natural in the following sense: if E is a KB-lineal of functions, then in many cases (in particular, for $E = C_Q$ or L^p_M $(1 \leq p < \infty)$) $E \mathbin{\widetilde{\otimes}} X$ is isometric to the space $E(X)$ of "the same" functions taking values in X, with the norm $\| x(\cdot) \|_{E(X)} = \| \| x(t) \|_X \|_E$; if E is a KB-lineal of real sequences in which the finite sequences form an everywhere dense set, then $E \mathbin{\widetilde{\otimes}} X$ is isometric to the space $E(X)$ of sequences $(x_k) \subset X$ for which $(\| x_k \|) \in E$, with the norm $\| (x_k) \|_{E(X)} = \| (\| x_k \|_X) \|_E$. Further $\Phi_E(X) = E \mathbin{\widetilde{\otimes}} X$ is a functor on any category

* The present translation incorporates suggestions made by the author.

1) See also [**8**], [**9**].

of Banach spaces, the morphisms being all continuous linear mappings.[2] A. S. Švarc in [3], [4] initiated the study of functors in categories of Banach spaces. In his survey article [5] he investigated, in particular, functors defined by special spaces of sequences and measurable functions, the so-called normed ideals (typical examples of normed ideals of sequences and functions are, respectively, the spaces l^p and $L^p_{[0,1]}$ for $p \geqq 1$). It turns out that the properties of these functors are determined by the order properties of the corresponding normed ideals. Thus, in the category of reflexive spaces the ring $\{\mathfrak{N} \to \mathfrak{N}\}$ of natural transformations of the functor $\mathfrak{N}$, generated by the normed ideal $\mathfrak{N}$, into itself is isomorphic to the ring of regular (in the sense of Kantorovič) operators of this normed ideal; for normed ideals of sequences $\mathfrak{n}$ an analogous result was proved in [5] for any category of Banach spaces. A study of the functor Φ_E gives us an insight into the reasons for this.

This paper consists of three sections. §1 contains preliminary information on tensor products of Banach spaces and also some auxiliary propositions.

In §2 we introduce the tensor product $E \widetilde{\otimes} X$ and study its properties. We prove that $E \widetilde{\otimes} X$ is isometric to $E(X)$ when E is a minimal normed ideal of sequences $\mathfrak{n}$ or of measurable functions $\mathfrak{N}$, the KB-lineal C_Q, or the KB-lineal L^p_M $(p \geqq 1)$, where $M = (S, \Sigma, \mu)$ is any measure space (Theorem 1). As a corollary we obtain that $L^p_{M_1} \widetilde{\otimes} L^p_{M_2} = L^p_{M_1 \times M_2}$ for any spaces M_1 and M_2 with σ-finite measures and any $p \geqq 1$. A complete description is given of when n_E coincides with the strong cross-norm γ and the weak cross-norm λ (Theorem 2). It turns out that $n_E = \gamma$ on $E \otimes X$ for any X if and only if E is an $\mathscr{L}$-space (a KB-lineal with additive norm), and $n_E = \lambda$ on $E \otimes X$ for any X if and only if E is an $\mathscr{M}$-space (a KB-lineal satisfying the condition $\|\sup(e_1, e_2)\| = \max(\|e_1\|, \|e_2\|)$ for any positive pair e_1, e_2). In §2 we also prove that the adjoint space $(E \widetilde{\otimes} X)'$ is isometric to the space $S(E, X')$ of summable mappings of E into X', endowed with a special norm (Theorem 3). A mapping $\alpha : E \to Y$ of a KB-lineal E to a Banach space Y is said to be summable if α maps any series $\sum_{k=1}^{\infty} e_k$ of E for which $\sum_{k=1}^{\infty} |e_k|$ is convergent to an absolutely convergent series $\sum_1^{\infty} \alpha e_k$ in Y. At the end of §2 we investigate the properties of permanence of the tensor product $E \widetilde{\otimes} X$ (Propositions 6 and 7 and Theorem 4). We prove that for any KB-lineals E and F of Banach spaces X and Y, the tensor product $\alpha \otimes \beta$ of a regular mapping $\alpha : E \to F$ and a continuous linear mapping $\beta : X \to Y$ extends to a continuous linear mapping of $E \widetilde{\otimes} X$ to $F \widetilde{\otimes} Y$. Further, if X is a closed subspace of Y, then $E \widetilde{\otimes} X$ is naturally isometric to a subspace of $E \widetilde{\otimes} Y$, as is $E \widetilde{\otimes} (Y/X)$ to $E \widetilde{\otimes} Y / E \widetilde{\otimes} X$.

§3 is devoted to the functor Φ_E. We prove that it is exact (Theorem 5); this result is apparently new even for $E = L^p$, $p > 1$. We consider the class of linear mappings $\alpha : X \to E$ that map the unit sphere into a bounded subset

[2] All spaces considered are assumed to be real. The action of Φ_E on the morphisms is defined in a natural way. That is, $\Phi_E(\alpha) = 1_E \otimes \alpha$ for any $\alpha : X \to Y$.

of E (with respect to the ordering); we call such mappings proper (when E is a K^+-space, they coincide with the *bo*-linear mappings of Kantorovič [**6**], [**7**]). A special norm is introduced on the space of proper linear mappings and we prove, under certain simple conditions on E, which are satisfied by most standard KB-lineals, that $\Phi_E(X') = E \widetilde{\otimes} X'$ is isometric to the closure of the set of finite-dimensional mappings $\alpha: X \to E$ in the space of all proper mappings; in addition, we consider the case when $\Phi_E(X')$ is isometric to the space of all proper mappings of X to E (Theorem 6). We show that the adjoint $\alpha^*: E' \to X'$ of a proper mapping $\alpha: X \to E$ is summable (Proposition 9). We characterize the space $\{\Phi_E \to \Phi_F\}$ of all natural transformations of Φ_E to Φ_F, when these functors are considered for a wide class of categories, including the categories of all reflexive and all Banach spaces (Theorems 7 and 8). Under simple conditions on E and F, which are satisfied by most standard KB-lineals, we establish an isometry between $\{\Phi_E \to \Phi_F\}$ and the space of regular mappings $\alpha: E \to F$ with the norm $\nu(\alpha) = \| \, |\alpha| \, \|$. By this result we obtain an affirmative answer to a question posed by Mitjagin and Švarc [**5**] on the ring of operators of the functor L^p $(p > 1)$. With a few further assumptions on E and F, which are also satisfied in the majority of standard cases, we obtain an isometry between $\{\Phi_E \to \Phi_F\}$ and the space of linear mappings of E to F, endowed with a special norm, which map any sequence in E that is bounded with respect to the ordering to a similar sequence in F. We characterize the functor $D\Phi_E$ dual to Φ_E in the sense of Švarc [**3**]; for any X, $D\Phi_E(X)$ is naturally isometric to $S(E, X)$ (Theorem 9). For functors defined by minimal normed ideals of functions this completely solves the problem raised in [**5**] of determining the dual functor in the case of the category of all Banach spaces. In addition, we show that[3)] in this category $D\mathfrak{N} \neq \mathfrak{N}^*$ for any $\mathfrak{N}$, so that the conjecture in [**5**] on the structure of the dual functor for the category of all Banach spaces is false (Proposition 11). In conclusion we prove a theorem (Theorem 10) on the nuclearity of certain product mappings $X \to E \to Y$ which includes as a special case the Hilbert-Schmidt theorem on the nuclearity of the product of two mappings. From Theorem 10 we derive, in particular, that the product $\beta\alpha$ of a completely continuous linear mapping $\alpha: X \to C$ and an absolutely summable mapping $\beta: C \to Y$ is nuclear for any two Banach spaces X and Y.

§1. Introductory information on tensor products and functors in categories of Banach spaces

We begin with some definitions and results from the theory of tensor products. Our exposition follows that in [**1**] and [**2**].

Let X and Y be Banach spaces. Following Schatten in [**1**] we say that a norm p defined on the algebraic tensor product $X \otimes Y$ is a *cross-norm* if it

[3)] $\mathfrak{N}^*$ is the functor generated by the normed ideal $\mathfrak{N}^*$ associated with $\mathfrak{N}$ [**5**].

satisfies $p(x \otimes y) = \|x\|_X \|y\|_Y$ for any $x \in X$, $y \in Y$. We denote the space $X \otimes Y$ with a cross-norm p by $X \otimes_p Y$, and its completion by $X \overline{\otimes}_p Y$.

There exists a *strongest cross-norm* γ; this is defined by

$$\gamma(z) = \inf \sum_{k=1}^{n} \|x_k\|_X \cdot \|y_k\|_Y,$$

where the lower bound is taken over all possible representations of $z \in X \otimes Y$ in the form $z = \sum_{k=1}^{n} x_k \otimes y_k$. The completion of $X \otimes_\gamma Y$ is called the *projective tensor product* of X and Y and is denoted by $X \hat{\otimes} Y$.

The spaces $X \otimes Y$ and $X' \otimes Y'$ form a dual pair with respect to the bilinear form

$$\left\langle \sum_{k=1}^{n} x_k \otimes y_k, \sum_{j=1}^{m} x_j' \otimes y_j' \right\rangle = \sum_{k=1}^{n} \sum_{j=1}^{m} \langle x_k, x_j' \rangle \langle y_k, y_j' \rangle .$$

Let p be a cross-norm on $X \otimes Y$; if for every $z' = \sum_{j=1}^{m} x_j' \otimes y_j' \in X' \otimes Y'$

$$p'(z') = \sup_{p(z) \leqq 1} |\langle z, z' \rangle| < \infty,$$

then p' is a norm on $X' \otimes Y'$ and is called the norm *associated with* p.

The weak cross-norm λ on $X \otimes Y$ is defined by

$$\lambda\left(\sum_{k=1}^{n} x_k \otimes y_k\right) = \sup_{\substack{\|x'\|_{X'} \leqq 1 \\ \|y'\|_{Y'} \leqq 1}} \left| \sum_{k=1}^{n} \langle x_k, x' \rangle \langle y_k, y' \rangle \right|;$$

the completion of $X \otimes_\lambda Y$ is called the *weak tensor product* of X with Y and is denoted by $X \hat{\hat{\otimes}} Y$. We note the following result.

SCHATTEN'S THEOREM ([**1**], p. 32). *Let p be a cross-norm on $X \otimes Y$. A necessary and sufficient condition for the existence of a cross-norm p' on $X' \otimes Y'$ associated with p is that $p \geqq \lambda$.*

Let p be a cross-norm on $X \otimes Y$. The elements of the space $(X \overline{\otimes}_p Y)'$ adjoint to $X \overline{\otimes}_p Y$ may be identified with the continuous linear transformations of X to Y'; more precisely, to every functional $f \in (X \overline{\otimes}_p Y)'$ there corresponds a mapping $L_f: X \to Y'$ defined by $\langle y, L_f x \rangle = \langle x \otimes y, f \rangle$. In addition,

$$\|L_f\|_{(X \overline{\otimes}_p Y)'} = \inf\{c : |\langle z, f \rangle| \leqq cp(z) \text{ for all } z \in X \otimes Y\}.$$

For $p \geqq \lambda$, $X' \overline{\otimes}_{p'} Y'$ is isometrically embedded in $(X \overline{\otimes}_p Y)'$; it is the closure in $(X \overline{\otimes}_p Y)'$ of the set of finite-dimensional mappings. The space $(X \hat{\otimes} Y)'$ consists of all continuous linear mappings of X into Y' and its norm is the usual operator norm.

The identity mapping of $X \otimes Y$ into itself extends to a continuous linear mapping $\omega: X \hat{\otimes} Y \to X \hat{\hat{\otimes}} Y$. We say that X satisfies *the one-to-one condition* if ω is one-to-one for any Banach space Y [**2**]. The following result holds.

GROTHENDIECK'S THEOREM ([2], CHAPTER 1, §5, pp. 164-165). *The following assertions are equivalent:*

1) *X satisfies the one-to-one condition.*

2) *Any completely continuous linear transformation of any Banach space Y into X is the limit with respect to the operator norm of a sequence of finite-dimensional linear mappings of Y into X.*

The tensor product $X' \otimes Y$ can be regarded as the space of all continuous finite-dimensional linear mappings of X into Y, by identifying $\sum_{k=1}^{n} x_k' \otimes y_k \in X' \otimes Y$ with the mapping $\sum_{k=1}^{n} \langle\, , x_k' \rangle y_k$. Under this correspondence the weak cross-norm λ corresponds to the operator norm, and $X' \hat{\hat{\otimes}} Y$ is isometric to the closure of the set of finite-dimensional linear mappings in the space $(X \to Y)$ of all continuous linear mappings of X into Y. The image of $X' \hat{\otimes} Y$ under the mapping $\omega : X' \hat{\otimes} Y \to X' \hat{\hat{\otimes}} Y$ consists of the nuclear mappings of X into Y.

Let $\alpha : X \to X_1$ and $\beta : Y \to Y_1$ be continuous linear mappings. We let $\alpha \otimes \beta$ denote the mapping $X \otimes Y \to X_1 \otimes Y_1$ defined by

$$(\alpha \otimes \beta) \sum_{k=1}^{n} x_k \otimes y_k = \sum_{k=1}^{n} \alpha x_k \otimes \beta y_k.$$

If $\alpha \otimes \beta$ extends to a continuous linear mapping $X \overline{\otimes}_p Y \to X_1 \overline{\otimes}_{p_1} Y_1$, then this extended mapping is also denoted by $\alpha \otimes \beta$. For any continuous mappings $\alpha : X \to X_1$ and $\beta : Y \to Y_1$ their tensor product is

$$\alpha \otimes \beta \in (X \hat{\otimes} Y \to X_1 \hat{\otimes} Y_1) \quad \text{and} \quad \alpha \otimes \beta \in (X \hat{\hat{\otimes}} Y \to X_1 \hat{\hat{\otimes}} Y_1).$$

If $X_1 = X/X_0$, $Y_1 = Y/Y_0$ and $\alpha : X \to X_1$, $\beta : Y \to Y_1$ are the canonical homomorphisms, then $X_1 \hat{\otimes} Y_1$ is isometric to the factor space of $X \hat{\otimes} Y$ and $\alpha \otimes \beta$ is the canonical homomorphism of $X \hat{\otimes} Y$ onto $X_1 \hat{\otimes} Y_1$. If $X_1 \subset X$ and $Y_1 \subset Y$ are complemented subspaces and $\alpha : X_1 \to X$ and $\beta : Y_1 \to Y$ are the identity isometric embeddings, then $\alpha \otimes \beta : X_1 \hat{\otimes} Y_1 \to X \hat{\otimes} Y$ is an isometric embedding (for arbitrary spaces this is not true). If $X_1 \subset X$ and $Y_1 \subset Y$ are any closed subspaces and $\alpha : X_1 \to X$ and $\beta : Y_1 \to Y$ are the identity isometric embeddings, then $\alpha \otimes \beta : X_1 \hat{\hat{\otimes}} Y_1 \to X \hat{\hat{\otimes}} Y$ is an isometric embedding.

We now pass on to the connections between tensor products and functors in categories of Banach spaces.[4)]

As a preliminary we recall the definition of a functor of type Σ. For every functor $\mathscr{F}$ a mapping i of the functor $\Sigma_{\mathscr{F}(I)}$ to $\mathscr{F}$ is defined by:

$$i = \{ i_X : X \in \mathscr{K} \}, \quad i_X \left(\sum_{k=1}^{n} a_k \otimes x_k \right) = \sum_{k=1}^{n} \mathscr{F}(\hat{x}_k) a_k,$$

[4)] For functors in categories of Banach spaces see [5].

where $\mathscr{K}$ is the category of Banach spaces under consideration, I is the space of real numbers with the usual modulus as norm, $a_k \in \mathscr{F}(I)$, $x_k \in X$, and $\hat{x}$ the mapping of I to $X: \hat{x}\lambda = \lambda x$, $\lambda \in I$. By this formula i_X is defined on $\mathscr{F}(I) \otimes X$; i_X can be uniquely extended by continuity to act on $\Sigma_{\mathscr{F}(I)}(x) = \mathscr{F}(I) \hat{\otimes} X$. A functor $\mathscr{F}$ is said to be of type Σ if $i_X(\mathscr{F}(I) \otimes X)$ is dense in $\mathscr{F}(X)$ for any $X \in \mathscr{K}$.

Lemma 1. *Let $\mathscr{F}$ be a functor on the category $\mathscr{K}$. Then for any $X \in \mathscr{K}$, i_X maps $\mathscr{F}(I) \otimes X$ one-to-one into $\mathscr{F}(X)$.*

Proof. Suppose that for some $z = \sum_{k=1}^n a_k \otimes x_k \in \mathscr{F}(I) \otimes X$

$$i_X z = \sum_{k=1}^n \mathscr{F}(\hat{x}_k) a_k = 0.$$

To every $x' \in X'$ there corresponds a mapping $\mathscr{F}(x') : \mathscr{F}(X) \to \mathscr{F}(I)$. Applying this to $i_X z$ we obtain $\mathscr{F}(x') i_X z = \sum_{k=1}^n \mathscr{F}(x') \mathscr{F}(\hat{x}_k) a_k = 0$. Further, since $\mathscr{F}(x')\mathscr{F}(\hat{x}_k) = \mathscr{F}(x'\hat{x}_k) = \langle x_k, x' \rangle 1_{\mathscr{F}(I)}$, we see that $\sum_{k=1}^n \langle x_k, x' \rangle a_k = 0$ for any $x' \in X'$. Without loss of generality we may assume that the elements $x_1, \cdots, x_n$ are linearly independent. Choosing x' so that $\langle x_j, x' \rangle = 1$ and $\langle x_k, x' \rangle = 0$ for $k \neq j$, we get that $a_j = 0$. Since this is true for $j = 1, \cdots, n$, we have $z = 0$ and the lemma is proved.

Thus we can assume that $\mathscr{F}(I) \otimes X$ is embedded in $\mathscr{F}(X)$. Accordingly we do not distinguish between the elements

$$z = \sum_{k=1}^n a_k \otimes x_k \in \mathscr{F}(I) \otimes X \quad \text{and} \quad i_X z = \sum_{k=1}^n \mathscr{F}(\hat{x}_k) a_k \in \mathscr{F}(X).$$

Proposition 1. *Let $\mathscr{F}$ be a functor of type Σ on some category $\mathscr{K}$. Then for any $X \in \mathscr{K}$, $\mathscr{F}(X)$ is isometric to the tensor product $\mathscr{F}(I) \overline{\otimes}_p X$, where p is a cross-norm and $p \geqq \lambda$. Further, $\mathscr{F}(\alpha) = 1_{\mathscr{F}(I)} \otimes \alpha$ for any $\alpha \in (X \to Y)$. If $\mathscr{G}$ is a functor on $\mathscr{K}$ and $\phi \in \{\mathscr{F} \to \mathscr{G}\}$, then $\phi_X = \phi_I \otimes 1_X$ for any $X \in \mathscr{K}$.*

Proof. By Lemma 1, $\mathscr{F}(I) \otimes X$ can be embedded in $\mathscr{F}(X)$, and since $\mathscr{F}$ is a functor of type Σ, it is everywhere dense there. We now show that the norm of $\mathscr{F}(X)$ is a cross-norm for $\mathscr{F}(I) \otimes X$. We have

$$\begin{aligned}\|a \otimes x\|_{\mathscr{F}(X)} = \|\mathscr{F}(\hat{x}) a\|_{\mathscr{F}(X)} &\leqq \|\mathscr{F}(\hat{x})\|_{(\mathscr{F}(I) \to \mathscr{F}(X))} \|a\| \\ &\leqq \|\hat{x}\|_{(I \to X)} \|a\| = \|a\| \, \|x\|.\end{aligned}$$

Further, from

$$\left\langle \sum_{k=1}^n a_k \otimes x_k, a' \otimes x' \right\rangle = \left\langle \mathscr{F}(x') \sum_{k=1}^n \mathscr{F}(\hat{x}_k) a_k, a' \right\rangle$$

it follows that $a' \otimes x' \in [\mathscr{F}(X)]'$ and

$$\|a' \otimes x'\|_{[\mathscr{F}(X)]'} = \|a' \circ \mathscr{F}(x')\|_{[\mathscr{F}(X)]'} \leqq \|a'\| \, \|\mathscr{F}(x')\|_{(\mathscr{F}(X) \to \mathscr{F}(I))} \leqq \|a'\| \cdot \|x'\|.$$

Then

$$\|a \otimes x\|_{\mathscr{F}(X)} = \sup_{g \in [\mathscr{F}(X)]', \|g\| \leqq 1} |\langle a \otimes x, g\rangle| \geqq \sup_{\|a' \otimes x'\|[\mathscr{F}(X)]' \leqq 1} |\langle a \otimes x, a' \otimes x'\rangle|$$

$$\geqq \sup_{\|a'\| \cdot \|x'\| \leqq 1} |\langle a, a'\rangle \langle x, x'\rangle| = \|a\| \, \|x\|.$$

Thus $\|a \otimes x\|_{\mathscr{F}(X)} = \|a\| \, \|x\|$, so that the norm on $\mathscr{F}(X)$ is a cross-norm on $\mathscr{F}(I) \otimes X$; we denote this by p. Then $\mathscr{F}(X) = \mathscr{F}(I) \overline{\otimes}_p X$. Further,

$$p\left(\sum_{k=1}^n a_k \otimes x_k\right) = \sup_{g \in [\mathscr{F}(X)]', \|g\| \leqq 1} \left|\left\langle \sum_{k=1}^n a_k \otimes x_k, g\right\rangle\right|$$

$$\geqq \sup_{\|a' \otimes x'\|[\mathscr{F}(X)]' \leqq 1} \left|\left\langle \sum_{k=1}^n a_k \otimes x_k, a' \otimes x'\right\rangle\right|$$

$$\geqq \sup_{\|a'\| \cdot \|x'\| \leqq 1} \left|\sum_{k=1}^n \langle a_k, a'\rangle \langle x_k, x'\rangle\right| = \lambda\left(\sum_{k=1}^n a_k \otimes x_k\right).$$

Now let $\alpha \in (X \to Y)$, $X, Y \in \mathscr{K}$. We have

$$\mathscr{F}(\alpha) \sum_{k=1}^n a_k \otimes x_k = \mathscr{F}(\alpha) \sum_{k=1}^n \mathscr{F}(\hat{x}_k) a_k = \sum_{k=1}^n \mathscr{F}(\alpha \hat{x}_k) a_k$$

$$= \sum_{k=1}^n \mathscr{F}(\alpha \hat{x}_k) a_k = \sum_{k=1}^n a_k \otimes \alpha x_k = (1_{\mathscr{F}(I)} \otimes \alpha) \sum_{k=1}^n a_k \otimes x_k,$$

that is, $\mathscr{F}(\alpha) = 1_{\mathscr{F}(I)} \otimes \alpha$. Further, let $\mathscr{G}$ be a functor on $\mathscr{K}$ and $\phi \in \{\mathscr{F} \to \mathscr{G}\}$. From the commutative diagram

$$\begin{array}{ccc} \mathscr{F}(I) & \xrightarrow{\ \phi_I\ } & \mathscr{G}(I) \\ \downarrow{\scriptstyle \mathscr{F}(\hat{x})} & & \downarrow{\scriptstyle \mathscr{G}(\hat{x})} \\ \mathscr{F}(X) & \xrightarrow{\ \phi_X\ } & \mathscr{G}(X) \end{array}$$

we see that

$$\phi_X\left(\sum_{k=1}^n a_k \otimes x_k\right) = \phi_X\left(\sum_{k=1}^n \mathscr{F}(\hat{x}_k) a_k\right) = \sum_{k=1}^n \phi_X \mathscr{F}(\hat{x}_k) a_k$$

$$= \sum_{k=1}^n \mathscr{G}(\hat{x}_k) \phi_I a_k = \sum_{k=1}^n \phi_I a_k \otimes x_k = (\phi_I \otimes 1_X) \sum_{k=1}^n a_k \otimes x_k;$$

that is, $\phi_X = \phi_I \otimes 1_X$. The proposition is proved.

COROLLARY. *The class $\{\mathscr{F} \to \mathscr{G}\}$ of all mappings from the functor $\mathscr{F}$ of type Σ to any functor $\mathscr{G}$ is a set, namely the subset of $(\mathscr{F}(I) \to \mathscr{G}(I))$ formed by those operators ϕ for which $\phi \otimes 1_X \in (\mathscr{F}(X) \to \mathscr{G}(X))$ for any $X \in \mathscr{K}$. In particular, the dual functor is defined for every functor of type Σ.*

Let $\mathscr{F}$ be a functor of type Σ on the category $\mathscr{K}$. For any $X \in \mathscr{K}$ the identity mapping of $\mathscr{F}(I) \otimes X$ onto itself extends by continuity to a mapping $k_X^{\mathscr{F}}: \mathscr{F}(X) \to \mathscr{F}(I) \hat{\hat{\otimes}} X$. We say that the functor *satisfies the one-to-one condition* if $k_X^{\mathscr{F}}$ is a one-to-one mapping for every $X \in \mathscr{K}$. For the functor Σ_A this condition reduces to the usual one-to-one condition for the space A.

Later we shall need the following lemma.

LEMMA 2. *Let $\mathscr{F}$ and $\mathscr{G}$ be functors of type Σ on some category $\mathscr{K}$, and let $\phi \in \{\mathscr{F} \to \mathscr{G}\}$. Then for any $X \in \mathscr{K}$ the following diagram is commutative*:

$$\begin{array}{ccc} \mathscr{F}(X) & \xrightarrow{\phi_I \otimes 1_X} & \mathscr{G}(X) \\ \downarrow k_X^{\mathscr{F}} & & \downarrow k_X^{\mathscr{G}} \\ \mathscr{F}(I) \hat{\hat{\otimes}} X & \xrightarrow{\phi_I \otimes 1_X} & \mathscr{G}(I) \hat{\hat{\otimes}} X \end{array}$$

PROOF. By the continuity of the mappings $k_X^{\mathscr{G}}(\phi_I \otimes 1_X)$ and $(\phi_I \otimes 1_X) k_X^{\mathscr{F}}$ it is sufficient to prove that they agree on the dense subspace $\mathscr{F}(I) \otimes X \subset \mathscr{F}(X)$; but this is obvious.

We need some concepts and results from the theory of partially ordered spaces. All results required can be found in [**6**] and [**7**]. We note that we use the definition of KB-lineal adopted in [**7**], and by a *conditionally complete* (respectively, *conditionally σ-complete*) KB-lineal we understand a KB-lineal that is a K-space (respectively, K_σ-space) if it is regarded as a K-lineal.

A norm on a KB-lineal E is said to be *additive* if

$$\|e_1 + e_2\| = \|e_1\| + \|e_2\| \text{ for any positive } e_1, e_2 \in E;$$

a KB-lineal with additive norm is said to be an *$\mathscr{L}$-space*. Typical examples are the spaces l^1 and L^1.

A KB-lineal E is said to be an *$\mathscr{M}$-space* if its norm satisfies the condition $\|\sup(e_1, e_2)\| = \max(\|e_1\|, \|e_2\|)$ for any positive $e_1, e_2 \in E$. Typical examples of μ-spaces are the spaces c_0, m and C.

In §2 we need Kakutani's theorems on the representation of $\mathscr{L}$- and $\mathscr{M}$-spaces. Before we state these theorems we define the KB-lineal $C[Q, (t_\alpha), (t'_\alpha), (\lambda_\alpha), A]$. Let Q be a compactum, A a set, $t_\alpha, t'_\alpha \in Q$, $0 \leqq \lambda_\alpha < 1$, where α runs over A; the KB-lineal $C[Q, (t_\alpha), (t'_\alpha), (\lambda_\alpha), A]$ consists of all continuous real functions on Q satisfying the condition $\phi(t_\alpha) = \lambda_\alpha \phi(t'_\alpha)$ for all $\alpha \in A$; its norm is defined by $\|\phi\| = \max_{t \in Q} |\phi(t)|$. When the set A is empty, we obtain the KB-lineal C_Q of all continuous real functions on Q.

KAKUTANI'S THEOREM I [**10**]. *Every $\mathscr{L}$-space is isometric and lattice isomorphic to the KB-lineal L_M^1, where $M = (S, \Sigma, \mu)$ is some measure space.*

KAKUTANI'S THEOREM II [**11**]. *Every $\mathscr{M}$-space is isometric and lattice isomorphic to a KB-lineal $C[Q, (t_\alpha), (t'_\alpha), (\lambda_\alpha), A]$.*

Normed ideals of sequences and measurable functions studied in [**5**] form a wide class of KB-lineals. Another example are the spaces L_M^p of functions summable to the pth power defined on a measure space $M = (S, \Sigma, \mu)$. The definition of the spaces L_M^p and also the corresponding spaces of vector functions $L_M^p(X)$ can be found in [**12**].

In §2 we need the following proposition,[5] which is of independent interest.

PROPOSITION 2. *For any Banach space X the space $L_M^1 \hat{\otimes} X$ is isometric to $L_M^1(X)$.*

PROOF. We consider the mapping $i_X: L_M^1 \hat{\otimes} X \to L_M^1(X)$ and show that i_X is an isometry. Let E_0 denote the subspace of integrable simple[6] functions. Then $i_X(E_0 \otimes X)$ is a subspace of integrable simple functions with values in X. Since $E_0 \otimes X$ is dense in $L_M^1 \hat{\otimes} X$ and since $i_X(E_0 \otimes X)$ is dense in $L_M^1(X)$, it is sufficient to show that $\gamma(z) = \| i_X z \|_{L_M^1(X)}$ for $z \in E_0 \otimes X$. Let $z = \sum_{k=1}^n f_k \otimes x_k$, where $f_k(s) = a_{ki}$ for $s \in Q_{ki}$, $i = 1, \cdots, m_k$, $Q_{ki} \in \Sigma$, $\bigcup_{i=1}^{m_k} Q_{ki} = S$ and $a_{ki} = 0$ for those i for which $\mu Q_{ki} = \infty$. We consider the set J of all possible choices of n indices $(i_1, \cdots, i_n)$, where $1 \leqq i_k \leqq m_k$, $k = 1, \cdots, n$, and the subset J_0 consisting of those sets of indices $(i_1, \cdots, i_n)$ for which $\bigcap_{k=1}^n Q_{ki_k}$ is nonempty and $\mu Q_{ki_k} < \infty$ for at least one k. We suppose that J_0 consists of m elements and number these with the natural numbers from 1 to m. For any $1 \leqq j \leqq m$ we put $P_j = \bigcap_{k=1}^n Q_{ki_k}(j)$ and $y_j = \sum_{k=1}^n a_{ki_k(j)} x_k$, where $i_k(j)$ is the kth member of $(i_1, \cdots, i_n) \in J_0$ with the suffix j. Clearly $P_j \in \Sigma$ and $\mu P_j < \infty$, so that[7] $\chi_{P_j} \in L_M^1$, $j = 1, \cdots, m$. Since $(i_1(j_1), \cdots, i_n(j_1)) \neq (i_1(j_2), \cdots, i_n(j_2))$ for $j_1 \neq j_2$, there exists k such that $i_k(j_1) \neq i_k(j_2)$. Thus P_{j_1} and P_{j_2} are disjoint, because they are contained in the disjoint sets $Q_{ki_k(j_1)}$ and $Q_{ki_k(j_2)}$. We consider the sets of indices

$$N_k = \{ i; \mu Q_{ki} < \infty \}, \quad A_{ki} = \{ j; i_k(j) = i \} = \{ j; P_j \subset Q_{ki} \}, \quad A_k = \bigcup_{i \in N_k} A_{ki},$$
$$i = 1, \cdots, m_k; \quad k = 1, \cdots, n.$$

We show that $Q_{ki} = \bigcup_{j \in A_{ki}} P_j$ for any Q_{ki} with finite measure. It is sufficient to show that $Q_{ki} \subset \bigcup_{j \in A_{ki}} P_j$, since the reverse inclusion is obvious. Let $s \in Q_{ki}$. For every $1 \leqq l \leqq n$, $l \neq k$, there exists a set Q_{li_l} containing s. Since $\mu Q_{ki} < \infty$, $(i_1, \cdots, i_{k-1}, i, i_{k+1}, \cdots, i_n) \in J_0$; suppose that it has the index j. Then $s \in P_j$ and, since $j \in A_{ki}$, the inclusion is proved. Thus $Q_{ki} = \bigcup_{j \in A_{ki}} P_j$. Hence distinct sets P_j are disjoint, and we obtain $\chi_{Q_{ki}} = \sum_{j \in A_{ki}} \chi_{P_j}$ for any Q_{ki} with finite measure.

---

[5] Grothendieck proved this proposition in the particular case when μ is a Radon measure on a locally compact space S (see [**2**], Chapter 1, §2, p. 59, Theorem 2). His proof does not carry over to the general case.

[6] A simple function is one that takes a finite set of values.

[7] χ_P denotes the characteristic function of a set P.

We have, further,

$$z = \sum_{k=1}^{n} f_k \otimes x_k = \sum_{k=1}^{n} \sum_{i \in N_k} a_{ki} \chi_{Q_{ki}} \otimes x_k = \sum_{k=1}^{n} \sum_{i \in N_k} \chi_{Q_{ki}} \otimes a_{ki} x_k$$

$$= \sum_{k=1}^{n} \sum_{i \in N_k} \sum_{j \in A_{ki}} \chi_{P_j} \otimes a_{ki} x_k = \sum_{k=1}^{n} \sum_{j \in A_k} \chi_{P_j} \otimes a_{ki_k(j)} x_k$$

$$= \sum_{j=1}^{m} \chi_{P_j} \otimes \left[\sum_{k=1}^{n} a_{ki\ (j)} x_k \right] = \sum_{j=1}^{m} \chi_{P_j} \otimes y_j .$$

Then

$$\gamma(z) \leqq \sum_{j=1}^{m} \| \chi_{P_j} \|_{L_M^1} \cdot \| y_j \| = \sum_{j=1}^{m} \mu P_j \cdot \| y_j \| = \int_S \left(\sum_{j=1}^{m} \chi_{P_j}(s) \| y_j \| \right) d\mu$$

$$= \int_S \left\| \sum_{j=1}^{m} \chi_{P_j}(s) y_j \right\| d\mu = \| i_X z \|_{L_M^1(X)} .$$

On the other hand,

$$\| i_X z \|_{L_M^1(X)} = \int_S \left\| \sum_{k=1}^{n} f_k(s) x_k \right\|_X d\mu \leqq \sum_{k=1}^{n} \int_S | f_k(s) | d\mu \cdot \| x_k \|$$

$$= \sum_{k=1}^{n} \| f_k \|_{L_M^1} \cdot \| x_k \| ,$$

and since this is true for any representation of $z = \sum_{k=1}^{n} f_k \otimes x_k$, $f_k \in L_M^1$, $x_k \in X$, we can take the greatest lower bound over all such representations on the right-hand side of the inequality. We obtain $\| i_X z \|_{L_M^1(X)} \leqq \gamma(z)$. Thus $\gamma(z) = \| i_X z \|_{L_M^1(X)}$, and the proposition is proved.

To conclude this section we consider some further conditions on a *KB*-lineal *E*.

A) If[8] $e_n \uparrow \infty$ and $e_n \geqq 0$, then $\| e_n \| \to \infty$ as $n \to \infty$.

B) If the sequence $e_n \geqq 0$ is monotonic increasing and is bounded above, then for every $\delta > 0$ there exists $e_\delta \in E$ such that $e_\delta \geqq e_n$ for all $n = 1, 2, \cdots$ and $\| e_\delta \| \leqq \lim_{n \to \infty} \| e_n \| + \delta$.

Condition A is satisfied by *KB*-spaces (in particular, L_M^p, $p \geqq 1$), the *KB*-lineal C_Q of continuous functions on a compactum Q, and maximal normed ideals of measurable functions (see Proposition 9 of [**5**]).

The space c_0 does not satisfy condition A.

Condition B is satisfied by *KB*-spaces, the *KB*-lineal C_Q, and normed ideals of sequences and functions.

When *E* is a conditionally σ-complete *KB*-lineal, B is equivalent to the following condition:

---

[8] The symbol $e_n \uparrow \infty$ means that the sequence e_n is monotonic increasing and unbounded above in the sense of the ordering.

B′) If $e_n \uparrow e$ and $e_n \geqq 0$, then $\|e_n\| \to \|e\|$ as $n \to \infty$.

For by B we have $e \leqq e_\delta$ for any $\delta > 0$; hence $\|e\| \leqq \inf_{\delta>0}\|e_\delta\| = \lim_{n\to\infty}\|e_n\|$. Then $\|e\| = \lim_{n\to\infty}\|e_n\|$, so that B implies B′ for any KB-lineal E. When E is a conditionally σ-complete KB-lineal, by condition B′ any sequence (e_n) has a least upper bound e, which can be taken to be e_δ for any $\delta > 0$, so that B′ implies B.

§2. Properties of the tensor product $E \tilde{\otimes} X$

Let E be a KB-lineal and X a Banach space. For any $z = \sum_{k=1}^n e_k \otimes x_k \in E \otimes X$ put

$$n_E(z) = \inf\left\{ \|u\|_E : u \geqq \left| \sum_{k=1}^n e_k \langle x_k, x' \rangle \right| \text{ for all } x' \in X', \|x'\| \leqq 1 \right\}.$$

Lemma 3. *n_E is a cross-norm on $E \otimes X$, and $n_E \geqq \lambda$.*

Lemma 4. *Let $z = \sum_{k=1}^n e_k \otimes x_k \in E \otimes X$ and suppose that the set*

$$\left\{ \left| \sum_{k=1}^n e_k \langle x_k, x' \rangle \right| : x' \in X', \|x'\| \leqq 1 \right\} \subset E$$

has an upper bound. Then

$$n_E(z) = \left\| \sup_{\|x'\| \leqq 1} \left| \sum_{k=1}^n e_k \langle x_k, x' \rangle \right| \right\|_E.$$

The proofs of Lemmas 3 and 4 are not difficult and we omit them.

Remark. The set $\{ |\sum_{k=1}^n e_k \langle x_k, x' \rangle| : x' \in X', \|x'\| \leqq 1 \}$ is bounded above by $\sum_{k=1}^n |e_k| \, \|x_k\| \in E$; therefore when E is a conditionally dense KB-lineal, we have $n_E(z) = \| \sup_{\|x'\| \leqq 1} | \sum_{k=1}^n e_k \langle x_k, x' \rangle | \|_E$ for every

$$z = \sum_{k=1}^n e_k \otimes x_k \in E \otimes X.$$

Lemma 5. *Let E be a conditionally σ-dense KB-lineal on the space C_Q. Then for any $z = \sum_{k=1}^n e_k \otimes x_k \in E \otimes X$ there exists*

$$\sup_{\|x'\| \leqq 1} \left| \sum_{k=1}^n e_k \langle x_k, x' \rangle \right| \in E$$

and consequently

$$n_E(z) = \left\| \sup_{\|x'\| \leqq 1} \left| \sum_{k=1}^n e_k \langle x_k, x' \rangle \right| \right\|_E.$$

Proof. To begin with let E be a conditionally σ-dense KB-lineal. We take $z = \sum_{k=1}^n e_k \otimes x_k \in E \otimes X$ and consider the subspace X_n of X spanned by the vectors $x_1, \cdots, x_n$. X_n is a finite-dimensional space; hence there exists a countable set $\{ f_j, j = 1, 2, \cdots \}$ that is dense in the unit sphere of the adjoint space X_n'. We extend each of the functionals f_j $(j = 1, 2, \cdots)$ to the whole of X with preservation of the norm. Let x_j' denote the extension of f_j to X,

$j=1,2,\cdots$. The countable set $\{|\sum_{k=1}^n e_k\langle x_k, x'_j\rangle|,\ j=1,2,\cdots\}$ is bounded above by $\sum_{k=1}^n |e_k|\ \|x_k\|$; therefore it has a least upper bound

$$u_0 = \sup_j \left| \sum_{k=1}^n e_k\langle x_k, x'_j\rangle \right| \in E.$$

Let $x' \in X'$ and $\|x'\| \leqq 1$. Denote by f the restriction of x' to X_n and choose a sequence f_{j_m} convergent to f. Then $\langle x_k, x'_{j_m}\rangle \to \langle x_k, x'\rangle$ as $m\to\infty$ for all $k=1,\cdots,n$ and $|\sum_{k=1}^n e_k\langle x_k, x'_{j_m}\rangle| \to |\sum_{k=1}^n e_k\langle x_k, x'\rangle|$ as $m\to\infty$. Since $u_0 \geqq |\sum_{k=1}^n e_k\langle x_k, x'_{j_m}\rangle|$, then $u_0 \geqq |\sum_{k=1}^n e_k\langle x_k, x'\rangle|$. On the other hand, if $u \geqq |\sum_{k=1}^n e_k\langle x_k, x'\rangle|$ for all $x'\in X'$, $\|x'\|\leqq 1$, then $u \geqq u_0$. We have also proved that $u_0 = \sup_{\|x'\|\leqq 1} |\sum_{k=1}^n e_k\langle x_k, x'\rangle|$.

Now let $E = C_Q$. Since the least upper bound in C_Q is defined pointwise, it is sufficient to prove that $\sup_{\|x'\|\leqq 1} |\sum_{k=1}^n e_k(t)\,\langle x_k, x'\rangle|$ is a continuous function of $t\in Q$ for any $z=\sum_{k=1}^n e_k(t)\otimes x_k \in C_Q\otimes X$. This follows from the equality $\sup_{\|x'\|\leqq 1} |\sum_{k=1}^n e_k(t)\,\langle x_k, x'\rangle| = \|\sum_{k=1}^n e_k(t)\, x_k\|_X$.

THEOREM 1. *If E is a minimal normed ideal of sequences $\mathfrak{n}$, of measurable functions $\mathfrak{N}$, of the KB-lineal L_M^p $(p\geqq 1)$, or the KB-lineal C_Q, then $E\widetilde{\otimes} X$ is isometric to the space $\mathfrak{n}(X)$, $\mathfrak{N}(X)$, $L_M^p(X)$, or $C_Q(X)$, respectively. For any normed ideal $\mathfrak{N}$, the tensor product $\mathfrak{N}\widetilde{\otimes} X$ is isometric to the closure in $\mathfrak{N}(X)$ of the set of functions of the form*

$$\sum_{k=1}^n \xi_k(t)\, x_k, \quad \textit{where} \quad \xi_k \in \mathfrak{N},\ x_k\in X, \quad k=1,\cdots,n.$$

PROOF. By Lemma 5, in the cases under discussion there exists $u_0 = \sup_{\|x'\|\leqq 1} |\sum_{k=1}^m e_k\langle x_k, x'\rangle| \in E$ for any $z=\sum_{k=1}^m e_k\otimes x_k \in E\otimes X$, and $n_E(z) = \|u_0\|_E$. We consider the mapping $i_X\colon E\otimes_{n_E} X \to E(X)$ $(E=\mathfrak{n},\mathfrak{N}, L_M^p, C_Q)$ and show that it isometrically embeds the first space in the second.

When $E=\mathfrak{n}$ $(e_k=(e_{kj})_j,\ k=1,\cdots,m)$, we have

$$i_X\left(\sum_{k=1}^m e_k\otimes x_k\right) = \sum_{k=1}^m n(\hat{x}_k)\, e_k = \left(\sum_{k=1}^m e_{kj}\, x_k\right)_j,$$

and since the least upper bound in a KB-lineal of sequences is defined coordinate-wise, we have

$$\begin{aligned} n_E\left(\sum_{k=1}^m e_k\otimes x_k\right) &= \left\| \sup_{\|x'\|\leqq 1} \left| \sum_{k=1}^m e_k\langle x_k, x'\rangle \right| \right\|_{\mathfrak{n}} \\ &= \left\| \left(\sup_{\|x'\|\leqq 1} \left| \left\langle \sum_{k=1}^m e_{kj}\, x_k, x' \right\rangle \right| \right)_j \right\|_{\mathfrak{n}} = \left\| \left(\left\| \sum_{k=1}^m e_{kj}\, x_k \right\|_X \right)_j \right\|_{\mathfrak{n}} \\ &= \left\| i_X\left(\sum_{k=1}^m e_k\otimes x_k\right) \right\|_{\mathfrak{n}(X)}. \end{aligned}$$

In the remaining cases $e_k = e_k(t)$ $(k=1,\cdots,m)$ we have $i_X(\sum_{k=1}^m e_k\otimes x_k) =$

$\sum_{k=1}^{m} e_k(t)\,x_k$, and since the upper bound in a KB-lineal of functions is defined pointwise, we have

$$\begin{aligned} n_E\left(\sum_{k=1}^{m} e_k \otimes x_k\right) &= \left\| \sup_{\|x'\| \leqq 1} \left| \sum_{k=1}^{m} e_k \langle x_k, x' \rangle \right| \right\|_E \\ &= \left\| \sup_{\|x'\| \leqq 1} \left| \left\langle \sum_{k=1}^{m} e_k(t)\,x_k, x' \right\rangle \right| \right\|_E = \left\| \left\| \sum_{k=1}^{m} e_k(t)\,x_k \right\|_X \right\|_E \\ &= \left\| i_X\left(\sum_{k=1}^{m} e_k \otimes x_k\right) \right\|_{E(X)} \qquad (E = \mathfrak{N}, L_M^p, C_Q). \end{aligned}$$

Since the functors $\mathfrak{n}$ and $\mathfrak{N}$ (for minimal normed ideals) and also L_M^p and C_Q are of type Σ, $i_X\ (E \otimes X)$ is dense in the corresponding $E(X)$, and the first part of the theorem is proved.

In the case of a normed ideal $\mathfrak{N}$ the mapping i_X can be extended by continuity to an isometric embedding of $\mathfrak{N} \widetilde{\otimes} X$ in $\mathfrak{N}(X)$, and the second assertion of the theorem follows from this.

From Theorem 1 and Fubini's Theorem ([**12**], p. 208) we deduce the

COROLLARY. *Let $M_S = (S, \Sigma_S, \mu_S)$ and $M_T = (T, \Sigma_T, \mu_T)$ be spaces with σ-finite measures. Then $L_{M_S}^p \widetilde{\otimes} L_{M_T}^p = L_{M_S \times M_T}^p$, where the equality sign denotes an isometry.*

THEOREM 2. *Let E be a KB-lineal.*

1) *A necessary and sufficient condition for $n_E = \gamma$ on $E \otimes X$ for any Banach space X is that E should be an $\mathscr{L}$-space.*

2) *A necessary and sufficient condition for $n_E = \lambda$ on $E \otimes X$ for any Banach space X is that E should be an $\mathscr{M}$-space.*

PROOF. 1) To prove the necessity of the condition we take $X = l^1$ and consider the element $e_1 \otimes \delta_1 + e_2 \otimes \delta_2$ in $E \otimes l^1$, where $e_1, e_2 \geqq 0$, $\delta_1 = (1, 0, 0, \cdots)$ and $\delta_2 = (0, 1, 0, \cdots)$. We have $n_E\,(e_1 \otimes \delta_1 + e_2 \otimes \delta_2) = \inf\{\|u\| : u \geqq |\lambda_1 e_1 + \lambda_2 e_2|$ for all λ_i, $|\lambda_i| \leqq 1$, $i = 1, 2\} = \|e_1 + e_2\|$, and $\gamma(e_1 \otimes \delta_1 + e_2 \otimes \delta_2) = \|e_1\| + \|e_2\|$, and hence $\|e_1 + e_2\| = \|e_1\| + \|e_2\|$. Since e_1 and e_2 are arbitrary positive elements of E, we see that E is an $\mathscr{L}$-space.

The sufficiency follows immediately from Kakutani's Theorem I (see §1) and Proposition 2.

2) NECESSITY. Take $X = c_0$ and consider the element $e_1 \otimes \delta_1 + e_2 \otimes \delta_2$ in $E \otimes c_0$, where $e_1, e_2 \geqq 0$, $\delta_1 = (1, 0, 0, \cdots)$ and $\delta_2 = (0, 1, 0, \cdots)$. Then

$$\begin{aligned} n_E(e_1 \otimes \delta_1 + e_2 \otimes \delta_2) &= \inf\{\|u\| : u \geqq |\lambda_1 e_1 + \lambda_2 e_2| \\ &\qquad \text{for all } \lambda_1, \lambda_2, |\lambda_1| + |\lambda_2| \leqq 1\} = \|\sup(e_1, e_2)\|, \\ \lambda(e_1 \otimes \delta_1 + e_2 \otimes \delta_2) &= \sup_{\|e'\| \leqq 1, |\lambda_1| + |\lambda_2| \leqq 1} |\langle e_1, e' \rangle \lambda_1 + \langle e_2, e' \rangle \lambda_2| \\ &= \sup_{|\lambda_1| + |\lambda_2| \leqq 1} \|\lambda_1 e_1 + \lambda_2 e_2\|_E = \max(\|e_1\|, \|e_2\|), \end{aligned}$$

and hence

$$\| \sup(e_1, e_2) \| = \max(\| e_1 \|, \| e_2 \|).$$

Since e_1 and e_2 are arbitrary positive elements of E, E is an $\mathscr{M}$-space.

SUFFICIENCY. By Kakutani's Theorem II we may assume that $E = C[Q, (t_\alpha), (t'_\alpha), (\lambda_\alpha), A]$ (see §1). For any $z = \sum_{k=1}^n e_k \otimes x_k \in E \otimes X$ we consider the set $\{ | \sum_{k=1}^n e_k \langle x_k, x' \rangle | \in E : x' \in X', \| x' \| \leqq 1 \}$. By Lemma 5 this set has in C_Q a least upper bound $\phi(t) = \| \sum_{k=1}^n e_k(t) x_k \|_X$. Further, since

$$\phi(t_\alpha) = \left\| \sum_{k=1}^n e_k(t_\alpha) x_k \right\|_X = \left\| \sum_{k=1}^n \lambda_\alpha e_k(t'_\alpha) x_k \right\|_X = \lambda_\alpha \left\| \sum_{k=1}^n e_k(t'_\alpha) x_k \right\|_X = \lambda_\alpha \phi(t'_\alpha)$$

for all $\alpha \in A$, it follows that $\phi \in E$ and $\phi = \sup_{\| x' \| \leqq 1} | \sum_{k=1}^n e_k \langle x_k, x' \rangle |$. Then from Lemma 4 we obtain

$$\begin{aligned} n_E(z) &= \| \phi \|_E = \max_{t \in Q} | \phi(t) | = \max_{t \in Q} \sup_{\| x' \| \leqq 1} \left| \sum_{k=1}^n e_k(t) \langle x_k, x' \rangle \right| \\ &= \sup_{\| x' \| \leqq 1} \max_{t \in Q} \left| \sum_{k=1}^n e_k(t) \langle x_k, x' \rangle \right| = \sup_{\| x' \| \leqq 1} \left\| \sum_{k=1}^n e_k \langle x_k, x' \rangle \right\|_E = \lambda(z). \end{aligned}$$

We consider the following condition on the KB-lineal E:

C) For every Banach space X and any $z = \sum_{k=1}^n e_k \otimes x_k \in E \otimes X$

$$n_E(z) = \inf \left\| \sum_{k=1}^n | e_k | \, \| x_k \| \right\|_E,$$

where the greatest lower bound is taken over all possible representations of z in the form $z = \sum_{k=1}^n e_k \otimes x_k$.

PROPOSITION 3. *Let E be a KB-lineal, $E_0 \subset E$ a linear manifold dense in E and such that for any finite set of elements $e_1, \cdots, e_n \in E_0$ there exists a set of pairwise disjoint*[9] *elements $\phi_1, \cdots, \phi_{N_n} \in E_0$ such that $e_1, \cdots, e_n$ can be expressed linearly in terms of them. Then E satisfies condition* C.

PROOF. Obviously, for any KB-lineal E

$$n_E(z) \leqq \inf \left\| \sum_{k=1}^n | e_k | \cdot \| x_k \| \right\|_E,$$

where the greatest lower bound is taken over all possible presentations of $z = \sum_{k=1}^n e_k \otimes x_k$. Further, it is easy to see that the expression

$$\inf \left\| \sum_{k=1}^n | e_k | \cdot \| x_k \| \right\|_E$$

has all the properties of a norm on $E \otimes X$. We show that this norm coincides with n_E on $E_0 \otimes X$. Let $z = \sum_{k=1}^n e_k \otimes x_k$, $e_k \in E_0$, $(k = 1, \cdots, n)$. There exists

[9] For the definition of disjointness see [6] or [7].

a set of pairwise disjoint elements $\phi_1, \cdots, \phi_{N_n}$ such that $e_1, \cdots, e_n$ can be expressed linearly in terms of them. Let $e_k = \sum_{j=1}^{N_n} \lambda_{kj} \phi_j$, $k = 1, \cdots, n$. Then z can be written in the form $z = \sum_{j=1}^{N_n} \phi_j \otimes y_j$, where $y_j = \sum_{k=1}^{n} \lambda_{kj} x_k$. Using the disjointness of the elements ϕ_j ($j = 1, \cdots, N_n$) we see that for this representation

$$\left\| \sum_{j=1}^{N_n} |\phi_j| \, \|y_j\| \right\|_E = \left\| \sup_{1 \leq j \leq N_n} |\phi_j| \, \|y_j\| \right\|_E$$

$$\leq \inf \left\{ \|u\|_E : u \geq \left| \sum_{j=1}^{N_n} \phi_j \langle y_j, x' \rangle \right| \text{ for all } x' \in X', \|x'\| \leq 1 \right\} = n_E(z);$$

hence both norms coincide on $E_0 \otimes X$. Since $E_0 \otimes X$ is dense in $E \widetilde{\otimes} X$, the proposition is proved.

COROLLARY. *Minimal normed ideals of sequences* $\mathfrak{n}$, *the measurable functions* $\mathfrak{N}$, *and the KB-lineals* L_M^p ($p \geq 1$) *satisfy condition* C.

PROOF. When $E = \mathfrak{n}$, we can take for E_0 the set of finite sequences. For $E = \mathfrak{N}$ the linear manifold E_0 consists of step functions, and for $E = L_M^p$, of simple integrable functions.

PROPOSITION 4. *The KB-lineal* C_Q *satisfies condition* C.

PROOF. Let $z = \sum_{k=1}^{n} e_k \otimes x_k \in C_Q \otimes X$. Given $\epsilon > 0$, choose a finite covering of the compactum Q by neighborhoods U_j, $j = 1, \cdots, N$, such that $|e_k(t') - e_k(t'')| \leq \epsilon / \sum_{k=1}^{n} \|x_k\|$ for any $t', t'' \in U_j$ ($j = 1, \cdots, N$) and for all $k = 1, \cdots, n$. For this covering we choose a partition of the identity; that is, a system of functions $\phi_j \in C_Q$, $j = 1, \cdots, N$, satisfying the conditions

1) $\phi_j(t) \geq 0$ ($j = 1, \cdots, N$);
2) $\phi_j(t) = 0$ for $t \notin U_j$, $j = 1, \cdots, N$;
3) $\sum_{j=1}^{N} \phi_j(t) \equiv 1$.

We consider a representation of z in the form

$$z = \sum_{j=1}^{N} \phi_j \otimes \left[\sum_{k=1}^{n} e_k(t_j) x_k \right] + \sum_{j=1}^{N} \sum_{k=1}^{n} \phi_j (e_k - e_k(t_j)) \otimes x_k,$$

where $t_j \in U_j$ ($j = 1, \cdots, N$). We denote the set of indices j for which $t \in U_j$ by $J(t)$. Then

$$\sum_{j=1}^{N} \phi_j(t) \left\| \sum_{k=1}^{n} e_k(t_j) x_k \right\| + \sum_{j=1}^{N} \sum_{k=1}^{n} \phi_j(t) |e_k(t) - e_k(t_j)| \, \|x_k\|$$

$$\leq \left(\max_{t \in Q} \left\| \sum_{k=1}^{n} e_k(t) x_k \right\| \right) \sum_{j=1}^{N} \phi_j(t) + \sum_{j \in J(t)} \sum_{k=1}^{n} \phi_j(t) \frac{\epsilon \|x_k\|}{\sum_{i=1}^{n} \|x_i\|}$$

$$= \max_{t \in Q} \left\| \sum_{k=1}^{n} e_k(t) x_k \right\| + \epsilon = n_{C_Q}(z) + \epsilon.$$

Since $\epsilon > 0$ is arbitrary, it follows that $\inf \| \sum_{i=1}^{m} |\psi_i(t)| \, \|y_i\| \, \|_{C_Q} \leq n_{C_Q}(z)$, where the greatest lower bound is taken over all possible representations

$z = \sum_{i=1}^{m} \psi_i \otimes y_i \in C_Q \otimes X$. On the other hand, since the reverse inequality is always true, the proposition is proved.

AUTHOR'S REMARK ADDED FOR THE TRANSLATION. After this article was published in Russian, Professor V. A. Heiler kindly called the author's attention to the fact that every KB-lineal satisfies condition B. Let us give an outline of Professor Heiler's proof of this fact.

Let $z = \sum_{k=1}^{n} e_k \otimes x_k \in E \otimes X$, where E is a KB-lineal and X is a Banach space. Take $f = \sum_{k=1}^{n} |e_k| \, \|x_k\| \in E$ and consider the sublineal $E_f = \{e \in E: \exists \lambda > 0, \lambda |e| \leqq f\}$. It follows from the theory of semiordered linear spaces that the K-lineal E_f is isomorphic to some C_Q, where the function $\equiv 1$ corresponds to the element f ([**7**], Russian p. 202; see also [**14**]). Taking into account this realization of E_f and using arguments analogous to the proof of Proposition 4, it is easy, for every $\epsilon > 0$, to give a presentation of z in the form $z = \sum_{k=1}^{m} f_k \otimes y_k$, where $f_k \in E_f$, $y_k \in X$ $(k = 1, \cdots, m)$ and $\| \sum_{k=1}^{m} |f_k| \, \|y_k\| \, \| \leqq n_E(z) + \epsilon$. From this it obviously follows that condition B is satisfied.

A mapping $\alpha \in (E \to Y)$ from a KB-lineal E to a Banach space Y is called *summable* if it maps any series $\sum_{k=1}^{\infty} e_k$ of E for which the series $\sum_{k=1}^{\infty} |e_k|$ is convergent[10)] to an absolutely convergent series $\sum_{k=1}^{\infty} \alpha e_k$ in Y. We recall that a series $\sum_{k=1}^{\infty} x_k$ in a Banach space X is said to be absolutely convergent if $\sum_{k=1}^{\infty} \|x_k\| < \infty$.

The class of summable mappings is very extensive; it contains, in particular, the absolutely summable mappings of Grothendieck and Pietsch (see [**2**] and [**13**]).

Let $B(E)$ be the space of sequences (e_k) of E for which the series $\sum_{k=1}^{\infty} |e_k|$ is convergent, with the norm $\| (e_k) \| = \| \sum_{k=1}^{\infty} |e_k| \, \|_E$.

LEMMA 6. *$B(E)$ is complete for any KB-lineal E.*

PROOF. Let $a_n = (e_k^{(n)})$ be a fundamental sequence in $B(E)$. This means that for any $\epsilon > 0$ there exists an $n_0 = n_0(\epsilon)$ such that $\| \sum_{k=1}^{\infty} |e_k^{(n)} - e_k^{(m)}| \, \|_E \leqq \epsilon$ for $m, n \geqq n_0$. Since $|e_k^{(n)} - e_k^{(m)}| \leqq \sum_{k=1}^{\infty} |e_k^{(n)} - e_k^{(m)}|$, we have

$$\| e_k^{(n)} - e_k^{(m)} \| = \| \, |e_k^{(n)} - e_k^{(m)}| \, \| \leqq \epsilon \text{ for } m, n \geqq n_0$$

for all $k = 1, 2, \cdots$; that is, $(e_k^{(n)})_n$ is a fundamental sequence in E for every $k = 1, 2, \cdots$. We consider $e_k = \lim_{n \to \infty} e_k^{(n)}$ and show that $a = (e_k) \in B(E)$. Take $n_0 = n_0(\epsilon/2)$ and consider the sequence $a_{n_0} - (e_k^{(n_0)}) \in B(E)$. Further, choose $N = N(\epsilon/2)$ such that $\| \sum_{k=N+1}^{N+p} |e_k^{(n_0)}| \, \| \leqq \epsilon/2$ for any p. For $m, n \geqq n_0$ we have $\| \sum_{k=N+1}^{N+p} |e_k^{(n)} - e_k^{(m)}| \, \| \leqq \epsilon/2$. Letting $m \to \infty$ we obtain

$$\left\| \sum_{k=N+1}^{N+p} |e_k^{(n)} - e_k| \right\| \leqq \frac{\epsilon}{2}$$

10) We mean convergence with respect to the norm.

for $n \geqq n_0$ and every p. We have, further,

$$\left\| \sum_{k=N+1}^{N+p} |e_k| \right\| \leqq \left\| \sum_{k=N+1}^{N+p} |e_k^{(n_0)}| \right\| + \left\| \sum_{k=N+1}^{N+p} |e_k^{(n_0)} - e_k| \right\| \leqq \frac{\epsilon}{2} + \frac{\epsilon}{2} = \epsilon$$

for any p. Thus $a = (e_k) \in B(E)$.

Letting $m \to \infty$ in the inequality $\| \sum_{k=1}^N |e_k^{(n)} - e_k^{(m)}| \| \leqq \epsilon$, which is true for $m, n \geqq n_0(\epsilon)$ and any N, we obtain $\| \sum_{k=1}^N |e_k^{(n)} - e_k| \| \leqq \epsilon$ for $n \geqq n_0(\epsilon)$ and any N. Thus $\| \sum_{k=1}^N |e_k^{(n)} - e_k| \| \leqq \epsilon$ for $n \geqq n_0(\epsilon)$; that is, a_n converges to a in $B(E)$, and the lemma is proved.

Let $\alpha : E \to Y$ be a summable mapping. Consider the mapping $\Psi_\alpha : B(E) \to l^1(Y)$ defined by $\Psi_\alpha(e_k) = (\alpha e_k)$.

Lemma 7. *The mapping Ψ_α is continuous for any KB-lineal E and Banach space Y.*

Proof. Suppose that $a_n = (e_k^{(n)}) \xrightarrow{B(E)} 0$ and $\Psi_\alpha a_n \xrightarrow{l^1(Y)} \bar{y} = (y_k)$ as $n \to \infty$. Then $e_k^{(n)} \xrightarrow{E} 0$, $\alpha e_k^{(n)} \xrightarrow{Y} y_k$ as $n \to \infty$, and $y_k = 0$ $(k = 1, 2, \cdots)$ by the continuity of α. At the same time we have proved that Ψ_α is a mapping with a closed graph. By Lemma 6 and Banach's closed graph theorem (see, for example, [**12**]) it follows that Ψ_α is continuous.

Let $\alpha : E \to Y$ be a summable mapping. We put

$$\nu_E(\alpha) = \sup \frac{\sum_{k=1}^n \| \alpha e_k \|_Y}{\| \sum_{k=1}^n |e_k| \|_E},$$

where the upper bound is taken over all possible finite sets of elements $e_1, \cdots, e_n \in E$. From Lemma 7 it follows that $\nu_E(\alpha) < \infty$ for every summable mapping α. For consider the sequence $a = (e_1, \cdots, e_n, 0, \cdots) \in B(E)$. We have

$$\sum_{k=1}^n \| \alpha e_k \|_Y = \| \Psi_\alpha a \|_{l^1(Y)} \leqq \| \Psi_\alpha \| \, \| a \|_{B(E)} = \| \Psi_\alpha \| \left\| \sum_{k=1}^n |e_k| \right\|_E,$$

from which it follows that $\nu_E(\alpha) \leqq \| \Psi_\alpha \| < \infty$. Clearly, ν_E is a norm on the space of summable mappings of E to Y.

We denote by $S(E, Y)$ the space of summable mappings of E to Y, with the norm ν_E.

Proposition 5. *$S(E, Y)$ is complete for any KB-lineal E and Banach space Y.*

Proof. First we show that $S(E, Y)$ is isometrically embedded in the space[11] $(B(E) \to l^1(Y))$. Consider the mapping $T : S(E, Y) \to (B(E) \to l^1(Y))$, defined by $T\alpha = \Psi_\alpha$. Given $\epsilon > 0$, we choose a sequence $a = (e_k) \in B(E)$ for which

[11] $(A \to B)$ denotes the space of all continuous linear mappings from A to B with the operator norm.

$$\frac{\|\Psi_\alpha a\|_{l^1(Y)}}{\|a\|_{B(E)}} \geqq \|\Psi_\alpha\| - \frac{\epsilon}{2}.$$

Further, we choose n such that

$$\sum_{k=n+1}^{\infty} \|\alpha e_k\| \leqq \frac{\epsilon\|a\|_{B(E)}}{2}.$$

Then

$$\frac{\sum_{k=1}^{n}\|\alpha e_k\|_Y}{\|\sum_{k=1}^{n}|e_k|\,\|_E} \geqq \frac{\sum_{k=1}^{\infty}\|\alpha e_k\|_Y - \sum_{k=n+1}^{\infty}\|\alpha e_k\|_Y}{\|\sum_{k=1}^{\infty}|e_k|\,\|_E}$$
$$= \frac{\|\Psi_\alpha a\|_{l^1(Y)}}{\|a\|_{B(E)}} - \frac{\sum_{k=n+1}^{\infty}\|\alpha e_k\|_Y}{\|a\|_{B(E)}} \geqq \|\Psi_\alpha\| - \epsilon.$$

Since $\epsilon > 0$ is arbitrary, it follows that $\nu_E(\alpha) \geqq \|\Psi_\alpha\|$, and since the reverse inequality is also true, T is an isometric embedding of $S(E, Y)$ in $(B(E) \to l^1(Y))$.

To complete the proof we show that $T[S(E, Y)]$ is closed in $(B(E) \to l^1(Y))$. Let $\Psi_{\alpha_n} \in T[S(E, Y)]$ and $\Psi_{\alpha_n} \to \Psi \in (B(E) \to l^1(Y))$. Then (α_n) is a fundamental sequence in $S(E, Y)$. From the definition of the norm ν_E it is clear that $\nu_E(\alpha) \geqq \|\alpha\|$ for any $\alpha \in S(E, Y)$; therefore (α_n) is fundamental in $(E \to Y)$ and converges there to some $\alpha \in (E \to Y)$. Let $a = (e_k) \in B(E)$. We set

$$a_k = (\underbrace{0, \cdots, 0, e_k}_{k}, 0, \cdots) \in B(E) \quad (k = 1, 2, \cdots),$$

and consider the series $\sum_{k=1}^{\infty} a_k$ in $B(E)$. Since

$$\left\| a - \sum_{k=1}^{n} a_k \right\|_{B(E)} = \left\| \sum_{k=N+1}^{\infty} |e_k| \right\|_E,$$

the series $\sum_{k=1}^{\infty} a_k$ converges in $B(E)$ and $\sum_{k=1}^{\infty} a_k = a$. We have, further, that $\Psi a = \sum_{k=1}^{\infty} \Psi a_k$, and since

$$\Psi a_k = \lim_{n\to\infty} \Psi_{\alpha_n} a_k = \lim_{n\to\infty} (\underbrace{0, \cdots, 0, \alpha_n e_k}_{k}, 0, \cdots) = (\underbrace{0, \cdots, 0, \alpha e_k}_{k}, 0, \cdots)$$
$$(k = 1, 2, \cdots),$$

it follows that $\alpha \in S(E, Y)$ and $\Psi = \Psi_\alpha$. The proposition is proved.

REMARK. It is not difficult to show that $S(E, E)$ is a Banach algebra, and for any $\alpha \in S(E, E)$ and $\beta \in (E \to E)$ we have $\beta\alpha \in S(E, E)$ and $\nu_E(\beta\alpha) \leqq \|\beta\| \nu_E(\alpha)$; that is, $S(E, E)$ is a left ideal in $(E \to E)$.

THEOREM 3. *Let the KB-lineal E satisfy condition* C *and let X be a Banach space. Then the space $(E \tilde{\otimes} X)'$ adjoint to $E \tilde{\otimes} X$ is isometric to the space $S(E, X')$ of summable mappings from E to X'.*

PROOF. Let $f \in (E \tilde{\otimes} X)'$. Consider the mapping $L_f : E \to X'$ corresponding to f (see §1). Since $\langle x, L_f e\rangle = \langle e \otimes x, f\rangle$, we have

$$\sum_{k=1}^{n} \|L_f e_k\|_{X'} = \sup_{\substack{\|x_k\| \leqq 1 \\ k=1,\cdots,n}} \sum_{k=1}^{n} \langle e_k \otimes x_k, f\rangle = \sup_{\substack{\|x_k\| \leqq 1 \\ k=1,\cdots,n}} \left\langle \sum_{k=1}^{n} e_k \otimes x_k, f \right\rangle$$

$$\leqq \|f\|_{(E \tilde{\otimes} X)'} \sup_{\substack{\|x_k\| \leqq 1 \\ k=1,\cdots,n}} n_E \left(\sum_{k=1}^{n} e_k \otimes x_k \right) \leqq \|f\|_{(E \tilde{\otimes} X)'} \left\| \sum_{k=1}^{n} |e_k| \right\|_E .$$

Since this inequality is true for any finite set of elements $e_1, \cdots, e_n \in E$, L_f is a *summable mapping*[12] and $\nu_E(L_f) \leqq \|f\|_{(E \tilde{\otimes} X)'}$.

Conversely let $\alpha \in S(E, X')$. Consider the linear form f on $E \otimes_{n_E} X$ defined by

$$\left\langle \sum_{k=1}^{n} e_k \otimes x_k, f \right\rangle = \sum_{k=1}^{n} \langle x_k, \alpha e_k \rangle .$$

We have

$$\left| \left\langle \sum_{k=1}^{n} e_k \otimes x_k, f \right\rangle \right| \leqq \sum_{k=1}^{n} \|x_k\| \, \|\alpha e_k\| = \sum_{k=1}^{n} \| \, \|x_k\| \alpha e_k \| \leqq \nu_E(\alpha) \left\| \sum_{k=1}^{n} \|x_k\| \, |e_k| \right\|_E .$$

Since this inequality is true for any representation of $z = \sum_{k=1}^{n} e_k \otimes x_k$ it follows from it and from condition C that

$$\left| \left\langle \sum_{k=1}^{n} e_k \otimes x_k, f \right\rangle \right| \leqq \nu_E(\alpha) n_E \left(\sum_{k=1}^{n} e_k \otimes x_k \right) .$$

Then $f \in (E \tilde{\otimes} X)'$ and $\|f\|_{(E \tilde{\otimes} X)'} \leqq \nu_E(\alpha)$. Further, it is obvious that $\alpha = L_f$; therefore $\|f\|_{(E \tilde{\otimes} X)'} \leqq \nu_E(L_f)$, and the theorem is proved.

To conclude this section we consider the property of permanence of the tensor product $E \tilde{\otimes} X$.

PROPOSITION 6. *Let E and F be KB-lineals, X and Y Banach spaces, $\alpha : E \to F$ a regular mapping* [**6**] *and $\beta \in (X \to Y)$. Then $\alpha \otimes \beta \in (E \tilde{\otimes} X \to F \tilde{\otimes} Y)$, and for any representation of α in the form $\alpha = \alpha_1 - \alpha_2$ where α_1 and α_2 are positive operators, we have $\|\alpha \otimes \beta\| \leqq \|\alpha_1 + \alpha_2\| \, \|\beta\|$. If, in addition, F is a conditionally complete KB-lineal,*[13] *then $\|\alpha \otimes \beta\| \leqq \| \, |\alpha| \, \| \, \|\beta\|$.*

Further, if F is a conditionally complete KB-lineal, E is a closed proper subspace [**6**] *of F, $\alpha : E \to F$ is the identity embedding and $\beta : X \to Y$ is an isometric embedding, then $\alpha \otimes \beta : E \tilde{\otimes} X \to F \tilde{\otimes} Y$ is an isometric embedding.*

PROOF. Let $\alpha = \alpha_1 - \alpha_2$ and let α_1, α_2 be positive mappings. Since $|(\alpha_1 - \alpha_2)e| \leqq (\alpha_1 + \alpha_2)|e|$ for any $e \in E$, we have

---

[12] Indeed, from the convergence of the series $\sum_{k=1}^{\infty} (e_k)$ and the inequality already proved, we obtain $\sum_{k=1}^{n} \|L_f e_k\|_{X'} \leqq \|f\|_{(E \tilde{\otimes} X)'} \| \sum_{k=1}^{\infty} |e_k| \|_E$ for any n, and consequently $\sum_{k=1}^{\infty} \|L_f e_k\|_{X'} < \infty$.

[13] $|\alpha|$ denotes the modulus of the regular mapping α (see [**6**]).

$$n_F\left[(\alpha\otimes\beta)\sum_{k=1}^{n}e_k\otimes x_k\right]$$
$$=\inf\left\{\|v\|_F: v\geqq\left|\sum_{k=1}^{n}\alpha e_k\langle\beta x_k, y'\rangle\right| \text{ for all } y'\in Y', \|y'\|\leqq 1\right\}$$
$$\leqq\inf\left\{\|v\|_F: v=(\alpha_1+\alpha_2)u, u\geqq\left|\sum_{k=1}^{n}e_k\langle x_k, \beta^* y'\rangle\right| \text{ for all } y'\in Y', \|y'\|\leqq 1\right\}$$
$$\leqq\|\alpha_1+\alpha_2\|\inf\left\{\|u\|_E: u\geqq\|\beta^*\|\left|\sum_{k=1}^{n}e_k\langle x_k, x'\rangle\right| \text{ for all } x'\in X', \|x'\|\leqq 1\right\}$$
$$\leqq\|\alpha_1+\alpha_2\|\,\|\beta^*\|\inf\left\{\|u\|_E: u\geqq\left|\sum_{k=1}^{n}e_k\langle x_k, x'\rangle\right| \text{ for all } x'\in X', \|x'\|\leqq 1\right\}$$
$$=\|\alpha_1+\alpha_2\|\,\|\beta\|n_E\left(\sum_{k=1}^{n}e_k\otimes x_k\right),$$

so that $\alpha\otimes\beta\in(E\widetilde{\otimes}X\to F\widetilde{\otimes}Y)$ and $\|\alpha\otimes\beta\|\leqq\|\alpha_1+\alpha_2\|\,\|\beta\|$.

If F is a conditionally dense KB-lineal, then by Kantorovič's theorem (see [**6**], p. 232) there exists a mapping $|\alpha|: E\to F$. Put $\alpha_1=\frac{1}{2}(|\alpha|+\alpha)$ and $\alpha_2=\frac{1}{2}(|\alpha|-\alpha)$. Clearly α_1 and α_2 are positive operators, $\alpha_1-\alpha_2=\alpha$ and $\alpha_1+\alpha_2=|\alpha|$, so that $\|\alpha\otimes\beta\|\leqq\|\,|\alpha|\,\|\,\|\beta\|$.

Now we prove the second part of the proposition. Since $\beta: X\to Y$ is an isometric embedding, the adjoint operator $\beta^*: Y'\to X'$ maps the unit sphere Y' onto the whole unit sphere X'. Then for any $z=\sum_{k=1}^{n}e_k\otimes x_k\in E\otimes X$

$$\sup_F\left\{\left|\sum_{k=1}^{n}\alpha e_k\langle\beta x_k, y'\rangle\right|: y'\in Y', \|y'\|\leqq 1\right\}$$
$$=\sup_F\left\{\left|\alpha\sum_{k=1}^{n}e_k\langle x_k, \beta^* y'\rangle\right|: y'\in Y', \|y'\|\leqq 1\right\}$$
$$=\sup_F\left\{\left|\alpha\sum_{k=1}^{n}e_k\langle x_k, x'\rangle\right|: x'\in X', \|x'\|\leqq 1\right\}.$$

From the fact that E is a proper subspace of F and $\alpha: E\to F$ is the identity mapping, it follows that

$$\sup_F\left\{\left|\alpha\sum_{k=1}^{n}e_k\langle x_k, x'\rangle\right|: x'\in X', \|x'\|\leqq 1\right\}$$
$$=\alpha\sup_E\left\{\left|\sum_{k=1}^{n}e_k\langle x_k, x'\rangle\right|: x'\in X', \|x'\|\leqq 1\right\}.$$

Finally we obtain

$$\sup_F\left\{\left|\sum_{k=1}^{n}\alpha e_k\langle\beta x_k, y'\rangle\right|: y'\in Y', \|y'\|\leqq 1\right\}$$
$$=\alpha\sup_E\left\{\left|\sum_{k=1}^{n}e_k\langle x_k, x'\rangle\right|: x'\in X', \|x'\|\leqq 1\right\},$$

and since α is an isometric embedding of E in F,

$$n_F[(\alpha\otimes\beta)z] = \left\| \sup_{\|y'\|\leqq 1} \left| \sum_{k=1}^n \alpha e_k \langle \beta x_k, y'\rangle \right| \right\|_F = \left\| \alpha \sup_{\|x'\|\leqq 1} \left| \sum_{k=1}^n e_k\langle x_k, x'\rangle \right| \right\|_F$$

$$= \left\| \sup_{\|x'\|\leqq 1} \left| \sum_{k=1}^n e_k\langle x_k, x'\rangle \right| \right\|_E = n_E(z).$$

By continuity we find that $n_F[(\alpha\otimes\beta)z] = n_E(z)$ for any $z\in E\tilde{\otimes} X$.

Proposition 7. *Let E be a KB-lineal, Y a Banach space, and X a closed subspace of Y. Then $E\tilde{\otimes} X$ is isometric to a closed subspace of $E\tilde{\otimes} Y$.*

Proof. Let β denote the identity embedding of X in Y. Then $\beta^*: Y'\to X'$ maps the unit sphere Y' onto the whole of the unit sphere X'. We obtain

$$n_E\left(\sum_{k=1}^n e_k\otimes\beta x_k\right) = \inf\left\{ \|u\| : u \geqq \left| \sum_{k=1}^n e_k\langle \beta x_k, y'\rangle \right| \text{ for all } y'\in Y', \|y'\|\leqq 1\right\}$$

$$= \inf\left\{ \|u\| : u \geqq \left| \sum_{k=1}^n e_k\langle x_k, \beta^* y'\rangle \right| \text{ for all } y'\in Y', \|y'\|\leqq 1\right\}$$

$$= \inf\left\{ \|u\| : u \geqq \left| \sum_{k=1}^n e_k\langle x_k, x'\rangle \right| \text{ for all } x'\in X', \|x'\|\leqq 1\right\}$$

$$= n_E\left(\sum_{k=1}^n e_k\otimes x_k\right).$$

From this it follows that $1_E\otimes\beta: E\tilde{\otimes} X\to E\tilde{\otimes} Y$ is an isometric embedding.

Theorem 4. *Let E be a KB-lineal satisfying condition* C, *X a Banach space, $X_0\subset X$ a closed subspace, and $Y = X/X_0$. Then $E\tilde{\otimes} Y$ is isometric to the factor space*[14] $E\tilde{\otimes} X/E\tilde{\otimes} X_0$.

Proof. Let β denote the canonical mapping of X onto Y. By Proposition 6, $1_E\tilde{\otimes}\beta\in(E\tilde{\otimes} X\to E\tilde{\otimes} Y)$. Consider the adjoint mapping $(1_E\otimes\beta)^*: (E\tilde{\otimes} Y)'\to(E\tilde{\otimes} X)'$. Let $f\in(E\tilde{\otimes} Y)'$, $g = (1_E\otimes\beta)^* f\in(E\tilde{\otimes} X)'$. By Theorem 3 the functionals f and g correspond to summable mappings $L_f: E\to Y'$, $L_g: E\to X'$ and

$$\sum_{k=1}^n \langle x_k, L_g e_k\rangle = \left\langle \sum_{k=1}^n e_k\otimes x_k, g\right\rangle = \left\langle \sum_{k=1}^n e_k\otimes\beta x_k, f\right\rangle$$

$$= \sum_{k=1}^n \langle \beta x_k, L_f e_k\rangle = \sum_{k=1}^n \langle x_k, \beta^* L_f e_k\rangle,$$

so that $L_g = \beta^* L_f$.

Moreover, since β^* is an isometric embedding of Y' in X', from this equation and from Theorem 3 it follows that $\|g\|_{(E\tilde{\otimes} X)'} = \nu_E(L_g) = \nu_E(L_f) = \|f\|_{(E\tilde{\otimes} Y)'}$; that is, $(1_E\otimes\beta)^*: (E\tilde{\otimes} Y)'\to(E\tilde{\otimes} X)'$ is an isometric embedding. Hence (see for example [**12**], p. 525) $1_E\otimes\beta$ maps $E\tilde{\otimes} X$ onto the whole of $E\tilde{\otimes} Y$. In addition $\|1_E\otimes\beta\| = \|(1_E\otimes\beta)^*\| = 1$. To complete the proof it remains

[14] We identify $E\tilde{\otimes} X_0$ with the subspace of $E\tilde{\otimes} X$ to which it is isometric by Proposition 7.

to show[15] that $E \tilde{\otimes} X_0$ is the kernel of the mapping $1_E \otimes \beta$. Obviously, $(1_E \otimes \beta)z = 0$ for any $z \in E \tilde{\otimes} X_0$. Suppose that there exists $z_0 \notin E \tilde{\otimes} X_0$ for which $(1_E \otimes \beta)z_0 = 0$. Then there exists a functional $g \in (E \tilde{\otimes} X)'$ such that $\langle z, g\rangle = 0$ for all $z \in E \tilde{\otimes} X_0$, and $\langle z_0, g\rangle \neq 0$. Let $e \in E$, $x \in X_0$. Then $z = e \otimes x \in E \tilde{\otimes} X_0$ and $\langle z, g\rangle = \langle e \otimes x, g\rangle = \langle x, L_g e\rangle = 0$. Since x is an arbitrary element of X_0, it follows that L_g maps E to $\beta^* Y'$; that is, $L_g = \beta^* L_f$, where $f \in (E \tilde{\otimes} Y)'$. Thus $g = (1_E \otimes \beta)^* f$ and $\langle (1_E \otimes \beta) z_0, f\rangle = \langle z_0, (1_E \otimes \beta)^* f\rangle = \langle z_0, g\rangle \neq 0$. On the other hand, $(1_E \otimes \beta) z_0 = 0$. This contradiction proves that $E \tilde{\otimes} X_0$ is the kernal of $1_E \otimes \beta$, and the proof of the theorem is complete.

§3. Properties of the functor Φ_E

Let E be a KB-lineal. Let $\Phi_E(X) = E \tilde{\otimes} X$ for any Banach space X and $\Phi_E(\alpha) = 1_E \otimes \alpha$ for any $\alpha \in (X \to Y)$.

LEMMA 8. *Φ_E is a functor of type Σ on any category $\mathscr{K}$ of Banach spaces.*

The proof reduces to a simple verification of the functor axioms [**5**]; the conditions $\Phi_E(\alpha) \in (\Phi_E(X) \to \Phi_E(Y))$ and $\|\Phi_E(\alpha)\| \leqq \|\alpha\|$ follow from Proposition 6. The rest is obvious.

From Theorem 1 it follows that the functor Φ_E includes as particular cases the functors defined by minimal normed ideals of sequences $\mathfrak{n}$ and of measurable functions $\mathfrak{N}$, and also the functors $L^p_M (p \geqq 1)$ and C_Q [**5**].

We show that for all the standard KB-lineals E, the functor Φ_E satisfies the one-to-one condition.

LEMMA 9. *Suppose that E and F are KB-lineals, that $\alpha : E \to F$ is a positive mapping and $\alpha \otimes 1_X : E \tilde{\otimes} X \to F \tilde{\otimes} X$ is one-to-one for every Banach space X. If Φ_F satisfies the one-to-one condition, then so does Φ_E.*

PROOF. By Lemma 2, the following diagram is commutative:

---

[15] For suppose that this has already been shown. Assume that the norm of $E \tilde{\otimes} Y$ does not coincide with the factor norm $\|w\|_1 = \inf\{n_E(z) : z \in E \tilde{\otimes} X, (1_E \otimes \beta)z = w\}$. Then there exists $w_0 \in E \tilde{\otimes} Y$ for which $n_E(w_0) \leqq 1$ and $\|w_0\|_1 > 1$. Since $n_E \leqq \|\cdot\|_1$, both norms are equivalent; consequently the convex set $U = \{w : \|w\|_1 \leqq 1\}$ is closed in $E \tilde{\otimes} Y$. Thus there exists $f \in (E \tilde{\otimes} Y)' : \langle w_0, f\rangle > 1$ and $\langle w, f\rangle \leqq 1$ for $w \in U$. Since U contains $-w$ together with each w, we have $|\langle w, f\rangle| \leqq 1$ for $w \in U$. Hence $\|f\| > 1$ and

$$\|(1_E \otimes \beta)^* f\| = \sup_{n_E(z) \leqq 1} |\langle (1_E \otimes \beta) z, f\rangle| = \sup_{w \in U} |\langle w, f\rangle| \leqq 1.$$

This contradicts the fact that $(1_E \otimes \beta)^*$ is an isometry. Thus the norm of $E \tilde{\otimes} Y$ coincides with the factor norm.

$$\begin{array}{ccc} \Phi_E(X) & \xrightarrow{\alpha\otimes 1_X} & \Phi_F(X) \\ \Big\downarrow k_X^{\Phi_E} & & \Big\downarrow k_X^{\Phi_F} \\ E\hat{\hat{\otimes}} X & \xrightarrow{\alpha\otimes 1_X} & F\hat{\hat{\otimes}} X \end{array}$$

Let $z\in\Phi_E(X)$ and $k_X^{\Phi_E}z=0$. From the diagram it follows that $k_X^{\Phi_F}(\alpha\otimes 1_X)z=0$. By the assumption of the lemma $k_X^{\Phi_F}(\alpha\otimes 1_X)$ is one-to-one; then $z=0$, and the lemma is proved.

Let $\mathfrak{N}$ be a normed ideal of measurable functions. The natural embedding $\alpha:\mathfrak{N}\to L^1$ gives rise to a mapping $\alpha_X:\mathfrak{N}(X)\to L^1(X)$ for any Banach space X.

Lemma 10. *The mapping α_X is one-to-one for any X.*

Proof. Consider $[\mathfrak{N}(X)]_0$, the subspace of $\mathfrak{N}(X)$ consisting of measurable functions $x(t)$ with values in X for which $\|x(\cdot)\|_X\in\mathfrak{N}$. By definition, $\mathfrak{N}(X)$ is the completion of $[\mathfrak{N}(X)]_0$ in the norm $\|\,\|x(t)\|_X\|_{\mathfrak{N}}$. Suppose that there exists $u\in\mathfrak{N}(X)$ for which $\alpha_Xu=0$. Choose a sequence $u_k=u_k(t)\in[\mathfrak{N}(X)]_0$, converging to u in $\mathfrak{N}(X)$. Then $\alpha_Xu_k=u_k(t)\to 0$ in $L^1(X)$; that is, $\int_0^1\|u_k(t)\|_X dt\to 0$ as $k\to\infty$. Consider the sequence $\|u_k(t)\|_X\in\mathfrak{N}$. Since $|\,\|u_k(t)\|_X-\|u_l(t)\|_X|\leqq\|u_k(t)-u_l(t)\|_X$, we obtain $\|\,\|u_k(t)\|_X-\|u_l(t)\|_X\|_{\mathfrak{N}}\leqq\|\,\|u_k(t)-u_l(t)\|_X\|_{\mathfrak{N}}=\|u_k-u_l\|_{\mathfrak{N}(X)}$, from which it follows that $(\|u_k(t)\|_X)$ is a fundamental sequence in $\mathfrak{N}$. Then there exists a function $\phi(t)\in\mathfrak{N}$ such that $\|u_k\|_X$ converges to $\phi(t)$ in $\mathfrak{N}$, and thus in L^1. But $\|u_k(t)\|_X\to 0$ in L^1; therefore $\phi(t)\equiv 0$, and so $\|\,\|u_k(t)\|_X\|_{\mathfrak{N}}\to 0$. Since $\|u_k\|_{\mathfrak{N}(X)}=\|\,\|u_k(t)\|_X\|_{\mathfrak{N}}$, this means that $u=0$.

Corollary. *The mapping $\alpha\otimes 1_X:\mathfrak{N}\tilde{\otimes}X\to L^1(X)$ is one-to-one for any X.*

Proposition 8. *Φ_E satisfies the one-to-one condition on any category $\mathscr{K}$ when E is a normed ideal of sequences or functions or is the KB-lineal C_Q. For $E=L_M^p$ $(p\geqq 1)$ the functor Φ_E satisfies the one-to-one condition on any category $\mathscr{K}'$ of Banach spaces X' that are adjoint to separable spaces.*

Proof. For $E=m$ the assertion is clear, since in this case $n_E=\lambda$ and so $k_X^{\Phi_E}=1_{m\hat{\hat{\otimes}}X}$. For any other normed ideal $\mathfrak{n}$ we have $\mathfrak{n}\tilde{\otimes}X\subset m\hat{\hat{\otimes}}X$ and the one-to-one condition follows from Lemma 9. For $E=L^1$ the assertion is true, because it becomes the Grothendieck one-to-one condition for L^1 (see §1), which is satisfied [**2**]. For any other normed ideal of functions $\mathfrak{N}$ the proposition holds by Lemma 9 and the Corollary to Lemma 10. The case $E=C_Q$ is trivial; since here $n_E=\lambda$ and $k_X^{C_Q}=1_{C_Q(X)}$

Let $E=L_M^p$, $X'\in\mathscr{K}'$. We identify the elements of $L_M^p\hat{\hat{\otimes}}X'$ with the completely continuous operators $\alpha:X\to L_M^p$ (see §1). Then the mapping $k_{X'}^{L_M^p}$ sends the function $u=u(s)\in L_M^p(X')$ to the operator $\tilde{u}=k_{X'}^{L_M^p}u$ acting by

$[\tilde{u}x](s) = \langle x, u(s)\rangle$. We have to prove that $\tilde{u} = 0$ implies $u(s) = 0$. We set

$$Q = \{s \in S : u(s) \neq 0\}, \quad Q_x = \{s \in S : \langle x, u(s)\rangle \neq 0\}.$$

Since $\tilde{u} = 0$, we have $\mu Q_x = 0$ for any $x \in X$. Let $\{x_n\}$ be a countable set that is everywhere dense in X. Then $Q = \bigcup_{n=1}^{\infty} Q_{x_n}$, and so $\mu Q = 0$. The proposition is proved.

THEOREM 5. *Let E be a KB-lineal satisfying condition* C, *and let Φ_E satisfy the one-to-one condition on the category of all Banach spaces. Then Φ_E is an exact functor; that is, any exact sequence $X \xrightarrow{\alpha} Y \xrightarrow{\beta} Z$ is mapped to an exact sequence $\Phi_E(X) \xrightarrow{\Phi_E(\alpha)} \Phi_E(Y) \xrightarrow{\Phi_E(\beta)} \Phi_E(Z)$.*

PROOF. Let $X \xrightarrow{\alpha} Y \xrightarrow{\beta} Z$ be exact; that is, $\alpha \in (X \to Y)$, $\beta \in (Y \to Z)$ and $\operatorname{Im}\alpha = \operatorname{Ker}\beta$. We show that

$$\Phi_E(X) \xrightarrow{\Phi_E(\alpha)} \Phi_E(Y) \xrightarrow{\Phi_E(\beta)} \Phi_E(Z)$$

is exact. Let $Y_0 = \operatorname{Im}\alpha = \operatorname{Ker}\tau$; Y_0, being the kernel of the continuous mapping β, is closed in Y. Let α_0 denote α, regarded as a mapping of X to Y_0. Then we have the commutative diagram

$$\begin{array}{ccccccccc} X & & \xrightarrow{\quad\alpha\quad} & & Y & & \xrightarrow{\quad\beta\quad} & & Z \\ & \searrow^{\alpha_0} & & \nearrow_{i} & & \searrow^{h} & & \nearrow_{j} & \\ & & Y_0 & & & & Y/Y_0 & & \end{array}$$

where i is the identity isometric embedding of Y_0 in Y, h is the canonical homomorphism of Y onto Y/Y_0, and j is the continuous embedding of Y/Y_0 in Z. Applying the functor Φ_E we obtain the commutative diagram

$$\begin{array}{ccccccccc} \Phi_E(X) & & \xrightarrow{\quad\Phi_E(\alpha)\quad} & & \Phi_E(Y) & & \xrightarrow{\quad\Phi_E(\beta)\quad} & & \Phi_E(Z) \\ & \searrow^{\Phi_E(\alpha_0)} & & \nearrow_{\Phi_E(i)} & & \searrow^{\Phi_E(h)} & & \nearrow_{\Phi_E(j)} & \\ & & \Phi_E(Y_0) & & & & \Phi_E(Y/Y_0) & & \end{array}$$

By Proposition 7, $\Phi_E(i)$ is an isometric embedding. We identify $\Phi_E(Y_0)$ with the subspace $\Phi_E(i)\Phi_E(Y_0)$ of $\Phi_E(Y)$ isometric to it. Then $\operatorname{Im}\Phi_E(\alpha) = \Phi_E(Y_0)$. Consider the diagram

$$\begin{array}{ccc} \Phi_E(Y/Y_0) & \xrightarrow{\Phi_E(j)} & \Phi_E(Z) \\ \Big\downarrow k_{Y/Y_0}^{\Phi E} & & \Big\downarrow k_Z^{\Phi E} \\ E \hat{\hat{\otimes}} Y/Y_0 & \xrightarrow{1_E \otimes j} & E \hat{\hat{\otimes}} Z \end{array}$$

It is clearly commutative. The mappings $k_{Y/Y_0}^{\Phi E}$ and $k_Z^{\Phi E}$ are one-to-one by the proposition. Further, from the fact that j is one-to-one it follows that $1_E \otimes j : E \hat{\hat{\otimes}} Y/Y_0 \to E \hat{\hat{\otimes}} Z$ is one-to-one. From the diagram we see that $\Phi_E(j)$ is also one-to-one. Thus $\operatorname{Ker}\Phi_E(\beta) = \operatorname{Ker}\Phi_E(h)$. By Theorem 4, $\Phi_E(Y/Y_0)$ is isometric to the factor space $\Phi_E(Y)/\Phi_E(Y_0)$, and also $\Phi_E(h)$ is the canonical homomorphism of $\Phi_E(Y)$ onto $\Phi_E(Y)/\Phi_E(Y_0)$, so that $\operatorname{Ker}\Phi_E(h) = \Phi_E(Y_0)$. Therefore $\operatorname{Im}\Phi_E(\alpha) = \operatorname{Ker}\Phi_E(\beta)$; that is, the sequence $\Phi_E(X) \xrightarrow{\Phi_E(\alpha)} \Phi_E(Y) \xrightarrow{\Phi_E(\beta)} \Phi_E(Z)$ is exact.

Let Φ_E satisfy the one-to-one condition on some category $\mathscr{K}$; then the mapping $k_{X'}^{\Phi_E}: \Phi_E(X') \to E \hat{\hat{\otimes}} X'$ establishes a one-to-one correspondence between $\Phi_E(X') = E \widetilde{\otimes} X'$ and a class of completely continuous operators $\alpha: X \to E$ for each $X' \in \mathscr{K}$. We describe this class of operators.

A mapping $\alpha \in (X \to E)$ that maps the unit sphere X to a bounded (in the sense of the ordering) subset of E is called *proper*.[16] We define a norm on the space of proper mappings by

$$n(\alpha) = \inf\{ \|u\|_E : u \geqq |\alpha x| \ \text{for all}\ x \in X, \|x\| \leqq 1\}.$$

THEOREM 6. *Suppose that the KB-lineal E satisfies condition* A *and that Φ_E satisfies the one-to-one condition on some category $\mathscr{K}$. Then for any $X' \in \mathscr{K}$ the space $\Phi_E(X')$ is isometric to the closure of the set of continuous finite-dimensional linear mappings of X to E in the space of proper mappings of X to E. If, in addition, there exists a sequence of finite-dimensional operators $v_n \in (E \to E)$ such that $v_n e \to e$ as $n \to \infty$ and $v_n e \leqq e$ for any positive $e \in E$, then $\Phi_E(X')$ is isometric to the space of all proper mappings X to E.*

PROOF. By the one-to-one condition every element $z \in \Phi_E(X')$ can be regarded as a completely continuous mapping of X to E. We show that this mapping is proper. Choose a sequence of elements $z_n \in E \otimes X'$ converging to z in $\Phi_E(X')$. Then $z_n \to z$ in $(X \to E)$. Let S_X denote the unit sphere of the space X. Since $z: X \to E$ is completely continuous, zS_X is separable. Choose a sequence $\{x_k\} \subset S_X$ for which $\{zx_k\}$ is dense in zS_X. Put

$$e_m = \sup_{1 \leqq k \leqq m} |zx_k|, \quad e_{mn} = \sup_{1 \leqq k \leqq m} |z_n x_k|, \quad m = 1, 2, \cdots; \ n = 1, 2, \cdots.$$

Clearly, $e_{mn} \to e_m$ as $n \to \infty$, for each $m = 1, 2, \cdots$. Let

$$z_n = \sum_{j=1}^{N_n} e_j^{(n)} \otimes x'_{jn}, \quad n = 1, 2, \cdots.$$

Then

$$n_E(z_n) = \inf\left\{ \|u\| : u \geqq \left| \sum_{j=1}^{N_n} e_j^{(n)} \langle x'_{jn}, x''\rangle \right| \ \text{for all}\ x'' \in X'', \|x''\| \leqq 1 \right\} \geqq \|e_{mn}\|$$

for $m = 1, 2, \cdots$, and since $z_n \to z$ in $\Phi_E(X')$, we see that $n_E(z_n) \leqq C < \infty$ $(n = 1, 2, \cdots)$. From this we obtain that $\|e_m\| \leqq C$ for all $m = 1, 2, \cdots$. Since $e_{m+1} \geqq e_m \geqq 0$ $(m = 1, 2, \cdots)$, it follows from condition A that the sequence $\{e_m\}$ is bounded above. Choose $e_0 \in E$ such that $e_m \leqq e_0$, $m = 1, 2, \cdots$. For any $e = zx \in zS_X$ there exists a sequence (x_{k_j}) such that $zx_{k_j} \to e$ as $j \to \infty$.

[16] When E is a K^+-space, proper mappings are the same as the *bo*-linear mappings of Kantorovič (see [6], p. 256 or [7], Russian p. 246).

Thus $|zx_{k_j}| \leqq e_{k_j} \leqq e_0$ for all $j = 1, 2, \cdots$, and so $|e| \leqq e_0$. We have proved that the mapping z is proper.

Now let $z = \sum_{k=1}^{n} e_k \otimes x'_k \in E \otimes X'$. Since the unit sphere of X is weakly dense in the unit sphere of X'', we see that

$$n_E(z) = \inf\left\{ \|u\| : u \geqq \left| \sum_{k=1}^{n} e_k \langle x'_k, x'' \rangle \right| \text{ for all } x'' \in X'', \|x''\| \leqq 1 \right\}$$
$$= \inf\left\{ \|u\| : u \geqq \left| \sum_{k=1}^{n} e_k \langle x, x'_k \rangle \right| \text{ for all } x \in X, \|x\| \leqq 1 \right\} = n(z).$$

Then by continuity $n_E(z) = n(z)$ for any $z \in \Phi_E(X')$, and the first part of the theorem is proved.

We now prove the second part. Let $\alpha : X \to E$ be a proper mapping and $|\alpha x| \leqq e_0$ for all $x \in X$, $\|x\| \leqq 1$. Then $|(1_E - v_n)\alpha x| \leqq (1_E - v_n)e_0$ for every $x \in X$, $\|x\| \leqq 1$, and $n(\alpha - v_n\alpha) = \inf\{ \|u\| : u \geqq |(1_E - v_n)\alpha x| \text{ for all } x \in X, \|x\| \leqq 1\} \leqq \|(1_E - v_n)e_0\|$, so that α is approximated with respect to the norm n by finite-dimensional mappings $v_n\alpha$, and consequently $\alpha \in \Phi_E(X')$.

REMARK. The sequence of operators (v_n) with the required properties exists for any KB-lineal of real sequences with the natural ordering in which the "unit vectors"

$$\delta_n = (\underbrace{0, \cdots, 0, 1}_{n}, 0, \cdots)$$

form a basis.

Proper mappings are connected with summable mappings, as the following proposition shows.

PROPOSITION 9.[17] *Let E be a KB-lineal, X a Banach space, and $\alpha : X \to E$ a proper mapping. Then its adjoint $\alpha^* : E' \to X'$ is a summable mapping and $\nu_{E'}(\alpha^*) \leqq n(\alpha)$.*

PROOF. Let $\sum_{k=1}^{\infty} e'_k$ be a series in E' for which the $\sum_{k=1}^{\infty} |e'_k|$ is convergent, and $e_0 \in E$ an element majorizing $|\alpha x|$ for all $x \in X$, $\|x\| \leqq 1$. Then

$$\sum_{k=1}^{n} \|\alpha^* e'_k\| = \sum_{k=1}^{n} \sup_{\|x_k\| \leqq 1} |\langle x_k, \alpha^* e'_k \rangle|$$
$$= \sum_{k=1}^{n} \sup_{\|x_k\| \leqq 1} |\langle \alpha x_k, e'_k \rangle| \leqq \sum_{k=1}^{n} \sup_{\|x_k\| \leqq 1} \langle |\alpha x_k|, |e'_k| \rangle$$
$$\leqq \sum_{k=1}^{n} \langle e_0, |e'_k| \rangle = \left\langle e_0, \sum_{k=1}^{n} |e'_k| \right\rangle \leqq \|e_0\| \left\| \sum_{k=1}^{n} |e'_k| \right\|.$$

Since for e_0 we can take any element majorizing $|\alpha x|$ for all $x \in X$, $\|x\| \leqq 1$, we obtain

[17] If E is a conditionally complete KB-lineal, then the equation $\nu_{E'}(\alpha^*) = n(\alpha)$ holds [15].

$$\sum_{k=1}^{n} \|\alpha^* e_k'\| \leqq \inf\left\{ \|e_0\| : e_0 \geqq |\alpha x| \text{ for all } x \in X, \|x\| \leqq 1 \right\} \left\| \sum_{k=1}^{n} |e_k'| \right\|$$

$$= n(\alpha) \left\| \sum_{k=1}^{n} |e_k'| \right\|.$$

Taking the limit as $n \to \infty$ we see that α^* is a summable mapping of E' to X'. If in the inequality

$$\sum_{k=1}^{n} \|\alpha^* e_k'\| \leqq n(\alpha) \left\| \sum_{k=1}^{n} |e_k'| \right\|$$

we take the least upper bound over all possible finite sets of elements $e_1', \cdots, e_n' \in E'$, we see that $\nu_{E'}(\alpha^*) \leqq n(\alpha)$.

Proposition 10. *Suppose that the KB-lineal E satisfies condition* C *and that* $\Phi_{E'}$ *satisfies the one-to-one condition. Then the natural embedding* $E' \otimes X' \to (E \widetilde{\otimes} X)'$ *extends to a continuous linear embedding* $E' \widetilde{\otimes} X' \to (E \widetilde{\otimes} X)'$.

Proof. We show first that E' satisfies condition A. Suppose the contrary. Then there exists a sequence of positive elements $e_n' \in E'$ such that $e_n' \uparrow \infty$ and $\|e_n'\| \leqq C < \infty$. For any positive $e \in E$ put $\langle e, e_0' \rangle = \lim_{n \to \infty} \langle e, e_n' \rangle$. Since (e_n') is a monotonic sequence, the limit exists and since $\langle e, e_n' \rangle \leqq \|e\| C$ for all $n = 1, 2, \cdots$, we see that $\langle e, e_0' \rangle \leqq \|e\| C$. Thus $e_0' \in E'$, and, since clearly $e_0' \geqq e_n'$ for $n = 1, 2, \cdots$, we get a contradiction. Thus E' satisfies condition A. From Theorem 6 it then follows that the elements of $E' \widetilde{\otimes} X'$ are proper mappings of X to E'. Consider the mapping f sending $\alpha \in E' \widetilde{\otimes} X'$ to the restriction to E of the adjoint operator $\alpha^* : E'' \to X'$. Clearly f is one-to-one. By Proposition 9, $\alpha^* \in S(E'', X')$ and $\nu_{E''}(\alpha^*) \leqq n_{E'}(\alpha)$; but then $f\alpha \in S(E, X')$ and $\nu_E(f\alpha) \leqq \nu_{E''}(\alpha^*) \leqq n_{E'}(\alpha)$. Since, by Theorem 3, $S(E, X')$ is isometric to $(E \widetilde{\otimes} X)'$, we see that f is a continuous embedding of $E' \widetilde{\otimes} X'$ into $(E \widetilde{\otimes} X)'$. We verify that the restriction of f to $E' \otimes X'$ is the canonical embedding of $E' \otimes X'$ in $(E \widetilde{\otimes} X)'$. For $\alpha = \sum_{k=1}^{n} e_k' \otimes x_k'$ we have $f\alpha = \sum_{k=1}^{n} \langle\ , e_k' \rangle x_k'$ and

$$\left\langle \sum_{j=1}^{m} e_j \otimes x_j, f\alpha \right\rangle = \sum_{j=1}^{m} \langle x_j, (f\alpha) e_j \rangle = \sum_{j=1}^{m} \left\langle x_j, \sum_{k=1}^{n} \langle e_j, e_k' \rangle x_k' \right\rangle$$

$$= \sum_{j=1}^{m} \sum_{k=1}^{n} \langle x_j, x_k' \rangle \langle e_j, e_k' \rangle.$$

The proposition is proved.

We proceed to a description of the space of mappings $\{\Phi_E \to \Phi_F\}$. In accordance with Proposition 1 we identify the elements of $\{\Phi_E \to \Phi_F\}$ with the mappings $E \to F$ by which they are generated. We recall that every element $\{\alpha_X\} \in \{\Phi_E \to \Phi_F\}$ is defined by some mapping $\alpha : E \to F$ according to the formula $\alpha_X = \alpha \otimes 1_X$ for every $X \in \mathscr{K}$, and this is a (one-to-one) correspondence (see Proposition 1).

Let E and F be KB-lineals. The mapping $\alpha \in (E \to F)$ is called *K-bounded* if it maps any bounded sequence (e_k) in the sense of the ordering of E to such a sequence $(\alpha e_k) \subset F$.

Let $A(E)$ be the space of sequences $(e_k) \subset E$ that are bounded in the sense of the ordering, with the norm $\|(e_k)\| = \inf\{\|u\| : u \geqq |e_k|,\ k = 1, 2, \cdots\}$.

Lemma 11. *The space* $A(E)$ *is complete for any KB-lineal E satisfying conditions* A *and* B.

Proof. Let $a_n = (e_k^{(n)})$ be a fundamental sequence in $A(E)$. This means that for any $\epsilon > 0$ there exists $n_0 = n_0(\epsilon)$ such that for $m, n \geqq n_0$

$$\inf\{\|u\| : u \geqq |e_k^{(m)} - e_k^{(n)}|,\ k = 1, 2, \cdots\} \leqq \epsilon.$$

Then $\|e_k^{(m)} - e_k^{(n)}\| = \|\,|e_k^{(m)} - e_k^{(n)}|\,\|$ for $m, n \geqq n_0$ and for all $k = 1, 2, \cdots$; that is, for each k the sequence $(e_k^{(n)})$ is fundamental in E. We take $e_k = \lim_{n\to\infty} e_k^{(n)}$ and show that $a = (e_k) \in A(E)$. We have $\|\sup_{1\leqq k\leqq m}|e_k|\,\| = \lim_{n\to\infty}\|\sup_{1\leqq k\leqq m}|e_k^{(n)}|\,\|$ for any $m = 1, 2, \cdots$, and since $\|\sup_{1\leqq k\leqq m}|e_k^{(n)}|\,\| \leqq \|a_n\|_{A(E)} \leqq C < \infty$ for all $n = 1, 2, \cdots$, we see that $\|\sup_{1\leqq k\leqq m}|e_k|\,\| \leqq C < \infty$ for any $m = 1, 2, \cdots$. Let $g_m = \sup_{1\leqq k\leqq m}|e_k|$; then (g_m) is a monotonic increasing sequence and $\|g_m\| \leqq C < \infty$. By condition A it follows that (g_m) is bounded above, and then $a = (e_k) \in A(E)$.

We have, further, $\|a - a_n\|_{A(E)} = \inf\{\|u\| : u \geqq |e_k - e_k^{(n)}|,\ k = 1, 2, \cdots\}$. Fix $n \geqq n_0(\epsilon)$ and let $\phi_N = \sup_{1\leqq k\leqq N}|e_k - e_k^{(n)}|$, $N = 1, 2, \cdots$. Then (ϕ_N) is a monotonic increasing sequence and $\|\phi_N\| = \lim_{m\to\infty}\|\sup_{1\leqq k\leqq N}|e_k^{(m)} - e_k^{(n)}|\,\| \leqq \epsilon$. By condition A it follows that (ϕ_N) is bounded above. Thus, by condition B, for any $\delta > 0$ there exists $u_\delta \in E$ such that $u_\delta \geqq \phi_N$, $N = 1, 2, \cdots$, and $\|u_\delta\| \leqq \epsilon + \delta$. Since $u_\delta \geqq |e_k - e_k^{(n)}|$ for all $k = 1, 2, \cdots$, we find that $\|a - a_k\|_{A(E)} \leqq \epsilon$; that is, $a = \lim_{n\to\infty} a_n$ and the lemma is proved.

Let E and F be KB-lineals, and $\alpha : E \to F$ a K-bounded mapping. Consider the mapping $\phi_\alpha : A(E) \to A(F)$ defined by $\phi_\alpha(e_k) = (\alpha e_k)$.

Lemma 12. *The mapping* ϕ_α *is continuous for any KB-lineals E and F satisfying conditions* A *and* B.

Proof. Let $a_n = (e_k^{(n)}) \xrightarrow{A(E)} 0$ and $\phi_\alpha a_n = (\alpha e_k^{(n)}) \xrightarrow{A(F)} b = (f_k)$ as $n \to \infty$. Thus, $\|e_k^{(n)}\| \leqq \|a_n\|_{A(E)} \to 0$ and $\|\alpha e_k^{(n)} - f_k\| \leqq \|\phi_\alpha a_n - b\|_{A(F)} \to 0$ as $n \to \infty$. By the continuity of α we have $f_k = 0$, $k = 1, 2, \cdots$, and so $b = 0$. We have also proved that ϕ_α is a mapping with a closed graph. Thus, by Banach's closed graph theorem, ϕ_α is continuous.

Suppose that E and F are KB-lineals satisfying conditions A and B, and that $\alpha : E \to F$ is a K-bounded mapping. Let

$$\|\alpha\|_0 = \sup \frac{\|\sup_{1\leqq k\leqq n}|\alpha e_k|\,\|_F}{\|\sup_{1\leqq k\leqq n}|e_k|\,\|_E},$$

where the least upper bound is taken over all finite sets of elements $e_1, \cdots, e_n \in E$.

By Lemma 12 it follows that $\|\alpha\|_0 < \infty$. For consider the sequence $a = (e_1, \cdots, e_n, 0, \cdots)$. We have $a \in A(E)$ and

$$\left\|\sup_{1 \leqq k \leqq n} |\alpha e_k|\right\|_F = \|\phi_\alpha a\|_{A(F)} \leqq \|\phi_\alpha\|_{(A(E)\to A(F))} \|a\|_{A(E)} = \|\phi_\alpha\|_{(A(E)\to A(F))} \left\|\sup_{1 \leqq k \leqq n} |e_k|\right\|_E.$$

Since this inequality is true for any finite set of elements $e_1, \cdots, e_n \in E$, we see that $\|\alpha\|_0 \leqq \|\phi_\alpha\|_{(A(E)\to A(F))} < \infty$. Clearly $\|\alpha\|_0$ has all the properties of a norm on the space of K-bounded mappings from E to F.

Let c_0^n denote the n-dimensional coordinate space of elements $x = (\xi_1, \cdots, \xi_n)$ with the norm $\|x\|_{c_0^n} = \max_{1 \leqq k \leqq n} |\xi_k|$.

Let l_n^1 denote the n-dimensional coordinate space of elements $x = (\xi_1, \cdots, \xi_n)$ with the norm $\|x\|_{l_n^1} = \sum_{k=1}^n |\xi_k|$.

LEMMA 13. *Let E and F be KB-lineals satisfying conditions* A *and* B, *and* $\mathscr{K}$ *a category of Banach spaces containing spaces* X_n *for which there exist isometric embeddings* $i_n: c_0^n \to X_n$ $(n = 1, 2, \cdots)$. *Further, let* $\alpha \in \{\Phi_E \to \Phi_F\}$, *where the functors* Φ_E, Φ_F *are considered on* $\mathscr{K}$. *Then* α *is a K-bounded mapping of E to F and* $\|\alpha\|_0 \leqq \|\alpha\|_{\{\Phi_E\to\Phi_F\}}$.

PROOF. Let $(e_k) \subset E$ be a bounded sequence with respect to the ordering. Consider the elements $z_n = \sum_{k=1}^n e_k \otimes i_n \delta_k^{(n)} \in E \otimes X_n$, $n = 1, 2, \cdots$, where

$$\delta_k^{(n)} = (\underbrace{0, \cdots, 0, 1}_{k}, 0, \cdots, 0) \in c_0^n.$$

By Proposition 7, $n_E(z_n) = n_E(\sum_{k=1}^n e_k \otimes \delta_k^{(n)})$. Further, since the space adjoint to c_0^n is isometric to l_n^1, we see that $n_E = (\sum_{k=1}^n e_k \otimes \delta_k^{(n)}) = \inf\{\|u\|_E : u \geqq |\sum_{k=1}^n \lambda_k e_k|,\ \sum_{k=1}^n |\lambda_k| \leqq 1\} = \|\sup_{1\leqq k \leqq n} |e_k|\,\|_E$. Thus $\|\sup_{1 \leqq k \leqq n}|e_k|\,\|_E = n_E(z_n)$. Similarly, $\|\sup_{1 \leqq k \leqq n} |\alpha e_k|\,\|_F = n_F[(\alpha \otimes 1_{X_n})_{z_n}]$. Furthermore, $\|\sup_{1 \leqq k \leqq n}|\alpha e_k|\,\|_F = n_F[(\alpha \otimes 1_{X_n}) z_n] \leqq \|\alpha\|_{\{\Phi_E\to\Phi_F\}} n_E(z_n) = \|\alpha\|_{\{\Phi_E\to\Phi_F\}} \|\sup_{1 \leqq k \leqq\leqq n} |e_k|\,\|_E$. Let $e_0 \in E$ be an element majorizing $|e_k|$ for all $k = 1, 2, \cdots$. Then $\|\sup_{1\leqq k \leqq n}|\alpha e_k|\,\|_F \leqq \|\alpha\|_{\{\Phi_E\to\Phi_F\}} \|e_0\|$ for all $n = 1, 2, \cdots$. By condition A the sequence $f_n = \sup_{1 \leqq k \leqq n} |\alpha e_k| \in F$ is bounded above, and hence so is $(|\alpha e_k|)$, so that the mapping α is K-bounded. In the inequality

$$\frac{\|\sup_{1\leqq k \leqq n}|\alpha e_k|\,\|_F}{\|\sup_{1 \leqq k \leqq n}|e_k|\,\|_E} \leqq \|\alpha\|_{\{\Phi_E\to\Phi_F\}}$$

taking the least upper bound over all finite sets of elements $e_1, \cdots, e_n \in E$, we find that $\|\alpha\|_0 \leqq \|\alpha\|_{\{\Phi_E\to\Phi_F\}}$.

LEMMA 14. *Let E and F be KB-lineals satisfying conditions* A *and* B, *and let* $\alpha: E \to F$ *be a K-bounded mapping. Then* $\alpha \in \{\Phi_E \to \Phi_F\}$ *for any category* $\mathscr{K}$ *and* $\|\alpha\|_{\{\Phi_E\to\Phi_F\}} \leqq \|\alpha\|_0$.

PROOF. Choose any $X \in \mathscr{K}$ and $z = \sum_{k=1}^n e_k \otimes x_k \in E \otimes X$. Consider a finite-dimensional subspace $X_n \subset X$ spanned by vectors $x_1, \cdots, x_n$ and choose a countably dense subset $\{f_j\}_{j=1}^\infty$ in the unit sphere of the adjoint space X_n'.

Extend each f_j with preservation of the norm to a continuous linear functional $x_j' \in X'$. Let $x' \in X'$, $\|x'\| \leqq 1$ and let f be the restriction of x' to X_n. Then there exists a sequence $f_{j_m} \to f$ as $m \to \infty$, and so $\langle x_k, x_{j_m}' \rangle \to \langle x_k, x' \rangle$ as $m \to \infty$ for all $k = 1, 2, \cdots$. From this it follows that

$$\left| \sum_{k=1}^{n} e_k \langle x_k, x_{j_m}' \rangle \right| \to \left| \sum_{k=1}^{n} e_k \langle x_k, x' \rangle \right| \quad \text{for } m \to \infty,$$

and if $u \geqq |\sum_{k=1}^{n} e_k \langle x_k, x_j' \rangle|$ for all $j = 1, 2, \cdots$, then $u \geqq |\sum_{k=1}^{n} e_k \langle x_k, x' \rangle|$ for any $x' \in X'$, $\|x'\| \leqq 1$. We obtain

$$\begin{aligned} n_F[(\alpha \otimes 1_X)z] &= \inf\left\{ \|u\| : u \geqq \left| \sum_{k=1}^{n} \alpha e_k \langle x_k, x' \rangle \right| \quad \text{for all } x' \in X', \|x'\| \leqq 1 \right\} \\ &= \inf\left\{ \|u\| : u \geqq \left| \sum_{k=1}^{n} \alpha e_k \langle x_k, x_j' \rangle \right| \quad \text{for any } j = 1, 2, \cdots \right\}. \end{aligned}$$

Put

$$g_m = \sup_{1 \leqq j \leqq m} \left| \sum_{k=1}^{n} \alpha e_k \langle x_k, x_j' \rangle \right|, \quad m = 1, 2, \cdots;$$

then $g_m \leqq g_0 = \sum_{k=1}^{n} |\alpha e_k| \, \|x_k\|$ for all $m = 1, 2, \cdots$. By condition B, for any $\delta > 0$ there exists $u_\delta \in F$ such that $u_\delta \geqq g_m$, $m = 1, 2, \cdots$, and $\|u_\delta\| \leqq \lim_{m \to \infty} \|g_m\| + \delta$. So we obtain

$$\begin{aligned} \inf\left\{ \|u\| : u \geqq \left| \sum_{k=1}^{n} \alpha e_k \langle x_k, x_j' \rangle \right| \text{ for all } j = 1, 2, \cdots \right\} &\leqq \inf\{ \|u_\delta\| : \delta > 0 \} \\ &= \lim_{m \to \infty} \|g_m\| = \lim_{m \to \infty} \left\| \sup_{1 \leqq j \leqq m} \left| \sum_{k=1}^{n} \alpha e_k \langle x_k, x_j' \rangle \right| \right\|. \end{aligned}$$

Moreover, from the definition of the norm $\|\alpha\|_0$ it follows that

$$\begin{aligned} \left\| \sup_{1 \leqq j \leqq m} \left| \sum_{k=1}^{n} \alpha e_k \langle x_k, x_j' \rangle \right| \right\|_F &\leqq \|\alpha\|_0 \left\| \sup_{1 \leqq j \leqq m} \left| \sum_{k=1}^{n} e_k \langle x_k, x_j' \rangle \right| \right\|_E \\ &\leqq \|\alpha\|_0 \inf\left\{ \|u\| : u \geqq \left| \sum_{k=1}^{n} e_k \langle x_k, x_j' \rangle \right| \text{ for all } j = 1, 2, \cdots \right\} = \|\alpha\|_0 n_E(z). \end{aligned}$$

Finally, we obtain $n_F[(\alpha \otimes 1_X)z] \leqq \|\alpha\|_0 n_E(z)$. Since this is true for any $z \in E \otimes X$, we see that $\alpha \in \{\Phi_E \to \Phi_F\}$ and $\|\alpha\|_{\{\Phi_E \to \Phi_F\}} \leqq \|\alpha\|_0$. The lemma is proved.

From Lemmas 13 and 14 we derive the following theorem.

THEOREM 7. *Let E and F be KB-lineals satisfying conditions* A *and* B, *and let $\mathcal{K}$ be a category of Banach spaces containing spaces X_n for which there exist isometric embeddings $i_n : c_0^n \to X_n$, $n = 1, 2, \cdots$. Then for the functors Φ_E and Φ_F considered on $\mathcal{K}$ the space $\{\Phi_E \to \Phi_F\}$ is isometric to the space of K-bounded mappings $\alpha : E \to F$, with the norm*

$$\|\alpha\|_0 = \sup \frac{\|\sup_{1 \leqq k \leqq n} |\alpha e_k|\,\|_F}{\|\sup_{1 \leqq k \leqq n} |e_k|\,\|_E},$$

where the least upper bound is taken over all finite sets of elements $e_1, \cdots, e_n \in E$.

When E is a conditionally complete KB-lineal, we can give another description of the space $\{\Phi_E \to \Phi_F\}$.

On the space $(E \to F)_r$ of regular mappings of E to F we introduce a norm ν by $\nu(\alpha) = \|\,|\alpha|\,\|$ for $\alpha \in (E \to F)_r$.

THEOREM 8. *Let E be a KB-lineal, F a conditionally complete KB-lineal satisfying conditions* A *and* B *and such that either F is separable, or any bounded set in E (with respect to the ordering) is separable. Further, let $\mathscr{K}$ be a category of Banach spaces containing spaces X_n for which there exist isometric embeddings $i_n : c_0^n \to X_n$, $n = 1, 2, \cdots$. Then for the functors Φ_E and Φ_F considered on $\mathscr{K}$ the space $\{\Phi_E \to \Phi_F\}$ is isometric to the space $(E \to F)_r$ of regular mappings of E to F, with the norm ν.*

PROOF. Let $\alpha : E \to F$ be a regular mapping. By Proposition 6, for any Banach space X

$$\alpha \otimes 1_X \in (\Phi_E(X) \to \Phi_F(X)) \quad \text{and} \quad \|\alpha \otimes 1_X\| \leqq \|\,|\alpha|\,\|;$$

then $\alpha \in \{\Phi_E \to \Phi_F\}$ and $\|\alpha\|_{\{\Phi_E \to \Phi_F\}} \leqq \nu(\alpha) = \|\,|\alpha|\,\|$.

Now let $\alpha \in (E \to F)$ be not regular. Then there exists in E a bounded set Q (with respect to the ordering) such that αQ is not bounded (with respect to the ordering) in F ([**6**], p. 234). Choose a sequence of elements $e_n \in Q$ whose images αe_n are dense in αQ. Then the sequence $g_n = \sup_{1 \leqq k \leqq n} |\alpha e_k|$ is not bounded above in F; applying condition A, we see that $\|g_n\| \to \infty$ as $n \to \infty$. Consider the elements

$$z_n = \sum_{k=1}^{n} e_k \otimes i_n \delta_k^{(n)} \in E \otimes X_n,$$

where

$$\delta_k^{(n)} = (\underbrace{0, \cdots, 0, 1}_{k}, 0, \cdots, 0) \in c_0^n, \quad n = 1, 2, \cdots.$$

Using Proposition 7, we get

$$n_E(z_n) = n_E\left(\sum_{k=1}^{n} e_k \otimes \delta_k^{(n)}\right) = \inf\left\{\|u\| : u \geqq \left|\sum_{k=1}^{n} \lambda_k e_k\right|, \sum_{k=1}^{n} |\lambda_k| \leqq 1\right\}$$
$$= \left\|\sup_{1 \leqq k \leqq n} |e_k|\,\|_E \leqq \|e_0\right\|_E,$$

where e_0 is an element of E majorizing the moduli of elements of Q. On the other hand,

$$n_F[(\alpha \otimes 1_{X_n}) z_n] = \left\|\sup_{1 \leqq k \leqq n} |\alpha e_k|\right\|_F = \|g_n\|_F \to \infty$$

as $n \to \infty$. Then $\|\alpha \otimes 1_{X_n}\| \to \infty$ as $n \to \infty$, and so $\alpha \notin \{\Phi_E \to \Phi_F\}$. Thus we have proved that $(E \to F)_r = \{\Phi_E \to \Phi_F\}$ and $\|\alpha\|_{\{\Phi_E \to \Phi_F\}} \leqq \nu(\alpha)$ for any $\alpha \in (E \to F)_r$.

Now we prove the reverse inequality. Let $\alpha \in (E \to F)_f$. Fix a positive element $e_0 \in E$ and choose a sequence of elements $e_n \in Q_{e_0} = \{e : |e| \leqq e_0\}$ whose images αe_n are dense in αQ_{e_0}. Put $u_n = \sup_{1 \leqq k \leqq n} |\alpha e_k|$. Then

$$\sup_{e \in Q_{e_0}} |\alpha e| = \sup_n u_n$$

and by condition B $\|\sup_{|e| \leqq e_0} |\alpha e|\| = \lim_{n \to \infty} \|u_n\|$. Next we have

$$\begin{aligned} \|u_n\|_F &= \left\| \sup_{1 \leqq k \leqq n} |\alpha e_k| \right\|_F = n_F \left(\sum_{k=1}^n \alpha e_k \otimes \delta_k^{(n)} \right) \\ &= n_F \left[\sum_{k=1}^n \alpha e_k \otimes i_n \delta_k^{(n)} \right] \leqq \|\alpha\|_{\{\Phi_E \to \Phi_F\}} n_E \left[\sum_{k=1}^n e_k \otimes i_n \delta_k^{(n)} \right] \\ &= \|\alpha\|_{\{\Phi_E \to \Phi_F\}} n_E \left(\sum_{k=1}^n e_k \otimes \delta_k^{(n)} \right) \\ &= \|\alpha\|_{\{\Phi_E \to \Phi_F\}} \left\| \sup_{1 \leqq k \leqq n} |e_k| \right\|_E \leqq \|\alpha\|_{\{\Phi_E \to \Phi_F\}} \|e_0\|. \end{aligned}$$

Taking the limit as $n \to \infty$, we obtain

$$\left\| \sup_{|e| \leqq e_0} |\alpha e| \right\| \leqq \|\alpha\|_{\{\Phi_E \to \Phi_F\}} \|e_0\|.$$

Since $|\alpha| e_0 = \sup_{|e| \leqq e_0} |\alpha e|$ (see [6] or [7]),

$$\nu(\alpha) = \| |\alpha| \| = \sup_{0 \leqq e_0 \in E} \frac{\| |\alpha| e_0 \|}{\|e_0\|} = \sup_{0 \leqq e_0 \in E} \frac{\| \sup_{|e| \leqq e_0} |\alpha e| \|}{\|e_0\|} \leqq \|\alpha\|_{\{\Phi_E \to \Phi_F\}},$$

and the theorem is completely proved.

REMARK. The condition imposed on $\mathscr{K}$ in Theorems 7 and 8 is satisfied by the categories of all Banach spaces, of all reflexive Banach spaces, and also any category containing the space c_0 or all finite-dimensional spaces.

COROLLARY 1. *Under the assumptions of Theorem 8 on the KB-lineals E and F the space $(E \to F)_r$ is a conditionally complete KB-lineal.*

COROLLARY 2. *Let $\mathfrak{N}$ be simultaneously a minimal and maximal normed ideal of measurable functions. Then the ring $\{\mathfrak{N} \to \mathfrak{N}\}$ of operators of the corresponding functor $\mathfrak{N}$, considered on a category $\mathscr{K}$ that satisfies the conditions of Theorem 8, is isometric to the ring of regular operators of the normed ideal $\mathfrak{N}$, with the norm ν.*

In [5] an analogous result was proved for the functor $\mathfrak{N}$ defined by a maximal normed ideal $\mathfrak{N}$ in any category of reflexive spaces $\mathscr{K}$ that contains the spaces l_n^1, $n = 1, 2, \cdots$, and together with every X contains its adjoint space X'. (Note that such categories satisfy the condition of Theorem 8.)

In the context of this result it was conjectured in [**5**] that for any normed ideal $\mathfrak{N}$ the ring of operators $\{\mathfrak{N} \to \mathfrak{N}\}$ is isomorphic to the ring of regular mappings into a normed ideal, provided that the functor $\mathfrak{N}$ is considered on the category of all Banach spaces. By Corollary 2 of Theorem 8 the conjecture is true for normed ideals that are simultaneously maximal and minimal. We note that this result is new even for the functors L^p, $p > 1$.

We pass on to a description of the dual functor $D\Phi_E$. In accordance with Proposition 1 we identify the elements of the space $D\Phi_E(X) = \{\Phi_E \to \Sigma_X\}$ with the mappings $E \to X$ by which they are generated.

LEMMA 15. *Let E be a KB-lineal satisfying condition* C, *X a Banach space and $\alpha: E \to X$ a summable mapping. Then α generates an element of $D\Phi_E(X)$ in any category $\mathscr{K}$ containing X, and $\|\alpha\|_{D\Phi_E(X)} \leqq \nu_E(\alpha)$.*

PROOF. Choose any $Y \in \mathscr{K}$ and

$$z = \sum_{k=1}^{n} e_k \otimes y_k \in E \otimes Y.$$

Then

$$\gamma[(\alpha \otimes 1_Y)z] \leqq \sum_{k=1}^{n} \|\alpha e_k\| \, \|y_k\| = \sum_{k=1}^{n} \| \alpha e_k \| \, y_k \| \, \|_X \leqq \nu_E(\alpha) \left\| \sum_{k=1}^{n} |e_k| \, \|y_k\| \right\|_E,$$

and since this is true for any representation $z = \sum_{k=1}^{n} e_k \otimes y_k$ and E satisfies condition C,

$$\gamma[(\alpha \otimes 1_Y)z] \leqq \nu_E(\alpha) \inf\left\{ \left\| \sum_{k=1}^{n} |e_k| \, \|y_k\| \right\|_E : z = \sum_{k=1}^{n} e_k \otimes y_k \right\} = \nu_E(\alpha) n_E(z).$$

Thus $\alpha \otimes 1_Y \in (E \tilde{\otimes} Y \to X \hat{\otimes} Y)$ and $\|\alpha \otimes 1_Y\| \leqq \nu_E(\alpha)$ for any $Y \in \mathscr{K}$; consequently $\alpha \in D\Phi_E(X)$ and $\|\alpha\|_{D\Phi_E(X)} \leqq \nu_E(\alpha)$.

LEMMA 16. *Let E be a KB-lineal, let $\mathscr{K}$ be a category of Banach spaces containing spaces X_n for which there exist isometric embeddings $i_n: l_n^1 \to X_n$, $n = 1, 2, \cdots, X \in \mathscr{K}$, and let $\alpha: E \to X$ generate an element of $D\Phi_E(X)$, where the functor Φ_E is considered on $\mathscr{K}$. Then α is a summable mapping and $\nu_E(\alpha) \leqq \|\alpha\|_{D\Phi_E(X)}$.*

PROOF. Let $\sum_{k=1}^{\infty} e_k$ be a series in E for which $\sum_{k=1}^{\infty} |e_k|$ is convergent. Consider the elements

$$z_n = \sum_{k=1}^{n} e_k \otimes i_n \delta_k^{(n)} \in E \otimes X_n,$$

where

$$\delta_k^{(n)} = (\underbrace{0, \cdots, 0, 1}_{k}, 0, \cdots, 0) \in l_n^1, \quad n = 1, 2, \cdots.$$

The subspace $i_n l_n^1 \subset X_n$ is finite dimensional and hence complemented in X_n; thus

$$\gamma[(\alpha\otimes 1_{X_n})z_n] = \left\| \sum_{k=1}^{n} \alpha e_k \otimes i_n\delta_k^{(n)} \right\|_{X\hat{\otimes} X_n}$$
$$= \left\| \sum_{k=1}^{n} \alpha e_k \otimes i_n\delta_k^{(n)} \right\|_{X\hat{\otimes} i_n l_n^1} = \left\| \sum_{k=1}^{n} \alpha e_k \otimes \delta_n^{(n)} \right\|_{X\hat{\otimes} l_n^1}.$$

By standard arguments it can be shown that

$$\left\| \sum_{k=1}^{n} \alpha e_k \otimes \delta_k^{(n)} \right\|_{X\hat{\otimes} l_n^1} = \sum_{k=1}^{n} \|\alpha e_k\|.$$

Since α generates an element of $D\Phi_E(X)$,

$$\sum_{k=1}^{n} \|\alpha e_k\|_X = \gamma[(\alpha\otimes 1_{X_n})z_n] \leqq \|\alpha\|_{D\Phi_E(X)} n_E(z_n).$$

Further, by Proposition 7

$$n_E(z_n) = n_E\left(\sum_{k=1}^{n} e_k \otimes \delta_k^{(n)}\right),$$

and since the space adjoint to l_n^1 is isometric to c_0^n we get

$$n_E(z_n) = n_E\left(\sum_{k=1}^{n} e_k \otimes \delta_k^{(n)}\right) = \inf\left\{ \|u\| : u \geqq \left|\sum_{k=1}^{n} \lambda_k e_k\right|, |\lambda_k| \leqq 1,\ k = 1,\cdots,n\right\}$$
$$\leqq \left\| \sum_{k=1}^{n} |e_k| \right\|_E \leqq \left\| \sum_{k=1}^{\infty} |e_k| \right\|_E.$$

Thus

$$\sum_{k=1}^{n} \|\alpha e_k\|_X \leqq \|\alpha\|_{D\Phi_E(X)} \left\| \sum_{k=1}^{\infty} |e_k| \right\|_E.$$

Taking the limit as $n\to\infty$ we see that

$$\sum_{k=1}^{\infty} \|\alpha e_k\|_X \leqq \|\alpha\|_{D\Phi_E(X)} \left\| \sum_{k=1}^{\infty} |e_k| \right\|_E,$$

so that α is a summable mapping. Taking the least upper bound over all finite sets of elements $e_1,\cdots,e_n\in E$ on the left-hand side of the inequality

$$\frac{\sum_{k=1}^{n}\|\alpha e_k\|_X}{\|\sum_{k=1}^{n}|e_k|\,\|_E} \leqq \|\alpha\|_{D\Phi_E(X)}$$

we obtain $\nu_E(\alpha) \leqq \|\alpha\|_{D\Phi_E(X)}$.

Lemmas 15 and 16 imply the following theorem.

THEOREM 9. *Let E be a KB-lineal satisfying condition* C, *and $\mathcal{K}$ a category of Banach spaces containing spaces X_n for which there exist isometric embeddings $i_n: l_n^1\to X_n$, $n = 1, 2, \cdots$. Then for the functor Φ_E considered on $\mathcal{K}$ and for any Banach space $X\in\mathcal{K}$, the space $D\Phi_E(X)$ is isometric to the space $S(E, X)$ of summable mappings of E to X.*

REMARK. Any category containing the space l^1 or all finite-dimensional spaces satisfies the condition of Theorem 8.

COROLLARY [5]. *For the functor $C = C_Q$, considered on the category of all Banach spaces, and any X, the space $DC(X)$ is isometric to the space of all absolutely summable mappings $\alpha: C \to X$ with the norm*

$$\|\alpha\|_{\text{a.s.}} = \inf\left\{ M : \sum_{k=1}^{n} \|\alpha\phi_k\| \leq M \sup_{|\epsilon_k| \leq 1} \left\| \sum_{k=1}^{n} \epsilon_k \phi_k \right\| \right.$$
$$\left. \textit{for any finite set } \phi_1, \cdots, \phi_n \in C \right\}.$$

PROOF.[18] By Proposition 4 the space C satisfies condition C, so that we can apply Theorem 9. Since in C the absolute convergence of the series $\sum_{k=1}^{\infty} \phi_k$ is equivalent to the convergence of the series of moduli $\sum_{k=1}^{\infty} |\phi_k|$, the classes of summable and absolutely summable mappings of C to X are the same. We prove the equality of the norms ν_C and $\|\cdot\|_{\text{a.s.}}$. Clearly

$$\|\alpha\|_{\text{a.s.}} = \sup \frac{\sum_{k=1}^{n} \|\alpha\phi_k\|_X}{\sup_{|\epsilon_k| \leq 1} \|\sum_{k=1}^{n} \epsilon_k \phi_k\|_C},$$

where the least upper bound is taken over all finite sets of elements $\phi_1, \cdots, \phi_n \in C$. Further,

$$\sup_{|\epsilon_k| \leq 1} \left\| \sum_{k=1}^{n} \epsilon_k \phi_k \right\|_C = \sup_{|\epsilon_k| \leq 1} \max_{t \in Q} \left| \sum_{k=1}^{n} \epsilon_k \phi_k(t) \right| = \max_{t \in Q} \sum_{k=1}^{n} |\phi_k(t)| = \left\| \sum_{k=1}^{n} |\phi_k| \right\|_C.$$

Substituting this expression in the formula for $\|\alpha\|_{\text{a.s.}}$, we obtain $\|\alpha\|_{\text{a.s.}} = \nu_C(\alpha)$.

In [5] it is proved that in the category of all reflexive Banach spaces for any minimal normed ideal $\mathfrak{N}$ the dual functor $D\mathfrak{N}$ is defined by an associated ideal $\mathfrak{N}^*$; that is, $D\mathfrak{N}(X)$ is naturally isometric to $\mathfrak{N}^*(X)$. In the context of this result the problem is raised in [5] of describing the functor $D\mathfrak{N}$ on the category of all Banach spaces, and it is conjectured that in this case $D\mathfrak{N} = \mathfrak{N}^*$ for any normed ideal $\mathfrak{N} \neq L^\infty$. We show that this conjecture is not true for every $\mathfrak{N}$.

PROPOSITION 11. *The identity embedding $\alpha: \mathfrak{N} \to L^1$ defines an element of $D\mathfrak{N}(L^1)$ that does not belong to $\mathfrak{N}^*(L^1)$.*

PROOF. Obviously $\alpha \in \{\mathfrak{N} \to L^1\}$, and since $L^1(X) = L^1 \hat{\otimes} X$ for any X we have $\alpha \in D\mathfrak{N}(L^1)$. We show that $\alpha \notin \mathfrak{N}^*(L^1)$. Suppose the contrary. By Lemma 10 the identity embedding $\beta: \mathfrak{N}^* \to L^1$ gives rise to an embedding $\beta_{L^1}: \mathfrak{N}^*(L^1) \to L^1(L^1)$. Let $\alpha(s,t)$ be the function of $L^1(L^1)$ corresponding to α under this embedding; that is, $\alpha(s,t) = \beta_{L^1}\alpha$. Consider the sequence of functions $u_n = u_n(s,t) \in [\mathfrak{N}^*(L^1)]_0$ convergent to α in $\mathfrak{N}^*(L^1)$. Then $u_n(s,t) \to \alpha(s,t)$ in $L^1(L^1)$; that is, $\int_0^1 \{\int_0^1 |u_n(s,t) - \alpha(s,t)|\,dt\}ds \to 0$ as $n \to \infty$. Obviously the integral $\int_0^1 \alpha(s,t)\phi(s)\,ds$ is defined for any function $\phi \in L^\infty$,

[18] In [5] this result is proved differently.

and its value lies in L^1. On the other hand, L^∞ is contained in $\mathfrak{N}$, so that the mapping α is defined on L^∞; we show that these two mappings coincide. Let $\phi \in L^\infty$. Then

$$[\alpha\phi](t) - \int_0^1 \alpha(s,t)\phi(s)\,ds = \lim_{n\to\infty} [u_n\phi](t) - \int_0^1 \alpha(s,t)\phi(s)\,ds$$
$$= \lim_{n\to\infty} \int_0^1 u_n(s,t)\phi(s)\,ds - \int_0^1 \alpha(s,t)\phi(s)\,ds$$
$$= \lim_{n\to\infty} \int_0^1 [u_n(s,t) - \alpha(s,t)]\phi(s)\,ds,$$

where the limit is taken in L^1. Now

$$\left\| [\alpha\phi](t) - \int_0^1 \alpha(s,t)\phi(s)\,ds \right\|_{L^1} = \lim_{n\to\infty} \left\| \int_0^1 [u_n(s,t) - \alpha(s,t)]\phi(s)\,ds \right\|_{L^1}$$
$$\leqq \lim_{n\to\infty} \left\| \int_0^1 |u_n(s,t) - \alpha(s,t)|\,ds \,\|\phi\|_{L^\infty} \right\|_{L^1}$$
$$= \lim_{n\to\infty} \int_0^1 \left\{ \int_0^1 |u_n(s,t) - \alpha(s,t)|\,ds \right\} dt\, \|\phi\|_{L^\infty}$$
$$= \lim_{n\to\infty} \int_0^1 \left\{ \int_0^1 |u_n(s,t) - \alpha(s,t)|\,dt \right\} ds\, \|\phi\|_{L^\infty} = 0,$$

so that $[\alpha\phi](t) = \int_0^1 \alpha(s,t)\phi(s)\,ds$ almost everywhere for any $\phi \in L^\infty$. On the other hand, by definition of the mapping $\alpha : \mathfrak{N} \to L^1$ we have $[\alpha\phi](t) = \phi(t)$. Since $L^1(L^1) = L^1 \hat{\otimes} L^1$ is isometrically embedded in $(L^\infty)' \hat{\otimes} L^1$, the integral operator with kernel $\alpha(s,t) \in L^1(L^1)$ is a nuclear mapping of L^∞ to L^1. This gives us a contradiction, since this operator is the identity embedding of L^∞ in L^1, which is not only not nuclear, but not even completely continuous.

For a minimal normed ideal $\mathfrak{N}$, Theorem 9 completely solves the problem of describing the functor $D\mathfrak{N}$.

In conclusion we prove a theorem that generalizes a well-known result of Hilbert-Schmidt on the nuclearity of the product of two operators.

THEOREM 10. *Let X and Y be Banach spaces, E a KB-lineal satisfying conditions* A *and* C *for which Φ_E satisfies the one-to-one condition, $\alpha : X \to E$ a proper mapping approximated by finite-dimensional mappings, and $u : E \to Y$ a summable mapping. Then the product $u \circ \alpha$ is a nuclear mapping of X to Y.*

PROOF. By Theorems 6 and 9 we know that $\alpha \in \Phi_E(X')$ and $u \in D\Phi_E(Y)$. Further, by Lemma 2 the following diagram is commutative:

$$\begin{array}{ccc}
\Phi_E(X') & \xrightarrow{\;u \otimes 1_{X'}\;} & Y \hat{\otimes} X' \\
\Big\downarrow k_{X'}^{\Phi_E} & & \Big\downarrow k_{X'}^{\Sigma_Y} \\
E \hat{\hat{\otimes}} X' & \xrightarrow{\;u \otimes 1_{X'}\;} & Y \hat{\otimes} X'
\end{array}$$

Let $\alpha_0 = \sum_{k=1}^{n} \langle\ , x'_k \rangle e_k$ be a finite-dimensional mapping of X to E, considered as an element of $\Phi_E(X')$. Then $(u \otimes 1_{X'}) k_{X'}^{\Phi E} \alpha_0 = (u \otimes 1_{X'}) \alpha_0 = \sum_{k=1}^{n} \langle\ , x'_k \rangle u e_k = u \circ \alpha_0$. Choose a sequence of finite-dimensional mappings $\alpha_n : X \to E$ convergent to α in $\Phi_E(X')$. Then $(u \otimes 1_{X'}) k_{X'}^{\Phi E} \alpha_n \to (u \otimes 1_{X'}) k_{X'}^{\Phi E} \alpha$, and thus $(u \otimes 1_{X'}) k_{X'}^{\Phi E} \alpha = u \circ \alpha$. From the diagram it follows that $u \circ \alpha = k_{X'}^{\check{\Sigma} Y} (u \otimes 1_{X'}) \alpha$ and $(u \otimes 1_{X'}) \alpha \in Y \hat{\otimes} X'$, so that $u \circ \alpha$ is a nuclear mapping of Z to Y.

REMARK. The Hilbert-Schmidt Theorem on the nuclearity of the product of two operators is a consequence of Theorem 10, if in the latter we put $E = l^2$ and take X and Y to be Hilbert spaces.

COROLLARY. *Let X and Y be Banach spaces, $\alpha : X \to C$ a completely continuous linear mapping, and $\beta : C \to Y$ an absolutely summable mapping. Then the product $\beta\alpha : X \to Y$ is nuclear.*

Received 10 NOV 1967

REFERENCES

[1] R. Schatten, *A theory of cross-spaces*, Ann. of Math. Studies, no. 26, Princeton Univ. Press, Princeton, N. J., 1950. MR **12**, 186.

[2] A. Grothendieck, *Produits tensoriels topologiques et espaces nucléaires*, Mem. Amer. Math. Soc. No. 16 (1955). MR **17**, 763.

[3] A. S. Švarc, *Duality of functors*, Dokl. Akad. Nauk SSSR **148** (1968), 288-291 = Soviet Math. Dokl. **4** (1963), 89-92. MR **29** #5887.

[4] ———, *Functors in categories of Banach spaces*, Dokl. Akad. Nauk SSSR **149** (1963), 44-47 = Soviet Math. Dokl. **4** (1963), 317-320. MR **27** #4046.

[5] B. S. Mitjagin and A. S. Švarc, *Functors in categories of Banach spaces*, Uspehi Mat. Nauk **19** (1964), no. 2 (116), 65-130 = Russian Math. Surveys **19** (1964), no. 2, 65-127. MR **29** #3866.

[6] L. V. Kantorovič, B. Z. Vulih and A. G. Pinsker, *Functional analysis in partially ordered spaces*, GITTL, Moscow, 1950. (Russian) MR **12**, 340.

[7] B. Z. Vulih, *Introduction to the theory of partially ordered spaces*, Fizmatgiz, Moscow, 1961; English transl., Noordhoff, Groningen, 1967. MR **24** #A3494; **37** #121.

[8] V. L. Levin, *Functors in categories of Banach spaces defined by KB-lineals*, Dokl. Akad. Nauk SSSR **162** (1965), 262-265 = Soviet Math. Dokl. **6** (1965), 648-651. MR **33** #4652.

[9] ———, *Tensor products and functors in Banach space categories defined by KB-lineals*, Dokl. Akad. Nauk SSSR **163** (1965), 1058-1060 = Soviet Math. Dokl. **6** (1965), 1059-1061. MR **33** #3077.

[10] S. Kakutani, *Concrete representation of abstract (L)-spaces and the mean ergodic theorem*, Ann. of Math. (2) **42** (1941), 523-537. MR **2**, 318.

[11] ———, *Concrete representation of abstract (M)-spaces. (A characterization of the space of continuous functions)*, Ann. of Math. (2) **42** (1941), 994-1024. MR **3**, 205.

[12] N. Dunford and J. T. Schwartz, *Linear operators*. I: *General theory*, Pure and Appl. Math., vol. 7, Interscience, New York, 1958; Russian transl., IL, Moscow, 1962. MR **22** #8302.

[13] A. Pietsch, *Nukleare lokalkonvexe Räume*, Akademie Verlag, Berlin, 1965; Russian transl., "Mir", Moscow, 1967. MR **31** #6114; **36** #4314.

[14] A. I. Veksler, *The concept of a linear lattice which is normal in itself, and certain applications of this concept to the theory of linear and normed linear lattices*, Izv. Vysš. Učebn. Zaved. Matematika **1966**, no. 4 (53), 13-22. (Russian) MR **34** #595.

[15] V. L. Levin, *Two classes of linear mappings which operate between Banach spaces and Banach lattices*, Sibirsk. Mat. Ž. **10** (1969), 903-909. (Russian) MR **40** #3268.

Translated by W. Stephenson

Trudy Moskov. Mat. Obšč.
Tom 20 (1969)

Trans. Moscow Math. Soc.
Vol. 20 (1969)

TENSORS WITH A GIVEN INVARIANCE GROUP AND SPHERICAL FUNCTIONS ON HOMOGENEOUS SPACES

P. K. RAŠEVSKIĬ

Contents

INTRODUCTION 79
§1. A fundamental lemma 81
§2. Lemma 3 (the contravariant case) 89
§3. Theorems A, A′ and A″ 93
§4. On spherical functions of homogeneous spaces 101
REFERENCES 107

Introduction

The following problem occupies a central position in this article. Suppose that we are given a system of tensors $\{z\}$ in the vector space $E^n(K)$ over the field K, and let w be some tensor admitting the invariance group of the system $\{z\}$. This means that w passes into itself under the action of any automorphism A of the space $E^n(K)$ that sends the tensors $\{z\}$ into themselves. We can then ask under what conditions the tensor w is expressible in terms of the tensors z, and in what way. A preliminary answer to this question is given in Theorem A (§3.2), which says in essence that w is expressible as the "quotient" of two comitants of the tensors $\{z\}$ and constant tensors (that is, tensors whose coordinates do not change when subjected to any tensorial transformation). We note that the tensors we consider are, generally speaking, relative, so that apart from δ^i_j, examples of such tensors are the n-vectors $\omega^{i_1\cdots i_n}$ of weight 1 ($\omega^{12\cdots n}=1$ in any basis), briefly written $\bar{\omega}$. By a comitant we understand a tensor that is formed from a given set of tensors by means of the standard tensorial operations. Without giving the full statement of the theorem here, we indicate some special cases of particular interest.

If z and w are contravariant tensors, then

$$wT(z, \bar{\omega}) = S(z, \bar{\omega});$$

and if z and w are absolute tensors, then

$$wT(z, \delta) = S(z, \delta);$$

if both properties hold, then

$$wT(z) = S(z).$$

The symbols T and S stand for comitants of the tensors in the brackets, and the left-hand side of each equation consists of the product of the tensor w with the comitant T ($\neq 0$).

It has been assumed here that the field K has characteristic zero, is algebraically closed, and contains "sufficiently many transcendentals" (for the exact formulation of this, see the beginning of §2).

The question arises as to whether Theorem A can be improved to the extent that the "fractional" character of the result is removed, and w is expressed simply as a comitant of z (and constant tensors). While this is impossible, in general, it can be done in the important special case where the invariance group of $\{z\}$ is irreducible.

The proof is based on Lemma 1, which is established in §1; it asserts that any nonzero contravariant tensor has a comitant of a certain specific structure, namely one which for a suitable basis $\{e_1, \cdots, e_n\}$ takes the form of the product of tensors

$$(*) \qquad e_1 \dots e_1 [e_1 e_2] \dots [e_1 e_2] [e_1 e_2 e_3] \dots [e_1 e_2 e_3 \dots e_n],$$

where $[e_1 e_2], [e_1 e_2 e_3], \cdots, [e_1 e_2 e_3 \cdots e_n]$ is a bivector, trivector, $\cdots$, n-vector, respectively, formed from the basis vectors $e_1, e_2, \cdots, e_n$.

Using Lemma 1, we reduce the comitant T which figures in Theorem A to this specific form (Theorem A$'$, §3.3); but then, in the case of an irreducible invariance group, the product (*) can only contain the n-vectors $[e_1 e_2 \cdots e_n]$, and this allows us to eliminate the comitant T and to express w simply in the form of a comitant of z and constant tensors, perhaps with a relative invariant factor (Theorem A$''$, §3.3, formulas (3.11) and (3.12)).

Finally, in §§3.7 and 3.8 we prove Theorems B and B$'$, which are specializations of Theorem A to the case when the invariance group of $\{z\}$ is a semisimple or a "slightly extended" semisimple group. Here we succeed in a certain sense in freeing the result from its "fractional" character.

Theorems B and B$'$ serve as a basis for the transition to §4, where we study spherical functions, first on complex-analytic homogeneous spaces $\mathfrak{G}/\mathfrak{H}$, where $\mathfrak{H}$ is a closed subgroup of the complex semisimple Lie group $\mathfrak{G}$. We investigate under what conditions all the spherical functions are expressible as polynomials in the spherical functions of a single system. It is a very curious fact that in some homogeneous spaces every spherical function is expressible as a polynomial in any system of spherical functions. This is the case if $\mathfrak{G}$ is a connected complex semisimple Lie group, $\mathfrak{H}$ a maximal subgroup and in some faithful linear representation of $\mathfrak{G}$ the connected component of the identity $\mathfrak{H}_0$ of $\mathfrak{H}$ is irreducible (of course, the representation of $\mathfrak{G}$ is a fortiori irreducible). These results are stated in the form of Theorems C, C$'$, C$''$ (§4.4).

Finally, in §§4.6 and 4.7 we transfer the results obtained to the case of compact homogeneous spaces $\mathfrak{G}/\mathfrak{H}$, with $\mathfrak{G}$ and $\mathfrak{H}$ compact. It turns out that any such space is naturally embeddable in its corresponding complex-analytic homogeneous space in such a way that the systems of spherical functions on the former are obtained by restriction from those on the latter (Proposition 3, §4.7). This allows us to transfer the results to the compact case.

An abbreviated version of the contents of §§1—3 was published in [**1**].

§1. A fundamental lemma

1. We consider a vector space $E^n(K)$ over a field K of characteristic zero, having a basis $\{e_1, \cdots, e_n\}$. By a tensor $a^{i_1\cdots i_k}_{j_1\cdots j_l}$ we mean a system of scalars that transform under a change of basis in accordance with the rule

$$a^{i'_1\cdots i'_k}_{j'_1\cdots j'_l} = A^{i'_1}_{i_1}\dots A^{i'_k}_{i_k} A^{j_1}_{j'_1}\dots A^{j_l}_{j'_l}\, a^{i_1\cdots i_k}_{j_1\cdots j_l}\,(\mathrm{Det}\,|\,A^i_{i'}\,|)^p,$$

where p is an integer (positive, negative, or zero) and the change of basis has the form

$$e_{i'} = A^i_{i'}e_i,$$

$\|A^i_{i'}\|$ being a nonsingular matrix $(A^i_{i'} \in K)$ and $\|A^{i'}_i\|$ is its inverse. The number p is called the *weight* of the tensor, and in the case $p = 0$ the tensor is called *absolute*. (Thus, by a tensor we mean, generally speaking, a relative tensor). In the usual way, we can perform the following operations on tensors: 1) addition, 2) multiplication by an element of K, 3) permutation of the indices, 4) multiplication of a tensor by a tensor, 5) contraction (with respect to one upper and one lower index).

Addition can only be performed in the case of tensors of the same structure (that is, with the same p, k and l); multiplication of tensors is associative but not commutative. In cases where it is unnecessary to specify the indices we simply write $a + b$ and ab for the sum and product of tensors, respectively. Tensors that can be obtained by means of the operations 1—5 from a given collection of tensors are called their comitants. Important examples are antisymmetrization (or alternation) and symmetrization: $a^{[i_1\cdots i_k]}$ and $a^{(i_1\cdots i_k)}$. (For us, division by $k!$ is always implicit, k being the number of indices taking part in the operation.) We write briefly $[a_1a_2\cdots a_k]$ for the tensor $a_1^{[i_1}a_2^{i_2}\cdots a_k^{i_k]}$ obtained by multiplication of the vectors $a_1, a_2, \cdots, a_k$ followed by antisymmetrization.

We introduce some notation: for the nonzero contravariant n-vector $\epsilon^{i_1i_2\cdots i_n}$ (an absolute tensor, skew-symmetric in all n indices) we simply write $\bar{\epsilon}$, and the analogous covariant n-vector $\epsilon_{i_1i_2\cdots i_n}$ is denoted by ϵ. They each have one essential coordinate: $\epsilon^{12\cdots n}$ (invariant of weight -1) and $\epsilon_{12\cdots n}$ (invariant of weight $+1$) and are defined to within an invariant factor $(\neq 0)$; we assume these factors to be chosen so that

$$\varepsilon^{12\dots n}\varepsilon_{12\dots n} = 1.$$

Putting

$$\omega^{i_1\cdots i_n} = \varepsilon^{i_1\cdots i_n}(\varepsilon^{12\ldots n})^{-1}, \quad \omega_{i_1\ldots i_n} = \varepsilon_{i_1\ldots i_n}(\varepsilon_{12\ldots n})^{-1},$$

we obtain relative n-vectors of weights 1 and -1, respectively, and, in any basis,

$$\omega^{12\ldots n} = \omega_{12\ldots n} = 1.$$

These are denoted briefly by $\bar{\omega}$ and ω, respectively. It is well known that $\bar{\omega}$ and its comitants exhaust the set of all constant contravariant tensors ([**2**], Russian p. 175; a tensor is called constant if the values of its coordinates are independent of the basis, the tensor δ^i_j, for example); an analogous result holds for ω.

In particular, we need the tensor

$$\Omega^{p_1\ldots p_k;\, q_1\ldots q_k;\, \ldots;\, t_1\ldots t_k},$$

having kn^k indices, divided up into $N = n^k$ "segments" containing k indices each, and skew-symmetric in all N segments; let its weight be kn^{k-1}. An ordered set of k indices (repetitions being allowed) is denoted by a single letter:

$$J = j_1 \cdots j_k,$$

so that the coordinates of Ω can be written in the form $\Omega^{J'_1\cdots J'_N}$, where $J'_1, \cdots, J'_N$ are arbitrary collections ("superindices"). It is obvious that Ω has only one essential coordinate, namely, that for which the $J'_1, \cdots, J'_N$ are all different and hence involve all the $N = n^k$ possible values of the superindex J. We denote these values by $J_1, \cdots, J_N$ in an arbitrary order. The weight kn^{k-1} of the tensor Ω is chosen so that the essential coordinate $\Omega^{J_1\cdots J_N}$ is the same in all bases; we assume that

$$\Omega^{J_1\ldots J_N} = 1.$$

Obviously, the tensor Ω reduces to $\bar{\omega}$ in case $k = 1$. Since Ω is a constant tensor, we have

$$\Omega = \Phi(\bar{\omega}),$$

where $\Phi(\bar{\omega})$ is a comitant of $\bar{\omega}$, homogeneous of degree kn^{k-1} relative to $\bar{\omega}$, as follows from comparing weights. The actual form of $\Phi(\bar{\omega})$ is of no interest to us.

Multiplying both sides of (1.1) by $(\epsilon^{12\cdots n})^{kn^{k-1}}$ we obtain on the left-hand side a certain tensor $\mathscr{E}$:

$$\mathscr{E}^{p_1\ldots p_k;\, q_1\ldots q_k;\, \ldots;\, t_1\ldots t_k}$$

of the same structure as Ω, but absolute (of weight 0); on the right-hand side $\bar{\omega}$ is replaced by $\bar{\epsilon}$:

$$\mathscr{E} = \Phi(\bar{\varepsilon}).$$

2. Lemma 1 (fundamental lemma). *Let $a^{i_1\cdots i_k} \neq 0$ be a given absolute contravariant tensor in $E^n(K)$, K being a field of characteristic zero. Then we*

can always construct a (nonzero) comitant c of the tensor a which in a suitable basis $\{e_1, \cdots, e_n\}$ *can be written as follows:*

$$c = e_1 \ldots e_1 [e_1e_2] \ldots [e_1e_2] \ldots [e_1e_2 \ldots e_n] \ldots [e_1e_2 \ldots e_n]; \tag{1.2}$$

that is, as a product of the tensors $e_1, [e_1e_2], [e_1e_2e_3], \cdots, [e_1 \cdots e_n]$, *each of which occurs with a specific (possibly zero) multiplicity.*

PROOF. 1) Let $x^{i_1 \cdots i_k}$ be an absolute contravariant tensor with indeterminate coordinates, and let $\overset{(1)}{y}_{i_1}, \overset{(2)}{y}_{i_1i_2}, \cdots, \overset{(n)}{y}_{i_1i_2\cdots i_n}$ be absolute covariant tensors, also with unspecified coordinates. Consider the set of all nonzero polynomials $J(x, y)$ (having zero constant term) in the coordinates of the tensors $x, \overset{(1)}{y}, \cdots, \overset{(n)}{y}$, that are absolute invariants of the tensorial transformations.

As is well known (see [**2**], Russian p. 180, for example), these invariants J are linear combinations of the polynomials P obtained by complete contraction:

$$P = x^{i_1 \ldots i_k} x^{i_{k+1} \ldots i_{2k}} \ldots x^{i_{sk-k+1} \ldots i_{sk}} \overset{(1)}{y}_{*} \ldots \overset{(1)}{y}_{*} \overset{(2)}{y}_{**} \ldots \overset{(2)}{y}_{**} \ldots \overset{(n)}{y}_{*\cdots *} \ldots \overset{(n)}{y}_{*\cdots *}, \tag{1.3}$$

where by the sequence of asterisks we mean an arbitrary permutation of the indices $i_1, \cdots, i_{sk}$.

By the first fundamental theorem of the theory of invariants we can choose a finite number of invariants of the form (1.3) [we call them $P_1, \cdots, P_N$] in such a way that all the remaining invariants are polynomials in them (having zero constant term):

$$P = Q(P_1, \ldots, P_N). \tag{1.4}$$

2) We now turn from indeterminate to specialized tensors x, y; by this we mean that the coordinates of x and y are subject to a definite tensor-invariant system of algebraic relations over K; in particular, x and y may be completely determined tensors.

Equation (1.4), being an identity in x and y, remains valid here (although we can no longer assert that the polynomials P given by (1.3) exhaust the set of all invariant polynomials in the coordinates of x and y).

3) In particular, we specialize x and y by putting

$$x^{i_1 \ldots i_k} = a^{i_1 \ldots i_k},$$

$$\overset{(1)}{y} = y, \; \overset{(2)}{y} = [yz], \; \overset{(3)}{y} = [yzu], \; \ldots, \overset{(n)}{y} = [yzu \ldots w], \tag{1.5}$$

where $a^{i_1 \cdots i_k}$ is the tensor in the conditions of the lemma, and $y, z, u, \cdots, w$ are n covectors with indeterminate coordinates. Any contraction (1.3) now assumes the form

$$P = a^{i_1 \ldots i_k} \ldots a^{i_{sk-k+1} \ldots i_{sk}} y_* \ldots y_* [yz]_{**} \ldots [yz]_{**} \ldots [yzu \ldots w]_{*\cdots *} \tag{1.6}$$

We wish to show that the contractions (1.6) are not all equal to zero, and hence that $P_i \neq 0$, where $1 \leqq i \leqq N$.

In fact, in the complete contraction (1.6), the antisymmetrizations $[yz] \cdots$

$[yz \cdots w]$ can be transferred to the corresponding indices in $a^{**\cdots *} \cdots a^{**\cdots *}$, the resulting tensor being then symmetrized with respect to the indices contracted with $y_* y_* \cdots y_*$, $z_* z_* \cdots z_*$, $\cdots$, $w_* w_* \cdots w_*$ in turn. As a result, the product $a^{**\cdots *} \cdots a^{**\cdots *}$ gives rise to a Young tensor (in symmetrized form), which will be contracted with $y_* \cdots y_* z_* \cdots z_* \cdots w_* \cdots w_*$; the result then coincides with the original contraction (1.6).

Let us suppose that the contractions in (1.6) are all equal to zero. Then it follows that any Young tensor obtained from the product $a^{**\cdots *} \cdots a^{**\cdots *}$ in accordance with the usual scheme (antisymmetrization followed by symmetrization) is equal to 0 in view of the indeterminacy of the covectors $y, z, \cdots, w$; but since any tensor decomposes into its Young components, it follows that

$$a^{**\cdots *} \cdots a^{**\cdots *} = 0.$$

This contradicts the hypothesis that $a \neq 0$.

4) Further, by $P_1, \cdots, P_n$ we mean the previous sequence, in which, however, only the nonzero P_i are retained.

As is clear from (1.6), any (nonzero) P is a homogeneous polynomial in the coordinates of each of the covectors $y, z, u, \cdots, w$. Let $\deg_P\{w\}$ be the degree of P with respect to w, and $\deg_P\{y, z, \cdots, w\}$ the sum of the degrees of P with respect to $y, z, \cdots, w$. We put

$$\theta_P = \deg_P\{w\}/\deg_P\{y, z, \ldots, w\}. \tag{1.7}$$

Suppose that θ_{P_i} $(i = 1, \cdots, N)$ assumes its maximum at $i = 1$, say. Then for any (nonzero) P

$$\theta_P \leqslant \theta_{P_1}. \tag{1.8}$$

For

$$P = Q(P_1, \ldots, P_N),$$

and on the right- and left-hand sides only monomials of fixed degrees in $y, z, \cdots, w$ occur; using any term $\alpha P_{i_1} \cdots P_{i_m}$ of the polynomial Q ($\alpha \in K$, $\alpha \neq 0$), we have

$$\deg_P\{w\} = \deg_{P_{i_1}}\{w\} + \ldots + \deg_{P_{i_m}}\{w\}$$

$$\leqslant \theta_{P_1}(\deg_{P_{i_1}}\{y, z, \ldots, w\} + \ldots + \deg_{P_{i_m}}\{y, z, \ldots, w\}) = \theta_{P_1} \deg_P\{y, z, \ldots, w\}.$$

This proves (1.8).

5) Writing the polynomial P_1 in the form (1.6) we eliminate on the right-hand side all the factors $y_* \cdots y_* [yz]_{**} \cdots [yz]_{**} \cdots$, except the highest in rank; that is, except for $[yz \cdots w]_{**\ldots *}$ which we retain unchanged. Then some of the indices in $a^{**\cdots *} \cdots a^{**\cdots *}$, become free, and instead of an invariant we obtain a certain (nonzero) tensor $b^{j_1 \cdots j_q}$ of the following form:

$$b^{j_1\ldots j_q} = \beta^{j_1\ldots j_q;\, \alpha_1\ldots\alpha_n;\, \ldots;\, \delta_1\ldots\delta_n} [yz \ldots w]_{\alpha_1\ldots\alpha_n} \cdots [yz \ldots w]_{\delta_1\ldots\delta_n}, \tag{1.9}$$

where the tensor β is a comitant of a, skew-symmetric in each of $\alpha_1 \cdots \alpha_n, \cdots$, $\delta_1 \cdots \delta_n$.

We now take the nth power of b (n being the dimension of our space) and perform antisymmetrization with respect to a set of indices chosen arbitrarily one from each factor:

$$b^{j_1 j_2 \cdots j_q} b^{j_{q+1} \cdots j_{2q}} \cdots b^{j_{nq-q+1} \cdots j_{nq}} [j_a j_b \cdots j_e], \tag{1.10}$$

where

$$1 \leqslant a \leqslant q;\ q+1 \leqslant b \leqslant 2q; \ldots ; nq - q + 1 \leqslant e \leqslant nq.$$

We consider two possibilities:

I. At least one of the tensors (1.10) is not zero. Then we contract it with $[yz \cdots w]_{j_a j_b \cdots j_e}$, and the resulting tensor is nonzero. We subject the remaining indices to the action of the Young symmetrizations (in the skew-symmetric variant); in at least one case the result is nonzero; we then contract the skew-symmetric groups from $1, 2, \cdots, n$ indices with $y_*, [yz]_*, \cdots$, $[yz \cdots w]_{** \cdots *}$, respectively, the result of which is again nonzero, by standard properties of Young tensors. As a result we obtain a certain nonzero polynomial P of the form (1.6), in which the percentage of contractions with $[yz \cdots w]$ is higher compared with P_1:

$$\theta_P > \theta_{P_1}.$$

But this contradicts (1.8), so that only the second possibility remains.

II. All the tensors (1.10) are zero. But this implies that the tensor $b^{** \cdots *}$ is of rank less than n; that is, in a suitable basis all its nonzero coordinates contain only the indices $1, 2, \cdots, n-1$, so that $b^{** \cdots *}$ can be regarded as a tensor in the subspace $E^{n-1}(K)$ spanned by $e_1, \cdots, e_{n-1}$.

In fact, we consider the set of all vectors arising from $b^{** \cdots *}$ by fixing all but one of its indices. Then the relations $(1.10) = 0$ mean that any n of these vectors are linearly dependent, and hence they all lie in some subspace $E^{n-1}(K)$; we choose the basis so that $e_1, e_2, \cdots, e_{n-1} \in E^{n-1}(K)$. Then, if at least one of the indices in the coordinates of $b^{** \cdots *}$ is equal to n, the coordinate is zero, and this implies *that the rank of b is less than n*.

6) We now prove our lemma by induction on n. For $n = 1$ the lemma is trivial; assume it to be proved for $n - 1$.

It is obvious that the tensor β of (1.9) can be presented in the form

$$\beta^{j_1 \cdots j_q; \alpha_1 \cdots \alpha_n; \cdots ; \delta_1 \cdots \delta_n} = \bar{b}^{j_1 \cdots j_q} [e_1 \ldots e_n]^{\alpha_1 \cdots \alpha_n} \ldots [e_1 \ldots e_n]^{\delta_1 \cdots \delta_n};$$

where b differs from $\bar{b}$ only by a scalar coefficient. Since $b \neq 0$, we see that $\bar{b} \neq 0$ and the rank of $\bar{b}$ is less than n; we can now apply our lemma and form a nonzero comitant $\bar{c}$ of $\bar{b}$ having the form (1.2) (with n replaced by $n - 1$). Further, we form a nonzero comitant c of β in such a way that $\bar{c}$ is the comitant formed from $\bar{b}$; here the factors $[e_1 \cdots e_n]$ are rewritten mechanically. But a comitant of β is together with β a comitant of a; at the same time it has the form (1.2). The lemma is proved.

3. Example. Let a be a tensor in $E^n(K)$ having the form $a^{i_1\cdots i_n} = v^{(i_1}u^{i_2}\cdots u^{i_k)}$ or, in brief notation; $a = (v\underbrace{u\cdots u}_{k-1})$, where $k \geqq 2$, and u and v are linearly independent vectors. Then for the comitant c we can take

$$c^{i_1\ldots i_{2k}} = a^{i_1\ldots i_{k-2}i_{2k-3}i_{2k-1}}a^{i_{k-1}\ldots i_{2k-4}i_{2k-2}i_{2k}}[i_{2k-3}, i_{2k-2}][i_{2k-1}, i_{2k}]$$

$$= -\frac{2}{k^2}u^{i_1}\ldots u^{i_{2k-4}}[uv]^{i_{2k-3}i_{2k-2}}[uv]^{i_{2k-1}i_{2k}},$$

or briefly

$$c = -\frac{2}{k^2}\underbrace{u\ldots u}_{2k-4}[uv][uv]$$

(the denominator is due to the fact that for us symmetrization always involves averaging). By choosing a basis in which $e_1 = u$ and $e_2 = v$, we obtain (1.2).

We note that the tensor a has no nonzero invariants (not even relative invariants), except in the case $k = n = 2$.

4. We need the subsidiary

Lemma 2. *Suppose that the absolute tensor $w^{i_1\cdots i_k}$ is the unique solution of the tensorial system of linear equations*

$$\sum_P \alpha_P^{(\lambda)} w^{\{i_1\ldots i_k}u_{(\lambda)}^{i_{k+1}\ldots i_{k+m}\}^P} = v_{(\lambda)}^{i_1\ldots i_{k+m}} \quad (\lambda = 1, 2, \ldots, L), \tag{1.11}$$

where P ranges over the set of all permutations of the indices $i_1, \cdots, i_{k+m}$; $\{i_1, \cdots, i_{k+m}\}^P$ is the result of the permutation P; $\alpha_P^{(\lambda)} \in K$ are certain coefficients; $u_{(\lambda)}$ and $v_{(\lambda)}$ $(\lambda = 1, 2, \cdots, L)$ are given absolute tensors. Then w satisfies the relation

$$wT(u, v) = S(u, v), \tag{1.12}$$

where on the left-hand side we have the product of w with some nonzero comitant T of $u_{(\lambda)}$, $v_{(\lambda)}$ $(\lambda = 1, \cdots, L)$, and S is also a comitant of $u_{(\lambda)}$, $v_{(\lambda)}$. These comitants, of course, are formed without contraction.

Proof. 1) (1.11) can obviously be rewritten in the form

$$\overset{(\lambda)}{U}{}^{i_1\ldots i_{k+m}}_{j_1\ldots j_k} w^{j_1\ldots j_k} = v_{(\lambda)}^{i_1\ldots i_{k+m}}, \tag{1.13}$$

where

$$\overset{(\lambda)}{U}{}^{i_1\ldots i_{k+m}}_{j_1\ldots j_k} = \sum_P \alpha_P^{(\lambda)}\delta^{\{i_1}_{j_1}\ldots\delta^{i_k}_{j_k}u_{(\lambda)}^{i_{k+1}\ldots i_{k+m}\}^P}; \tag{1.14}$$

in (1.13) contractions are performed with respect to $j_1, \cdots, j_k$. Since the system of linear equations (1.13) in n^k unknown $w^{j_1\cdots j_k}$ has one and only one solution, it must contain a subsystem of n^k equations having nonzero determinant. In what follows we consider only those values of λ for which we obtain, for suitable collections $i_1, \cdots, i_{k+m}$, equations in this subsystem. If we allow the tensors $u_{(\lambda)}$, $v_{(\lambda)}$ and the coefficients $\alpha_P^{(\lambda)}$ to be identical for different λ's, we may assume that for each λ only one set of indices $i_1, \cdots, i_{k+m}$ yields an equation in the subsystem; thus λ takes the n^k values

$$\lambda = 1, \ldots, n^k.$$

2) For brevity we denote an arbitrary ordered set of k or $k+m$ indices by a single letter ("superindex"):

$$J = j_1 \ldots j_k; \quad I = i_1 \ldots i_{k+m}. \tag{1.15}$$

Then (1.13) takes the form

$$\overset{(\lambda)}{U}{}^I_J w^J = v^I_{(\lambda)} \quad (\lambda = 1, \ldots, n^k). \tag{1.16}$$

Let $J_1, J_2, \cdots, J_N$ $(N = n^k)$ be the set of all values of the superindex J in some order, and let $I_1, I_2, \cdots, I_N$ be any values of I. Multiplying the equation

$$\overset{(1)}{U}{}^{I_1}_J w^J = v^{I_1}_{(1)}$$

termwise by $\overset{(2)}{U}{}^{I_2}_{J_2} \cdots \overset{(N)}{U}{}^{I_N}_{J_N}$ and antisymmetrizing with respect to the pairs $\begin{bmatrix} I_1; & I_2; & \cdots; & I_N \\ (1) & (2) & & (N) \end{bmatrix}$, we obtain on the left-hand side the same result as that produced by the antisymmetrization $[J, J_2, \cdots, J_N]$; hence it follows that for the contraction it is sufficient to take the value $J = J_1$. So we obtain

$$w^{J_1} \overset{(1)}{U}{}^{I_1}_{[J_1} \overset{(2)}{U}{}^{I_2}_{J_2} \ldots \overset{(N)}{U}{}^{I_N}_{J_N]} = v^{I_1}_{(1)} \overset{(2)}{U}{}^{I_2}_{J_2} \ldots \overset{(N)}{U}{}^{I_N}_{J_N} \begin{bmatrix} I_1; & I_2; & \ldots; & I_N \\ (1) & (2) & & (N) \end{bmatrix}. \tag{1.17}$$

By our assumption the coefficient of w^{J_1} (for at least one set $I_1, \cdots, I_N$) is different from 0.

In a more symmetrical form we can write

$$w^J \overset{(1)}{U}{}^{I_1}_{[J_1} \ldots \overset{(N)}{U}{}^{I_N}_{J_N]} = v^{I_1}_{(1)} \delta^J_{[J_1} \overset{(2)}{U}{}^{I_2}_{J_2} \ldots \overset{(N)}{U}{}^{I_N}_{J_N]} \begin{bmatrix} I_1 & ; \ldots ; & I_N \\ (1) & & (N) \end{bmatrix}, \tag{1.18}$$

where J is any set $j_1 \cdots j_k$. In fact, for $J = J_1$ equation (1.18) reduces to (1.17) and hence is valid; by renumbering the indices $J_1, \cdots, J_N$ we obtain the same result for any J_α.

Finally, (1.18) can be written in the form

$$w^J \overset{(1)}{U}{}^{I_1}_{[J'_1} \ldots \overset{(N)}{U}{}^{I_N}_{J'_N]} = v^{I_1}_{(1)} \delta^J_{[J'_1} \overset{(2)}{U}{}^{I_2}_{J'_2} \ldots \overset{(N)}{U}{}^{I_N}_{J'_N]} \begin{bmatrix} I_1 & ; \ldots ; & I_N \\ (1) & & (N) \end{bmatrix}, \tag{1.19}$$

where $J'_1, \cdots, J'_N$ are completely arbitrary sets J; in fact, if they are all distinct, we recover (1.18); if not, both sides of (1.19) are equal to 0.

We now contract (1.19) termwise with the tensor $\Omega^{J'_1 \cdots J'_N}$ (see §1.1); in accordance with (1.1),

$$\Omega = \Phi(\bar{\omega}), \tag{1.20}$$

where Φ is a homogeneous comitant of $\bar{\omega}$ of degree kn^{k-1}. Since all the lower indices in (1.19), after evaluation of superindices, are lower indices of the factors δ_j^i and they are all contracted with indices of the comitant $\Phi(\bar{\omega})$, we have

$$wT(\bar{\omega}, u) = S(\bar{\omega}; u, v), \tag{1.21}$$

where T and S are certain comitants of the tensors in parentheses ($T \neq 0$).

3) We first consider the case when the system of tensors $\{u_{(\lambda)}, v_{(\lambda)}\}$ has full rank ($= n$). We form the set of all comitants of the form (1.10) with only that difference that instead of the unique tensor b we make use of the tensors $u_{(\lambda)}$ and $v_{(\lambda)}$ in all possible combinations. Not all these comitants can be 0, or else the rank of the system $\{u_{(\lambda)}, v_{(\lambda)}\}$ would be less than n (this is seen just as in the case of (1.10)). We take a nonzero comitant of the form (1.10) formed from the tensors $u_{(\lambda)}$ and $v_{(\lambda)}$. It can obviously be presented in the form

$$c^{i_1 \cdots i_m} = \omega^{i_1 \cdots i_n} d^{i_{n+1} \cdots i_m},$$

after antisymmetrization of the n indices in the factor $\bar{\omega}$. We now multiply both sides of (1.21) by d as many times as the degree of (1.21) contains the coordinates of the tensor $\bar{\omega}$ (that is, kn^{k-1} times). By adjoining each factor $\bar{\omega}$ to one of the factors d, we replace $\bar{\omega}$ in (1.21) by a comitant c of the tensors $u_{(\lambda)}$ and $v_{(\lambda)}$. As a result we obtain (1.21), and the lemma is proved.

4) Let the system $\{u_{(\lambda)}, v_{(\lambda)}\}$ have rank $\nu < n$. This implies that, for a suitable basis, only those coordinates of $u_{(\lambda)}$ and $v_{(\lambda)}$ having all indices $\leqq \nu$ are different from 0. Then the solution w has the same property, for otherwise we could replace its remaining coordinates by zero and obtain another solution of (1.11) (in fact, if $i_1, \cdots, i_{k+m}$ are all less than or equal to ν, (1.11) is satisfied, since w is a solution, while if at least one of these indices is greater than ν, we obtain $0 = 0$). But since there cannot be two solutions, w has the required property.

Thus (1.11) can be regarded in a lower dimension (ν instead of n), for which the condition of fullness of rank holds. A repetition of all the preceding constructions yields (1.12), the values of the indices being restricted to $1, 2, \cdots, \nu$. But since otherwise (1.12) assumes the form $0 = 0$, the lemma is proved.

REMARK. It is obvious that the relation (1.12) is not violated if the tensors w, $u_{(\lambda)}$ and $v_{(\lambda)}$ are multiplied by a nonzero element of K in such a way that equality is preserved in (1.11). For the subsystem with nonzero determinant remains as before, and the form of the comitants T and S is not changed, since all the arguments and formulas can be repeated literally.

COROLLARY. *Lemma 2, which has been proved for absolute tensors, is also true for relative tensors.*

For let us replace the relative tensors w, $u_{(\lambda)}$ and $v_{(\lambda)}$, which satisfy the conditions of the lemma, by absolute tensors $\mathring{w}$, $\mathring{u}_{(\lambda)}$ and $\mathring{v}_{(\lambda)}$ that coincide with them in some fixed basis. In an arbitrary basis w, $u_{(\lambda)}$ and $v_{(\lambda)}$ differ from $\overset{\circ}{w}$, $\overset{\circ}{u}_{(\lambda)}$ and $\overset{\circ}{v}_{(\lambda)}$ only by nonzero factors, and both sets of tensors satisfy the system (1.11), the former by the hypotheses of the lemma and the latter because of the invariant character of the relation (1.11). By Lemma 2, $\overset{\circ}{w}$, $\overset{\circ}{u}_{(\lambda)}$ and $\overset{\circ}{v}_{(\lambda)}$ satisfy (1.12), and the above remark shows that the same is true of w, $u_{(\lambda)}$ and $v_{(\lambda)}$.

§2. Lemma 3 (the contravariant case)

1. At this point we assume that the field K (of characteristic 0) has the following properties:

(A) K is algebraically closed;

(B) for any elements $x_1, \cdots, x_N \in K$, there is an element $y \in K$ which is transcendental over the subfield $Q(x_1, \cdots, x_N)$ of K formed by adjoining the elements $x_1, \cdots, x_N$ to the field Q of rational numbers.

It is obvious that the field C of complex numbers has these properties.

Let $\{z\}$ be any system (finite set) of contravariant tensors $z^{i_1\cdots i_k}$. The invariance group $\mathfrak{G}$ of the system $\{z\}$ is the set of all automorphisms of the vector space $E^n(K)$ that carry the tensors in $\{z\}$ into themselves. More precisely, we fix some basis $\{e_1^0, \cdots, e_n^0\}$ so that each automorphism can be written as a nonsingular linear transformation of the coordinates of a vector in the given basis:

$$y^i = A^i_j x^j \quad (A^i_j \in K; \quad \operatorname{Det} |A^i_j| \neq 0). \tag{2.1}$$

Then the invariance group of the system $\{z\}$ consists of the matrices $\|A^i_j\|$ satisfying the conditions

$$\underset{0}{z}^{i_1\ldots i_k} = A^{i_1}_{j_1} \ldots A^{i_k}_{j_k} \underset{0}{z}^{j_1\ldots j_k} (\operatorname{Det} | A^i_j |)^{-p} \tag{2.2}$$

for every tensor in $\{z\}$. Here p is the weight of z, and the suffix 0 indicates that the coordinates of z are taken in the fixed basis $\{e_i^0\}$.

We say that a tensor admits a given group of automorphisms if the group is contained in its invariance group.

2. LEMMA 3. *Suppose that a contravariant tensor $w \neq 0$ of weight q admits the invariance group of the system $\{z\}$ of contravariant tensors. Then w satisfies the relation*

$$wT(z, \bar{\omega}) = S(z, \bar{\omega}), \tag{2.3}$$

where T and S are certain nonzero comitants of the tensors z and $\bar{\omega}$ (formed, of course, without contractions), the left-hand side being the product of the tensors w and T. In case w and z are absolute tensors, the result can be simplified: $\bar{\omega}$ does not appear in (2.3).

PROOF. 1) Let z_0 and w_0 be the coordinates of the tensors z and w in the fixed basis $\{e_i^0\}$; we form the subfield $Q(z_0, w_0)$ of K obtained by adjoining the coordinates z_0 and w_0 to the field Q of rational numbers. Let $\underline{A}^i_j$ be a matrix whose entries are in K and are transcendental over $Q(z_0, w_0)$. The existence of such a matrix is guaranteed by property (B) of the field K. Denote by $\underline{z}$ and $\underline{w}$ the coordinates of the tensors z and w in the basis $\{e_i\}$ related to the fixed basis by the transformation

$$e_i^0 = \underline{A}_i^j \underline{e}_j. \tag{2.4}$$

Thus

$$\underline{z}^{i_1\cdots i_k} = \underline{A}^{i_1}_{j_1}\cdots \underline{A}^{i_k}_{j_k} z_0^{j_1\cdots j_k}(\mathrm{Det}\,|\underline{A}^i_j|)^{-p},$$
$$\underline{w}^{\alpha_1\cdots\alpha_m} = \underline{A}^{\alpha_1}_{\beta_1}\cdots \underline{A}^{\alpha_m}_{\beta_m} w_0^{\beta_1\cdots\beta_m}(\mathrm{Det}\,|\underline{A}^i_j|)^{-q}. \tag{2.5}$$

Further, we understand by $\underline{w}$ an arbitrarily chosen *single* coordinate $\underline{w}^{\alpha_1\cdots\alpha_m}$ (for a fixed set $\alpha_1,\cdots,\alpha_m$), but $\underline{z}$ denotes *all* coordinates of all the tensors z in the basis $\{\underline{e}_i\}$. We also regard the system (2.5) as having the same restriction. We are interested in the transition from $Q(z_0, w_0, \underline{z}, \underline{w})$ to its extension field $Q(z_0, w_0, \underline{w}, \{\underline{A}^i_j\}) = Q(z_0, w_0, \{\underline{A}^i_j\})$. This can be achieved by consecutive adjunctions—some transcendental, some algebraic—of the elements $\underline{A}^i_j$ in some definite order.

2) We show that $\underline{w}$ cannot be transcendental over $Q(z_0, w_0, \underline{z})$; otherwise we take any other element $\underline{\underline{w}} \in K$, also transcendental over $Q(z_0, w_0, \underline{z})$. The field $Q(z_0, w_0, \underline{z}, \underline{\underline{w}})$ is naturally isomorphic to $Q(z_0, w_0, \underline{z}, \underline{w})$, and so the process of successive adjunction of the elements $\underline{A}^i_j$ to $Q(z_0, w_0, \underline{z}, \underline{w})$ can be copied isomorphically in the form of successive adjunction of certain elements $\underline{\underline{A}}^i_j$ to $Q(z_0, w_0, \underline{z}, \underline{\underline{w}})$. By virtue of this isomorphism, the elements $\underline{\underline{A}}^i_j$, like the $\underline{A}^i_j$, are transcendental over $Q(z_0, w_0)$ and satisfy (2.5) with $\underline{w}$ replaced by $\underline{\underline{w}}$.

In addition, by properties (A) and (B) of K, the successive adjunction of the $\underline{\underline{A}}^i_j$ can always be realized in K; hence $\underline{\underline{A}}^i_j \in K$.

We proceed to derive a contradiction. In the bases $\{\underline{e}_i\}$ and $\{\underline{\underline{e}}_i\}$ (where the basis $\underline{\underline{e}}_i$ is analogous to (2.4), but by means of $\underline{\underline{A}}^i_j$), the values of the coordinates of z are identical: $z = \underline{z}$, while those of w are not: $w = \underline{w}$ and $w = \underline{\underline{w}}$. The automorphism of $E^n(K)$ taking $\{\underline{e}_i\}$ to $\{\underline{\underline{e}}_i\}$ thus preserves the tensors z, but alters w, in conflict with the hypothesis of the lemma.

The coordinate $\underline{w}$ is thus algebraic over $Q(z_0, w_0, \underline{z})$ and is a root of a certain algebraic equation with coefficients in $Q(z_0, w_0, z)$:

$$\mathscr{P}(\underline{w}) = 0. \tag{2.6}$$

3) We assert that the equation (2.6) is linear; that is, it has the form

$$\underline{w}T(z_0, w_0, \underline{z}) - S(z_0, w_0, \underline{z}) = 0, \tag{2.7}$$

where T $(\neq 0)$ and S are rational expressions (which may be assumed to be polynomials) in z_0, w_0, $\underline{z}$ with rational coefficients.

For suppose that the irreducible polynomial $\mathscr{P}(w)$ has degree greater than 1 and consider a root $\underline{\underline{w}} \neq \underline{w}$; since K is algebraically closed, $\underline{\underline{w}} \in K$.

Since $Q(z_0, w_0, \underline{z}, \underline{w})$ is naturally isomorphic to $Q(z_0, w_0, \underline{z}, \underline{\underline{w}})$, a repetition of the construction in 2) leads to a contradiction in exactly the same way.

Thus $\underline{w}$ satisfies (2.7):

(2.8) $$\underline{w}T(z_0, w_0, \underline{z}) = S(z_0, w_0, \underline{z}).$$

In what follows we do not write z_0 and w_0 in the symbols for T and S, and regard $T(\underline{z})$ and $S(\underline{z})$ as polynomials over $Q(z_0, w_0)$. We recall that for us $\underline{w}$ stands for $\underline{w}^{\alpha_1 \cdots \alpha_m}$, $\alpha_1, \cdots, \alpha_m$ being an arbitrarily fixed set of indices, so that T and S depend, of course, on this set:

$$T = T^{\alpha_1 \ldots \alpha_m}(\underline{z}), \quad S = S^{\alpha_1 \ldots \alpha_m}(\underline{z})$$

and (2.8) takes the form

(2.9) $$\underline{w}^{\alpha_1 \ldots \alpha_m} T^{\alpha_1 \ldots \alpha_m}(\underline{z}) = S^{\alpha_1 \ldots \alpha_m}(\underline{z}); \quad T^{\alpha_1 \ldots \alpha_m}(\underline{z}) \neq 0.$$

4) The coordinates $\underline{w}$ and $\underline{z}$ are obtained from w_0 and z_0 by means of the transformation (2.5), where the $\underline{A}^i_j$ are transcendental over $Q(z_0, w_0)$. If in (2.5) we replace the matrix $\|\underline{A}^i_j\|$ by its product $\|C^i_j\|$ with an arbitrary nonsingular matrix $\|B^i_j\|$, the elements

(2.10) $$C^i_j = B^i_k \underline{A}^k_j$$

remain transcendental over $Q(z_0, w_0)$, and so all the above arguments can be repeated word for word, and (2.9) remains true.

But the change from $\underline{A}^i_j$ to C^i_j in (2.5) involves the tensorial transformation of the coordinates $\underline{w}$ and $\underline{z}$ by the matrix $\|B^i_j\|$. Thus the relation (2.9) remains true *after any tensorial transformation of* $\underline{w}, \underline{z}$. Hence we conclude at once that for every set $\alpha_1, \cdots, \alpha_m$, (2.9) yields a *tensorial* relation between w and z, linear in w.

5) For the fixed set $\alpha_1, \cdots, \alpha_m$ the expressions $T^{\alpha_1 \cdots \alpha_m}(\underline{z})$ and $S^{\alpha_1 \cdots \alpha_m}(\underline{z})$ are completely determined polynomials in the coordinates $\underline{z}$. We multiply the individual terms of these polynomials by suitable powers of $\omega^{12 \cdots n}$ to make the weights of all the terms in the polynomial

(2.11) $$\underline{w}^{\alpha_1 \ldots \alpha_m} T^{\alpha_1 \ldots \alpha_m}(\underline{z}) - S^{\alpha_1 \ldots \alpha_m}(\underline{z})$$

equal to the maximal of these weights (where the weight of a term is understood to mean the sum of the weights of the tensors whose coordinates appear as factors in that term). We denote the resulting polynomial by $\Lambda(\underline{w}, \underline{z}, \bar{\omega})$. We recall (§1.1) that $\bar{\omega}$ has weight 1 and in any basis $\omega^{12 \cdots n} = 1$, so that the meaning of the relation (2.9) is not changed if we rewrite it it the form

(2.12) $$\Lambda(\underline{w}, \underline{z}, \bar{\omega}) = 0,$$

At the same time, in view of the equality of the weights of all the terms, it remains true after a tensorial transformation of $\underline{w}$, $\underline{z}$ and $\bar{\omega}$ by the arbitrary matrix $\|B^i_j\|$, *and in this case, when weights are ignored,* it is assumed to be precisely equal to zero.

We emphasize that in the case of absolute tensors w and z, it is unnecessary to multiply by $\omega^{12 \cdots n}$.

6) First we take for $\| B_j^i \|$ a diagonal matrix with arbitrary nonzero elements $\lambda_1, \cdots, \lambda_n \in K$ on the leading diagonal. This has the effect of multiplying each term of Λ by $\lambda_1^{p_1} \cdots \lambda_n^{p_n}$, where p_1 is the number of indices equal to 1 in the given term, $\cdots$, p_n is the number of indices equal to n. Since (2.12) remains true for any $\lambda_1, \cdots, \lambda_n$, it follows that the sums of all the terms in Λ having a fixed number of indices equal to $1, \cdots, n$ are equal to 0 individually. We now take Λ to mean any one of these sums (subject to the condition that it actually contains $\underline{w}$).

Thus Λ can now be regarded as a certain coordinate $\Lambda^{i_1^0 \cdots i_s^0}$ of some comitant $\Lambda^{i_1 \cdots i_s}$ of $\underline{w}$, $\underline{z}$ and $\bar{\omega}$ (linear relative to $\underline{w}$), and (2.12) takes the form

$$\Lambda^{i_1^0 \ldots i_s^0} = 0, \tag{2.13}$$

where the indices $i_1^0, \cdots, i_s^0$ are fixed. For the sake of definiteness, let

$$i_1^0 \ldots i_s^0 = \underbrace{11 \ldots 1}_{a} \; \underbrace{22 \ldots 2}_{b} \ldots \underbrace{nn \ldots n}_{f}.$$

But (2.13) remains true after any tensorial transformation; that is,

$$B_{i_1}^1 \ldots B_{i_a}^1 B_{i_{a+1}}^2 \ldots B_{i_{a+b}}^2 \ldots B_{i_{s-f+1}}^n \ldots B_{i_s}^n \Lambda^{i_1 \ldots i_s} = 0. \tag{2.14}$$

Assuming B_j^i to be transcendental, we may equate to zero the coefficients of the polynomial corresponding to B_j^i. This obviously yields

$$\Lambda^{(i_1 \ldots i_a)\,(i_{a+1} \ldots i_{a+b}) \ldots (i_{s-f+1} \ldots i_s)} = 0. \tag{2.15}$$

Here the left-hand side is a comitant of $\underline{w}$, $\underline{z}$, $\bar{\omega}$, linear in $\underline{w}$ and obtained from Λ by the symmetrizations indicated; in contrast to (2.13), here not one but all the coordinates of the comitant turn out to be 0. In view of the tensorial character of the result, the underlining in the symbols (indicating a particular basis) can now be omitted, and we simply write w and z (for any basis).

7) Thus for any set $\alpha_1, \cdots, \alpha_m$ we obtain a relation of the form (2.15); we consider the system of these relations as a special case of (1.11) in the conditions of Lemma 2, where in our case the tensors $u_{(\lambda)}$ and $v_{(\lambda)}$ are comitants of the tensors z and $\bar{\omega}$. The uniqueness condition is satisfied, because if in (2.15) we put

$$i_1 = i_2 = \ldots = i_a = 1; \; i_{a+1} = \ldots = 2; \ldots ; i_{s-f+1} = \ldots = i_s = n$$

and again specialize the basis $e_i = \underline{e_i}$, we recover (2.13) or, what is the same thing, (2.12):

$$\underline{w}^{\alpha_1 \ldots \alpha_m} T^{\alpha_1 \ldots \alpha_m} (\underline{z}) - S^{\alpha_1 \ldots \alpha_m} (\underline{z}) = 0; \; T \neq 0.$$

We see that the coordinate $\underline{w}^{\alpha_1 \cdots \alpha_m}$ is defined in a unique way, and since the set $\alpha_1, \cdots, \alpha_m$ is arbitrary, the tensor w satisfying the system of relations (2.15) is unique.

Applying Lemma 2 we obtain (1.12), which clearly means in our case that

$$wT(z, \bar{\omega}) = S(z, \bar{\omega}); \; T(z, \bar{\omega}) \neq 0. \tag{2.16}$$

In the case of absolute tensors w and z, it is not necessary to introduce $\bar{\omega}$ (as was mentioned in 5. above), and we have

$$wT(z) = S(z); \quad T(z) \neq 0. \tag{2.17}$$

Lemma 3 is proved. Of course, it is also true for covariant tensors with $\bar{\omega}$ replaced by ω.

§3. Theorems A, A′ and A″

1. Suppose now that the hypotheses of Lemma 3 are fulfilled by a tensor w and a system of tensors z *of arbitrary structure*. By means of the constant tensor $\bar{\omega}$ we can "raise" every lower index of every tensor in accordance with the scheme

$$\{z_i\} \to \{z^{i_1\ldots i_{n-1}}\}, \quad z^{i_1\ldots i_{n-1}} = \omega^{i_1\ldots i_{n-1}i} z_i, \tag{3.1}$$

thereby changing this index into a skew-symmetric group of upper indices. Conversely, a skew-symmetric group of upper indices can always be "lowered" in accordance with the scheme

$$\{z^{i_1\ldots i_{n-1}}\} \to \{z_i\}, \quad z_i = \frac{1}{(n-1)!} \omega_{i_1\ldots i_{n-1}i} z^{i_1\ldots i_{n-1}}, \tag{3.2}$$

thereby changing it into a single lower index. An elementary calculation shows that these operations are inverse to one another.

Therefore, no matter what kind of tensors w and z are, they can always be converted to contravariant form by means of (3.1); under such an operation, the invariance group is not changed by virtue of the fact that $\bar{\omega}$ is a constant tensor (as is the tensor ω which provides the inverse transformation). Applying Lemma 3 to the contravariant forms of w and z, we arrive at (2.16), in which the symbols T and S can be regarded as having the same form as z, since the presence of $\bar{\omega}$ allows us to convert to contravariant form *under the symbols T and S*.

Thus Lemma 3 remains true for arbitrary tensors, provided that in (2.16) w is replaced by its contravariant form; but the original form of w can be recovered by lowering those indices of w which had to be raised (let the number of these be m_1) and performing the corresponding operation on S. Thus we obtain

$$wT(z, \bar{\omega}) = \downarrow S(z, \bar{\omega}), \tag{3.3}$$

where the arrow denotes the lowering of the m_1 skew-symmetric groups of $n-1$ indices by means of the constant tensor ω.

Of course, the dual form of (3.3) is also true: simply interchange the roles of the upper and lower indices, and replace $\bar{\omega}$ by ω, and vice versa.

The weights of the individual terms in (3.3) are all equal; for otherwise (3.3) would split into a number of relations, each homogeneous with respect to weight, and we could simply take one of these. (Actually, homogeneity with respect to weight is already guaranteed by the constructions in the proof of Lemma 3.)

We consider the case of *absolute* tensors w and z separately; then only the tensors $\bar{\omega}$ and ω contribute to the weights of the terms in (3.3) (the latter occur only on the right-hand side and appear each time an index is lowered; that is, to the power m_1). The homogeneity with respect to weight implies that $\bar{\omega}$ occurs homogeneously in T (to the power l, say) and in S (to the power $l+m_1$). After termwise multiplication of (3.3) l times by ω we can group the factors ω and $\bar{\omega}$ in pairs, and replace each product by using the standard formula

$$\omega_{j_1\dots j_n}\omega^{i_1\dots i_n} = n!\, \delta^{[i_1}_{j_1} \dots \delta^{i_n]}_{j_n}. \tag{3.4}$$

As a result, (3.3) takes the form

$$wT(z, \delta) = S(z, \delta),$$

where T and S are comitants of z and $\delta = \{\delta^i_j\}$.

2. Let us summarize all that has been said. We have been dealing with a vector space $E^n(K)$, where K is an algebraically closed field of characteristic 0 having the following property: for any $x_1, \cdots, x_N \in K$, K contains an element y that is transcendental over the subfield $Q(x_1, \cdots, x_N)$, Q being the field of rational numbers.

THEOREM A. *Suppose that a tensor w admits the invariance group of the system of tensors $\{z\}$. Then w satisfies the relation*

$$wT(z, \bar{\omega}) = \downarrow S(z, \bar{\omega}), \tag{3.5}$$

where the arrow denotes the lowering (*in accordance with the scheme* (3.2)) *of m_1 skew-symmetric groups of $n-1$ indices, m_1 being the number of lower indices in w. In particular, when z and w are contravariant*

$$wT(z, \bar{\omega}) = S(z, \bar{\omega}). \tag{3.6}$$

In case w and z are absolute, the relation (3.5) *takes the form*

$$wT(z, \delta) = S(z, \delta), \tag{3.7}$$

and the relation (3.6) *takes the form*

$$wT(z) = S(z). \tag{3.8}$$

In all cases, T and S are symbols for comitants of the corresponding tensors, and $T \neq 0$; in (3.5) (and of course, in (3.6)), these comitants are contravariant tensors. (The dual propositions with covariance and contravariance interchanged are also true, of course.)

3. We multiply both sides of (3.5) by T a sufficient number of times so that Lemma 1 (§1.2) can be applied to construct a comitant of the form (1.2) from the product of factors T on the left-hand side. But Lemma 1 was proved for absolute tensors, whereas T in general is relative; therefore the replacement of the comitant of T by (1.2) is only permissible for a fixed basis; to pass to an arbitrary basis we must be sure of the right weight (say p); this can be done by introducing the additional factor $(\epsilon^{12\cdots n})^{-p}$.

Repeating these operations on the right-hand side (in which the lowered indices obviously do not participate), we obtain

$$we_1 \ldots e_1 [e_1e_2] \ldots [e_1e_2] \ldots [e_1e_2 \ldots e_n] \ldots [e_1e_2 \ldots e_n] (\varepsilon^{12\ldots n})^{-p} = \downarrow S_1(z, \bar{\omega}), \tag{3.9}$$

where $S_1(z, \bar{\omega})$ is a new comitant formed from S and T in the manner described.

If z and w are contravariant tensors, then there is no lowering $\downarrow$ of indices on the right-hand side.

Finally, if z and w are absolute contravariant tensors, then we start from (3.8) (instead of (3.5)) and obtain

$$we_1 \ldots e_1 [e_1e_2] \ldots [e_1e_2] \ldots [e_1e_2 \ldots e_n] \ldots [e_1e_2 \ldots e_n] = S_1(z). \tag{3.10}$$

Of course, some or even all of the factors $e_1, \cdots, [e_1e_2 \cdots e_n]$ may be absent. We summarize our results.

THEOREM A′. *In Theorem* A *the relations* (3.5) *and* (3.8) *can be expanded to* (3.9) *and* (3.10) *by means of the specific construction on the comitants* $R(z, \bar{\omega})$ *and* $T(z)$, *respectively.*

If the invariance group of the system $\{z\}$ is irreducible, then of the polyvectorial factors on the left-hand side of (3.9) only $[e_1e_2 \cdots e_n] = \bar{\epsilon}$ can occur. Let there be q such factors. We contract each of these completely with the tensor $\epsilon = \{e_{i_1 \cdots i_n}\}$, which yields the factor $n!$ each time (we recall that $\epsilon^{12\cdots n} = (\epsilon_{12\cdots n})^{-1}$ by the convention adopted in §1.1). Performing these contractions on the right-hand side we substitute for ϵ by means of the formula

$$\varepsilon = (\varepsilon^{12\ldots n})^{-1} \omega$$

to obtain

$$w = \lrcorner \downarrow S_1(z, \bar{\omega}) \, (\varepsilon^{12\ldots n})^{-p-q}, \tag{3.11}$$

where $\lrcorner \downarrow$ now denotes both sets of complete contractions with ω—the previous ones and those just described. (The factor $(n!)^{-q}$ has been absorbed into S_1.)

In the case of (3.10) we perform the contraction with ϵ directly on the right-hand side and obtain

$$w = S_1(z) \, \varepsilon\varepsilon \ldots \varepsilon, \tag{3.12}$$

where $\epsilon\epsilon \cdots \epsilon$ is completely contracted with part of the upper indices in $S_1(z)$.

THEOREM A″. *If the invariance group of the system* $\{z\}$ *is irreducible, then the relations* (3.5) *and* (3.8) *assume the forms* (3.11) *and* (3.12), *respectively.*

4. EXAMPLE 1. We take in $E^n(K)$ $(n \geqq 2)$ a system of two absolute contravariant tensors:

$$z_{(1)} = (\underbrace{vuu\ldots u}_{p}), \quad z_{(2)} = (\underbrace{vuu\ldots u}_{q}), \quad p > q$$

(see §1.3 for the notation). If we choose a basis in which $e_1 = u$, $e_2 = v$, then the invariance group of the system consists of the automorphisms of $E^n(K)$ of the following form:

$$e_1 \to \lambda e_1, \quad e_2 \to \mu e_2, \quad \text{where } \lambda^p\mu = 1, \quad \lambda^q\mu = 1,$$

that is,

$$\lambda^{p-q} = 1, \quad \mu = \lambda^{-p} \,(= \lambda^{-q});$$

the remaining e_i going to arbitrary vectors (e_1 and e_2 are linearly independent, of course).

We take for w the tensor $\underbrace{e_1 e_1 \cdots e_1}_{p-q}$; it admits the given invariance group, and therefore must satisfy a relation of the form (3.10). For we can take the comitant T formed from $z_{(2)}$ just as c was formed from a in §1.3:

$$T^{i_1\ldots i_{2q+2}} = z_{(2)}^{i_1\ldots i_{q-1}\, i_{2q-1}\, i_{2q+1}}\, z_{(2)}^{i_q\ldots i_{2q-2}\, i_{2q}\, i_{2q+2}}\, [i_{2q-1} i_{2q}]\, [i_{2q+1} i_{2q+2}].$$

Hence we obtain

$$T = -\frac{2}{(q+1)^2}\underbrace{e_1 \ldots e_1}_{2q-2}\, [e_1 e_2]\, [e_1 e_2].$$

Multiplying w by T, we obtain the tensor

$$S = -\frac{2}{(q+1)^2}\underbrace{e_1 \ldots e_1}_{p+q-2}\, [e_1 e_2]\, [e_1 e_2],$$

which is a comitant of $z_{(1)}$ and $z_{(2)}$:

$$S^{i_1\ldots i_{p+q+2}} = \sigma z_{(1)}^{i_1\ldots i_{p-1}\, i_{p+q-1}\, i_{p+q+1}}\, z_{(2)}^{i_p\ldots i_{p+q-2}\, i_{p+q}\, i_{p+q+2}}\, [i_{p+q-1} i_{p+q}]\, [i_{p+q+1} i_{p+q+2}],$$

where σ is a certain rational coefficient.

The invariance group in this example is reducible, and (3.12) does not hold; w is expressible as the "quotient" of the two comitants S and T, but not simply as a comitant of z.

EXAMPLE 2. Let g^{ij} be a symmetric, twice contravariant, nondegenerate, absolute tensor in $E^n(C)$. Its invariance group in a suitable basis is the orthogonal group $O(n, C)$. We form from g^{ij} a nonzero comitant $V^{i_1\cdots i_{2p}}$ (of degree p relative to g and denoted briefly by V_p). The invariance group of V_p can be larger than $O(n, C)$; either (I) it contains $SL(n, C)$, or (II) it is an extension of $O(n, C)$ by scalar matrices.

The scalar matrix λI belongs to the invariance group of V_p if and only if $\lambda^{2p} = 1$. Thus, all the comitants V_p of type (II) having the given valency $2p$ have one and the same invariance group, which is in fact irreducible, and by (3.12) each of them is a comitant of another, contracted with $\epsilon\epsilon \cdots \epsilon$. The number of factors here is always even (for otherwise reflection would change the sign of the right-hand side).

Let d be the greatest common divisor of p and n; then

$$d + kn = lp \quad (k, l \text{ are positive integers}).$$

We form the comitants of $\{g^{ij}\}$:

$$V^{(1)} = \underbrace{gg \ldots g}_{d} \underbrace{G \ldots G}_{k},$$

where

$$g = \{g^{ij}\};\ G = \{g^{i_1 j_1} \ldots g^{i_n j_n} [i_1 \ldots i_n]\};$$

$$V^{(2)} = \underbrace{V_p V_p \ldots V_p}_{l},$$

and where V_p is any comitant of g of type (II) of degree p. Then, as has been pointed out, $V^{(1)}$ is expressible in terms of $V^{(2)}$. For the sake of greater clarity we restrict ourselves to those bases for which $\det|g^{ij}| = 1$. Then G and $\epsilon\epsilon$ have constant coordinates and we obtain: *if d is the greatest common divisor of p and n, then the products $g^{i_1 j_1} \ldots g^{i_d j_d}$ are expressible as polynomials in the coordinates of an (arbitrary) comitant V_p of type* (II).

As regards the comitants of type (I), these are arbitrary comitants of the tensor G, as follows once again from (3.12) (here the contraction with $\epsilon \cdots \epsilon$ can be discarded without loss of generality).

5. We wish to discuss an important special case of Theorem A (namely, Theorem B in subsection 7). First we recall some facts. Let $\mathfrak{H}$ be a complex semisimple connected linear Lie group, which therefore acts on the vector space $E^n(C)$ as a subgroup of its automorphism group $GL(n, C)$.

As is well known, $\mathfrak{H}$ is always an algebraic linear group and is thus defined by the condition of invariance (up to multiplication by a numerical coefficient) of some system of absolute tensors $\{z_{\mathfrak{H}}\}$, which may be assumed to be contravariant (see the Appendix in [**3**], for example). But as a consequence of the connectedness and semisimplicity of $\mathfrak{H}$ this coefficient is equal to 1, for otherwise we could find an ideal of codimension 1 in the Lie algebra H of $\mathfrak{H}$, which in the semisimple case is impossible. Thus, *$\mathfrak{H}$ is the invariance group of some system of absolute contravariant tensors.*

If the condition of connectedness of $\mathfrak{H}$ is replaced by the condition of *finitely many connected components,* then the connected component $\mathfrak{H}_0$ of the identity is algebraic, and it follows in an obvious way that $\mathfrak{H}$ is also; the tensors in the corresponding system $\{z_{\mathfrak{H}}\}$ remain unchanged under the action of $\mathfrak{H}_0$ (for the same reason), and the action of any of the remaining components $\mathfrak{H}_\alpha$ results in the multiplication of these tensors by a coefficient which is necessarily an m_αth root of unity, where $(\mathfrak{H}_\alpha)^{m_\alpha} \subset \mathfrak{H}_0$. Let m be a common multiple of all the m_α. We form the new system of tensors $\{\tilde{z}_{\mathfrak{H}}\}$, consisting of all m-fold products of the tensors of $\{z_{\mathfrak{H}}\}$. It is obvious that the tensors $\tilde{z}_{\mathfrak{H}}$ are invariant under the action of $\mathfrak{H}$. Conversely, the invariance of the tensors $\tilde{z}_{\mathfrak{H}}$ implies the invariance of their factors (the tensors $z_{\mathfrak{H}}$) up to a numerical coefficient. This coefficient is common to all the $z_{\mathfrak{H}}$, since the invariance of the products $z_{\mathfrak{H}}$ is not infringed if the arbitrary factor $\tilde{z}_{\mathfrak{H}}$ is replaced by any other factor $z'_{\mathfrak{H}}$. Thus $\mathfrak{H}$ is characterized by the invariance of the system $\{\tilde{z}_{\mathfrak{H}}\}$. So we obtain

PROPOSITION 1. *Every complex semisimple linear Lie group* $\mathfrak{H} \subset GL(n, C)$ *with a finite number of connected components is the invariance group of some finite system of absolute contravariant tensors* $\{z_{\mathfrak{H}}\}$ *(we omit the tilde).*

PROPOSITION 2. *Suppose that a complex linear Lie group* $\mathfrak{H} \subset SL(n, C)$ *has an irreducible connected component* $\mathfrak{H}_0$ *of the identity. Then* $\mathfrak{H}_0$ *is semisimple and* $\mathfrak{H}$ *(also semisimple) has a finite number of connected components.*

PROOF. It follows from the irreducibility and the connectedness of $\mathfrak{H}_0$ that it either is semisimple or is obtained from a semisimple group by the adjunction of a one-dimensional group of scalar matrices; but since $\mathfrak{H} \subset SL(n, C)$, the latter possibility cannot occur. Thus $\mathfrak{H}_0$ is semisimple. Each of the remaining components $\mathfrak{H}_\alpha$ induces some connected component $(\operatorname{Aut}\mathfrak{H}_0)_\alpha$ of the automorphism group $\operatorname{Aut}\mathfrak{H}_0$ of $\mathfrak{H}_0$, whereby $\mathfrak{H}_0$ itself, as is well known, induces the component of the identity, $(\operatorname{Aut}\mathfrak{H}_0)_0 = \operatorname{Int}\mathfrak{H}_0$ (the group of inner automorphisms). If $(\operatorname{Aut}\mathfrak{H}_0)_\alpha = \operatorname{Int}\mathfrak{H}_0$, then $\mathfrak{H}_\alpha$ contains a matrix that commutes with $\mathfrak{H}_0$; that is, a scalar matrix, since $\mathfrak{H}_0$ is irreducible. But there are only n such matrices in $SL(n, C)$. Therefore there are only a finite number $m \leqq n$ of components $\mathfrak{H}_\alpha$ for which $(\operatorname{Aut}\mathfrak{H}_0)_\alpha = \operatorname{Int}\mathfrak{H}_0$, and if the $\mathfrak{H}_\alpha$ are partitioned in systems with the same $(\operatorname{Aut}\mathfrak{H}_0)_\alpha$, then there are m components in each system. On the other hand, it is known that for connected semisimple $\mathfrak{H}_0$ the number of distinct components of $\operatorname{Aut}\mathfrak{H}_0$ is finite, and so the number of distinct $\mathfrak{H}_\alpha$ is finite.

COROLLARY. *Under the conditions of Proposition* 2, $\mathfrak{H}$ *is the invariance group of some system of absolute contravariant tensors.*

6. The space $E^n(C)$ on which $\mathfrak{H}$ acts ($\mathfrak{H}$ is the same as at the beginning of subsection 5) is known to be completely reducible: it decomposes into invariant basic planes $E^{n_\lambda}_{(\lambda)}$ on each of which $\mathfrak{H}$ acts irreducibly:

$$E^n = E^{n_1}_{(1)} + \dots + E^{n_L}_{(L)}.$$

(If this decomposition is not unique, we select a particular one.) To each plane we can assign its "direction polyvector"

$$\bar{\varepsilon}_{(\lambda)} = [e_{(\lambda)1} \dots e_{(\lambda)n_\lambda}] \quad (\lambda = 1, \dots, L), \tag{3.13}$$

where $\{e_{(\lambda)1}, \cdots, e_{(\lambda)n_\lambda}\}$ is a basis for $E^{n_\lambda}_{(\lambda)}$.

7. THEOREM B. *Let* $\mathfrak{H}$ *be a connected complex semisimple linear Lie group and* $\mathfrak{H} \subset \mathfrak{H}_1 \subset SL(n, C)$, *where* $\mathfrak{H}_1$ *is the invariance group of a system of absolute contravariant tensors* $\{z\}$. *Then every absolute contravariant tensor* w *admitting the group* $\mathfrak{H}_1$ *has the form*

(3.14) $$w = S(z, \bar{\varepsilon}_{(1)}, \ldots, \bar{\varepsilon}_{(L)}; \varepsilon) = (S_1(z).\Pi)\,\varepsilon\varepsilon\ldots\varepsilon,$$

where the comitant S is formed in the following way: a certain comitant $S_1(z)$ is multiplied by a tensor Π of the form

$$\Pi = \bar{\varepsilon}_{(\mu_1)} \ldots \bar{\varepsilon}_{(\mu_k)},$$

and in the result obtained we contract some of the indices of S and all the indices of Π (all these indices are upper) with all the indices of the product $\epsilon\epsilon\cdots\epsilon$ (all these indices are lower, $\epsilon = \{\epsilon_{i_1\cdots i_n}\}$).

We emphasize that this structure of w is necessary but not sufficient for its $\mathfrak{H}_1$-invariance.

PROOF. By Theorem A′, (3.10),

(3.15) $$we_1\ldots e_1[e_1e_2]\ldots[e_1e_2]\ldots[e_1e_2\ldots e_n]\ldots[e_1e_2\ldots e_n] = S_1(z),$$

where $e_1,\cdots,e_n$ is some basis for the space $E^n(C)$ (in general, not related to the basis $\{e_{(\lambda)i}\}$). Since z and w certainly admit the group $\mathfrak{H}$, so does the product $e_1\cdots[e_1e_2\cdots e_n]$, and hence so do the individual factors $[e_1e_2\cdots e_s]$. For it is clear that these factors are invariant up to multiplication by a number; but this number can only be 1, in view of the semisimplicity of $\mathfrak{H}$ and its consequent unimodularity on the invariant plane $\{e_1,e_2,\cdots,e_s\}$.

We successively extend the invariant plane $\{e_1,e_2,\cdots,e_s\}$ by adjoining to it step by step the irreducible invariant planes $E^{n_\lambda}_{(\lambda)}$; at each step the next $E^{n_\lambda}_{(\lambda)}$ either lies in the plane already constructed or intersects it in the zero vector only. The plane generated by the totality of the $E^{n_\lambda}_{(\lambda)}$ of the latter type (for $\lambda = \lambda_1,\lambda_2,\cdots,\lambda_m$, say) is complementary to $\{e_1,e_2,\cdots,e_s\}$; the polyvector corresponding to it is obviously $[\bar{\epsilon}_{(\lambda_1)},\bar{\epsilon}_{(\lambda_2)},\cdots,\bar{\epsilon}_{(\lambda_m)}]$, where antisymmetrization with respect to all the vectorial factors occurring in $\bar{\epsilon}_{(\lambda_1)},\cdots,\bar{\epsilon}_{(\lambda_m)}$ is implicit.

We now multiply each factor $[e_1e_2\cdots e_s]$ of the left-hand side of (3.15) by the corresponding polyvector

(3.16) $$[\bar{\varepsilon}_{(\lambda_1)}\ldots\bar{\varepsilon}_{(\lambda_m)}]$$

and completely contract each of the resulting tensors with $\epsilon = \{\epsilon_{i_1}\cdots\epsilon_{i_n}\}$. We obtain a nonzero constant, since

$$[e_1\ldots\varepsilon_s\bar{\varepsilon}_{(\lambda_1)}\ldots\bar{\varepsilon}_{(\lambda_m)}] \neq 0$$

in view of the fact that the planes in question are mutually complementary.

We perform the corresponding multiplications and contractions on the right-hand side of (3.15), and obtain the required comitant S. To obtain Π, we collect all the products of the form (3.16) taking part in the construction (that is, one for each factor $[e_1\cdots e_s]$ on the right-hand side of (3.15)). We omit the antisymmetrization in (3.16): in view of the contraction with ϵ, this takes place automatically. As to the left-hand side, there remains only the tensor w (with a nonzero numerical coefficient). The theorem is proved.

8. We consider an important special case of Theorem B: we assume that $\mathfrak{H}_1$ sends the planes $E^{n_\lambda}_{(\lambda)}$ into themselves. Then under the action of $h \in \mathfrak{H}_1$ the polyvectors $\bar{\epsilon}_{(\lambda)}$ are multiplied by numbers:

$$\bar{\varepsilon}_{(\lambda)} \rightarrow \sigma_\lambda (h) \cdot \bar{\varepsilon}_{(\lambda)}, \tag{3.17}$$

and hence Π in (3.14) can only be multiplied by a number. But if this number were not 1, we would have a contradiction to the $\mathfrak{H}_1$-invariance of the tensors, w, z, ϵ.

Thus the product Π in (3.14) is $\mathfrak{H}_1$-invariant; but among the totality of $\mathfrak{H}_1$-invariant products of the form $\Pi = \epsilon_{(\mu_1)} \cdots \epsilon_{(\mu_k)}$, we can find a finite number of generators

$$\Pi_1,\ \Pi_2,\ \ldots,\ \Pi_N, \tag{3.18}$$

such that all the remaining Π are products of these, up to the order of the factors $\epsilon_{(\mu)}$ in Π.

Theorem B′. *If under the conditions of Theorem* B *the basic $\mathfrak{H}$-invariant planes $E^{n_\lambda}_{(\lambda)}$ are also $\mathfrak{H}_1$-invariant, then*

$$w = (S_1 (z)\, \Pi_{\alpha_1} \Pi_{\alpha_2} \ldots \Pi_{\alpha_t})\, \varepsilon\varepsilon \ldots \varepsilon, \tag{3.19}$$

where contraction is understood, taking as upper indices all the indices y of all the Π_α and some of the indices y of S_1, and as lower indices all the indices y of all the ϵ.

Here $\Pi_{\alpha_1}, \cdots, \Pi_{\alpha_t}$ are part of the generators (3.18), possibly with repetitions. That w has this structure is not only necessary, but is obviously also *sufficient* for the $\mathfrak{H}_1$-invariance of w.

Note that we do not indicate here the possible permutation of the factors $\epsilon_{(\mu)}$ in the product $\Pi_{\alpha_1} \cdots \Pi_{\alpha_t}$, since when this product is fully contracted with $\epsilon\epsilon \cdots \epsilon$ this permutation can always be replaced by a permutation of the indices of $\epsilon\epsilon \cdots \epsilon$; for the same reason it can always be assumed that $\alpha_1 \leqq \alpha_2 \leqq \cdots \leqq \alpha_t$.

9. **Example.** Let $\mathfrak{H} = SL(p, C) \times SL(q, C) \subset SL(n, C)$, where $p + q = n$ and $\mathfrak{H}$ consists of $n \times n$ matrices having two unimodular blocks of types $p \times p$ and $q \times q$ on the leading diagonal. It is assumed that $n \geqq 3$. Obviously $\mathfrak{H}$ is determined in $GL(n, C)$ by the condition of invariance of the tensors $\bar{\epsilon}_{(1)} = [e_1 \cdots e_p]$ and $\bar{\epsilon}_{(2)} = [e_{p+1} \cdots e_n]$. We distinguish the subgroup $\mathfrak{H}_1$ of $GL(n, C)$ by the condition of invariance of the product $\bar{\epsilon}_{(1)}\bar{\epsilon}_{(2)}$ of these tensors. $\mathfrak{H}_1$ is obtained from $\mathfrak{H}$ by the adjunction of the matrices of the following form: scalar matrices λI_p and μI_q in the first and second blocks, respectively, with the condition that $\lambda^p \mu^q = 1$. The conditions of Theorem B are obviously satisfied, and so all the contravariant tensors w admitting the group $\mathfrak{H}_1$ are comitants of the product $\bar{\epsilon}_{(1)}\bar{\epsilon}_{(2)}$, possibly subject to contraction with the tensor $\epsilon = \{\epsilon_{i_1 \cdots i_n}\}$. We illustrate this result in a trivial case, (true, this

is formally illegal because $\mathfrak{H}$ is not semisimple): let $p = q = 1$ and $n = 2$. Then $\mathfrak{H} = \{e\}, \mathfrak{H}_1 \ni \begin{Vmatrix} \lambda & 0 \\ 0 & \lambda^{-1} \end{Vmatrix}$ and all the tensors admitting $\mathfrak{H}_1$ are comitants of $e_1 e_2$; that is, they have nonzero coordinates only when the indices 1 and 2 appear the same number of times (contraction with ϵ does not yield anything new in this case).

§4. On spherical functions of homogeneous spaces

1. Let $\mathfrak{G}$ be a connected complex semisimple Lie group, $\mathfrak{H}$ a closed complex Lie subgroup of $\mathfrak{G}$, and $\mathfrak{G}/\mathfrak{H}$ the corresponding complex-analytic homogeneous space (possibly having nonfaithful $\mathfrak{G}$-action).

A finite-dimensional complex linear space $\{\sigma(x)\}$ of complex analytic functions $\sigma(x)$ on $\mathfrak{G}/\mathfrak{H}$, invariant under translation of the functions $\sigma(x)$ by the transformations in $\mathfrak{G}$, is called a *space of spherical functions*, and a basis $\{\sigma^1(x), \cdots, \sigma^N(x)\}$ of this space is called a *system of spherical functions*. A system is called *irreducible* if the corresponding space has no nontrivial invariant subspaces.

Let $x \in \mathfrak{G}/\mathfrak{H}$ and $g \in \mathfrak{G}$; then

$$\sigma^i(gx) = A^i_j(g)\,\sigma^j(x), \tag{4.1}$$

and we obtain a matrix representation of $\mathfrak{G}$: $g \to \| A^i_j(g) \|$ (this representation is contragradient to the representation of $\mathfrak{G}$ naturally induced in the space $\{\sigma(x)\}$ by translation of the functions $\sigma(x)$ by the transformations $g \in \mathfrak{G}$).

In this section we want to investigate the situation when *all* the spherical functions are expressible as *polynomials* in the spherical functions of *some definite system*; it is of particular interest that for some homogeneous spaces we can take *any* system of spherical functions.

2. We regard $\mathfrak{G}$ in a faithful analytic linear representation:

$$\varphi_0 : \mathfrak{G} \subset GL(n, C)$$

(in fact, here, by the semisimplicity of $\mathfrak{G}, \mathfrak{G} \subseteq SL(n, C)$), where $GL(n, C)$ is the automorphism group of the vector space $E^n(C)$. Now suppose that we are given an arbitrary irreducible analytic linear representation ϕ of the same group $\mathfrak{G}$. *We assert that it can always be realized in a suitable finite-dimensional space T_ϕ of tensors over $E^n(C)$.* For the connected semisimple complex group $\mathfrak{G}$ is contained in the semisimple (complex) group $SL(n, C)$; by appealing to their compact forms (see §4.6) it is not difficult to show that any irreducible analytic linear representation $\phi(\mathfrak{G})$ can be embedded in a suitable irreducible analytic representation $\tilde{\phi}$ of $SL(n, C)$; more precisely, ϕ is obtained from $\tilde{\phi}$ by restricting it to $\mathfrak{G} \subseteq SL(n, C)$ and restricting the representation space T_ϕ to one of its invariant planes (if such planes arise after restriction of $\tilde{\phi}$ to $\mathfrak{G}$). But all the $\tilde{\phi}(SL(n, C))$ are known to be realizable in suitable spaces $T_{\tilde{\phi}}$ of absolute (if necessary, contravariant) tensors, for instance, in spaces generated by Young symmetrizers. It follows that $\phi(\mathfrak{G})$ can be realized in some subspace T_ϕ of $T_{\tilde{\phi}}$, which proves our assertion.

3. Returning to the space $\mathfrak{G}/\mathfrak{H}$ (with the hypotheses of subsections 1 and 2), let w be some (absolute and contravariant) tensor in $E^n(C)$ admitting the group $\mathfrak{H}$. Under translation by the automorphisms of $\mathfrak{G}$ $(\subset SL(n, C))$, the tensor w describes a certain complex analytic manifold T_w in the space T of all contravariant absolute tensors of the given valency. This manifold serves in an obvious way as a model for the homogeneous space $\mathfrak{G}/\mathfrak{H}$ (perhaps not faithful if $\mathfrak{H}$ does not exhaust the invariance group of w). In some fixed basis the coordinates of w under translation by $\mathfrak{G}$ clearly form a system of spherical functions on $\mathfrak{G}/\mathfrak{H}$ (it being assumed here that of the coordinates of w we only take a maximal number of linearly independent ones).

Conversely, every irreducible system of spherical functions on $\mathfrak{G}/\mathfrak{H}$ under translation by the transformations $g \in \mathfrak{G}$ affords in accordance with (4.1) a certain (irreducible) representation $\phi(\mathfrak{G})$ and so can be interpreted as the set of "flow coordinates" on some model (possibly not faithful) of $\mathfrak{G}/\mathfrak{H}$ in E_ϕ of the representation ϕ. But, as has been explained, E_ϕ can always be regarded as some tensor space over $E^n(C)$, and we recover the original situation. Here w plays the role of the tensor corresponding to a point on the model $\mathfrak{G}/\mathfrak{H}$, whose stabilizing group coincides with $\mathfrak{H}$.

Thus, any $\mathfrak{H}$-invariant tensor w over $E^n(C)$ gives rise to a system of spherical functions on $\mathfrak{G}/\mathfrak{H}$, and any irreducible system of spherical functions can be obtained in this way.

Hence any system can be so obtained if instead of individual tensors we allow finite systems of tensors.

4. THEOREM C. *Let $\mathfrak{H} \subset \mathfrak{G}$ be connected complex semisimple Lie groups, provided with a faithful analytic linear representation ϕ_0. To increase the generality we also consider extensions $\mathfrak{H}_1$ of $\mathfrak{H}$ (which may be neither semisimple nor connected) subject to the conditions:*

1) $\mathfrak{H}_1 \subset \mathfrak{G}$;

2) *the planes $E^{n_\lambda}_{(\lambda)}$ (corresponding to $\mathfrak{H}$) of some basic system are also invariant relative to $\mathfrak{H}_1$ (and hence the corresponding polyvectors $\bar{\epsilon}_{(1)}, \cdots, \bar{\epsilon}_{(L)}$ are invariant up to multiplication by a number);*

3) *$\mathfrak{H}_1$ is determined as a subgroup of $\mathfrak{G}$ by the condition of invariance of some system of tensors $\{z\}$ (for $\mathfrak{H}_1 = \mathfrak{H}$ this condition is automatically satisfied; see* §3.5).

Then any spherical function f on $\mathfrak{G}/\mathfrak{H}_1$ is expressible as a polynomial in the spherical functions $\{z, \Pi_1, \cdots, \Pi_N\}$ (that is, in the coordinates of the $\mathfrak{H}_1$-invariant tensors $z, \Pi_1, \cdots, \Pi_N$ under translation by $\mathfrak{G}$; the tensors Π are those of (3.18)).

PROOF. An arbitrary spherical function on $\mathfrak{G}/\mathfrak{H}_1$ can be regarded as a linear combination of the coordinates of certain tensors $\{w\}$ admitting $\mathfrak{H}_1$ and translated by $\mathfrak{G}$ (see subsection 3). The group $\mathfrak{H}_1$ is distinguished in $\mathfrak{G}$ by the condition of invariance of the tensors $\{z\}$, and hence in $GL(n,C)$ by the condition of invariance of the system $\{z\} \cup \{z_{\mathfrak{G}}\}$, where the condition of invariance of $\{z_{\mathfrak{G}}\}$ distinguishes $\mathfrak{G}$ in $GL(n,C)$ (see §3.5). Then the assertion of the theorem follows from (3.19) (Theorem B$'$). Note that under translation by the operations in $\mathfrak{G}$ the tensors $\{z_{\mathfrak{G}}\}$ and ϵ go over into themselves, their coordinates remaining constant, and thus, using (3.19), we do not take them into account in the final result.

We mention the important special case where the group $\mathfrak{H}_1$ cannot be extended without violating the properties 1 and 3; in this case $\mathfrak{H}_1$ is called a conditionally maximal subgroup of $\mathfrak{G}$. Then the original system of tensors $\{z\}$ can be replaced by any $\mathfrak{H}_1$-invariant (but not $\mathfrak{G}$-invariant) tensor w, whose invariance group must necessarily intersect $\mathfrak{G}$ in $\mathfrak{H}_1$ (since it cannot be larger). Insofar as any irreducible system of spherical functions on $\mathfrak{G}/\mathfrak{H}_1$ can be constructed from the coordinates of a suitable $\mathfrak{H}_1$-invariant tensor w, Theorem C takes the form

THEOREM C$'$. *Under the assumptions of Theorem* C, *if $\mathfrak{H}_1$ is a conditionally maximal subgroup of $\mathfrak{G}$, then any spherical function f on $\mathfrak{G}/\mathfrak{H}_1$ is expressible as a polynomial in the spherical functions of any irreducible system congruent to the "standard" spherical functions* $\Pi_1, \cdots, \Pi_N$.

If in Theorem C the group $\mathfrak{H}$ is irreducible, then condition 2 disappears and the spherical functions $\Pi_1, \cdots, \Pi_N$ are absent. The group $\mathfrak{H}_1$ and by the same token its connected component of the identity $\mathfrak{H}_{10}$ ($\supseteq \mathfrak{H}$) are also irreducible. Then by §3.5, $\mathfrak{H}_{10}$ is semisimple and $\mathfrak{H}_1$ contains only a finite number of connected components and is the invariance group of some system of absolute contravariant tensors $\{z\}$ [so that condition 3) also becomes superfluous]. For the same reason, any extension of $\mathfrak{H}_1$ contained in $\mathfrak{G}$ (in general, contained in $SL(n,C)$) is an invariance group. Therefore, if $\mathfrak{H}_1$ is conditionally maximal in $\mathfrak{G}$, then it is a maximal Lie subgroup of $\mathfrak{G}$, so that for irreducible $\mathfrak{H}$ there is no necessity to consider conditional maximality. In addition, we can assume in this case that $\mathfrak{H}$ coincides with $\mathfrak{H}_{10}$, without affecting the degree of generality of the homogeneous space $\mathfrak{G}/\mathfrak{H}_1$.

Thus, for irreducible $\mathfrak{H}$, Theorem C$'$ can be restated in the following form:

THEOREM C$''$. *Let $\mathfrak{G}$ be a connected complex semisimple linear Lie group, and $\mathfrak{H}_1$ a maximal subgroup of $\mathfrak{G}$ whose connected component of the identity $\mathfrak{H}$ $(= \mathfrak{H}_{10})$ is irreducible (and so semisimple). Then any spherical function f on $\mathfrak{G}/\mathfrak{H}_1$ is expressible as a polynomial in the spherical functions of any irreducible (and of any) system.*

5. EXAMPLE 1. Let $\mathfrak{G} = SL(n,C)$, $\mathfrak{H} = SO(n,C)$ and $\mathfrak{H}_1 = \mathfrak{H}$ $(n \geqq 3)$; for $\{z\}$ we take the metric tensor $g_0^{ij} = \begin{cases} 1 \text{ for } i = j \\ 0 \text{ for } i \neq j \end{cases}$. The conditions of Theorem

C are satisfied, and any spherical function on $SL(n,C)/SO(n,C)$ is an arbitrary polynomial in the functions g^{ij}, which in this space can be evaluated as follows: the coordinates of $z = \{g_0^{ij}\}$ after the transformation of z by an arbitrary unimodular matrix $a = \|a_j^i\| \in \mathfrak{G}$ become equal to

$$g^{ij} = \sum_k a_k^i a_k^j,$$

and these values are functions of the coset $a\mathfrak{H}$, and conversely, are completely determined by it; the space of 2-contravariant symmetric tensors g^{ij}, $\det|g^{ij}| = 1$, can be naturally identified with $SL(n,C)/SO(n,C)$.

EXAMPLE 2. By modifying the last example we can apply the more powerful Theorem C″. Namely, for $\mathfrak{H}_1$ we take $\mathfrak{H}$ extended by the unimodular similarity transformations (that is, with coefficients ρ, where $\rho^n = 1$). Now we take for $\{z\}$ the tensor $g_0^{ij} g_{pq}^0$, where g_{pq}^0 is the covariant form of the same metric tensor (if we wish to avoid lower indices, we can always raise them by means of $\epsilon = \{\epsilon^{i_1 \cdots i_n}\}$, §1.1).

Now $\mathfrak{H}_1$ is a maximal subgroup of $\mathfrak{G} = SL(n,C)$ and is irreducible; and so any spherical function on $\mathfrak{G}/\mathfrak{H}_1$ expressed as a polynomial either in the functions $g^{ij}g_{pq}$ or in the functions of any irreducible (or a fortiori reducible) spherical system.

The geometric interpretation of $\mathfrak{G}/\mathfrak{H}_1$ is that it is the space of all nondegenerate cones of the second order in $E^n(C)$.

EXAMPLE 3. Let $\mathfrak{G} = SL(n,C)$ and $\mathfrak{H} = SL(p,C) \times SL(q,C)$, $p+q=n$, and let $\bar{\epsilon}_{(1)}$ and $\bar{\epsilon}_{(2)}$ be as in the example of §3.9. Putting $\mathfrak{H}_1 = \mathfrak{H}$ and $\{z\} = \{\bar{\epsilon}_{(1)}, \bar{\epsilon}_{(2)}\}$, we see that the conditions of Theorem C are satisfied, so that any spherical function f on $\mathfrak{G}/\mathfrak{H}$ is expressible as a polynomial in the coordinates of the tensors $\bar{\epsilon}_{(1)}$, $\bar{\epsilon}_{(2)}$ (which play the roles of the tensors z and Π). Explicitly, let $a = \|a_j^i\| \in \mathfrak{G}$; then these spherical functions are

$$a_1^{[i_1} a_2^{i_2} \cdots a_p^{i_p]}, \quad a_{p+1}^{[i_{p+1}} \cdots a_n^{i_n]}$$

(where $i_1, \cdots, i_p$; $i_{p+1}, \cdots, i_n$ are arbitrary combinations of p and $n-p$ elements in $1, \cdots, n$).

EXAMPLE 4. Let $\mathfrak{G}$ and $\mathfrak{H}$ be as in Example 3; form $\mathfrak{H}_1$ by adjoining to $\mathfrak{H}$ the matrices having $\lambda I_{(p)}$ in the first block and $\mu I_{(q)}$ in the second (by the condition of unimodularity $\lambda^p \mu^q = 1$). Then $\mathfrak{H}_1$ is the invariance group of the tensor $z = \bar{\epsilon}_{(1)}\bar{\epsilon}_{(2)}$, which at the same time plays the role of the unique generator Π_1. Therefore, by Theorem C, all the spherical functions on $\mathfrak{G}/\mathfrak{H}_1$ are expressible as polynomials in the coordinates of the tensor z.

Geometrically $\mathfrak{G}/\mathfrak{H}_1$ is the space of ordered pairs of mutually complementary planes of dimensions p and q, respectively, in $E^n(C)$.

6. So far we have been concerned only with complex analytic homogeneous spaces.

We now show that our results can be adapted to the case of compact homogeneous spaces. To this end we state some preliminary facts.

PROPOSITION 1 (COROLLARY OF THE PETER-WEYL THEOREM). *Let $\mathfrak{H}_K \subset \mathfrak{G}_K$ be compact Lie groups (not necessarily connected). Then any irreducible linear representation $\psi(\mathfrak{H}_K)$ on the complex vector space $E_\psi(C)$ can be obtained from a suitable irreducible linear representation $\phi(\mathfrak{G}_K)$ by a double restriction: of $\phi(\mathfrak{G}_K)$ to $\phi(\mathfrak{H}_K)$, and then of $E_\phi(C)$ to one of its $\phi(\mathfrak{H}_K)$-invariant subspaces $E_\psi(C)$ (should $\phi(\mathfrak{H}_K)$ be reducible).*

PROPOSITION 2. *Let $\phi(\mathfrak{G}_K)$ be a faithful linear representation of the connected semisimple compact Lie group $\mathfrak{G}_K$ on the space $E_\phi(C)$. Then any irreducible linear representation $\psi(\mathfrak{G}_K)$ can be realized on a suitable space T of tensors over $E_\phi(C)$. (This means that under the transformations of the tensors of T by the $\phi(\mathfrak{G}_K)$ the space T passes into itself and the resulting representation is equivalent to $\psi(\mathfrak{G}_K)$.)*

PROOF. In view of the semisimplicity and compactness of $\mathfrak{G}_K$ the representation $\phi(\mathfrak{G}_K)$ is always unitary and unimodular in a suitable basis of $E_\phi(C)$: $\mathfrak{G}_K \subseteq SU(n_\phi, C)$, where $n_\phi = \dim E_\phi(C)$. Using Proposition 1 we obtain $\psi(\mathfrak{G}_K)$ by the double restriction of some irreducible representation of $SU(n_\phi, C)$; but all such representations, as is well known, are realizable on spaces of tensors (for instance, by the action of the Young symmetrizer). Thus, $\psi(\mathfrak{G}_K)$ is realized on some subspace of this space of tensors.

7. Let $\mathfrak{H}_K \subset \mathfrak{G}_K$ be compact Lie groups, with $\mathfrak{G}_K$ connected and semisimple, and let $\mathfrak{G}_K/\mathfrak{H}_K$ be the corresponding homogeneous space (obviously connected). We make the usual assumption that $\mathfrak{H}_K$ does not contain a normal subgroup of $\mathfrak{G}_K$. It is well known that $\mathfrak{G}_K$ can be extended[1)] to a complex Lie group $\mathfrak{G} \supset \mathfrak{G}_K$ in such a way that $\mathfrak{H}_K$ is extended to $\mathfrak{H} \supset \mathfrak{H}_K$ (with $\mathfrak{H} \subset \mathfrak{G}$). Since $\mathfrak{G}_K$ is compact and since $\mathfrak{H}$ can be shown to be closed in $\mathfrak{G}$, we see that $\mathfrak{G}_K \cap \mathfrak{H}$ is compact, and hence

$$\mathfrak{G}_K \cap \mathfrak{H} = \mathfrak{H}_K$$

in view of the maximality of the compact subgroup $\mathfrak{H}_K$ in $\mathfrak{H}$. Every coset $g_K\mathfrak{H}_K$ $(\in \mathfrak{G}_K/\mathfrak{H}_K)$ can be extended unambiguously to a coset $g_K\mathfrak{H}$ $(\in \mathfrak{G}/\mathfrak{H})$, and the mapping is a monomorphism: for if $g_K\mathfrak{H} = g'_K\mathfrak{H}$, then $g_K^{-1}g'_K \in \mathfrak{H}$, and hence $g_K^{-1}g'_K \in \mathfrak{G}_K \cap \mathfrak{H} = \mathfrak{H}_K$, so that $g_K\mathfrak{H}_K = g'_K\mathfrak{H}_K$.

Thus

$$\mathfrak{G}_K/\mathfrak{H}_K \subset \mathfrak{G}/\mathfrak{H}. \tag{4.2}$$

Spherical functions on $\mathfrak{G}_K/\mathfrak{H}_K$ are defined as in subsection 1, only now we do not, of course, require them to be complex analytic (real-analyticity is obtained automatically).

PROPOSITION 3. *Any system of spherical functions $\{\sigma_K^1(x), \cdots, \sigma_K^N(x)\}$ on*

1) This means that 1) $\mathfrak{G}_K$ is a maximal compact subgroup of the complex Lie group $\mathfrak{G}$, and 2) the Lie algebra of $\mathfrak{G}$ can be obtained by complexification of the Lie algebra of $\mathfrak{G}_K$ embedded in it.

$\mathfrak{G}_K/\mathfrak{H}_K$ can be obtained by the restriction to $\mathfrak{G}_K/\mathfrak{H}_K$ (in accordance with (4.2)) of some system of spherical functions $\{\sigma^1, \cdots, \sigma^N\}$ on $\mathfrak{G}/\mathfrak{H}$, and $\{\sigma^i\}$ is uniquely determined by $\{\sigma_K^i\}$.

PROOF. Obviously, as in subsection 3 the spherical functions $\{\sigma_K^i\}$ can always be regarded as flow coordinates ξ^i in the model $\mathfrak{G}_K/\mathfrak{H}_K$ in the space E_ϕ of some suitable linear representation $\phi(\mathfrak{G}_K)$, namely, that which transforms $\sigma_K^1, \cdots, \sigma_K^N$ among themselves in accordance with (4.1). The equations of the (possibly not faithful) model then have the form

$$\xi^i = \sigma_K^i(x_K) \quad (i = 1, \ldots, N); \quad x_K \in \mathfrak{G}_K/\mathfrak{H}_K. \tag{4.3}$$

Let $x_{K0} \in \mathfrak{G}_K/\mathfrak{H}_K$ be a point with stabilizing group $\mathfrak{H}_K$; the corresponding vector $\xi_0 \in E_\phi$ is invariant under the action of $\mathfrak{H}_K$.

The representation $\phi(\mathfrak{G}_K)$ can be regarded as a homomorphism $\mathfrak{G}_K \to SU(N)$; but under the extension of $\mathfrak{G}_K$ and $SU(N)$ to $\mathfrak{G}$ and $SL(N, C)$, this homomorphism is extended uniquely to a complex analytic homomorphism $\mathfrak{G} \to SL(N, C)$; that is, we extend the representation $\phi(\mathfrak{G}_K)$ to the representation $\phi(\mathfrak{G})$. Here the vector ξ_0 also remains invariant relative to $\phi(\mathfrak{H})$ $(\mathfrak{H} \subset \mathfrak{G})$. Indeed, $\mathfrak{H} = \mathfrak{H}_K \cdot \mathfrak{H}_0$, where $\mathfrak{H}_0$ is the connected component of the identity of $\mathfrak{H}$; ξ is invariant relative to $\phi(\mathfrak{H}_K)$, and the connected linear Lie group $\phi(\mathfrak{H}_0)$ is completely determined by its Lie algebra $\phi(\mathfrak{H})$, but the latter is the "complexified" Lie algebra $\phi(H_K)$, and hence leaves the vector ξ_0 fixed (by the definition of the extension of $\mathfrak{H}_K$ to $\mathfrak{H}$, H is obtained from H_K by complexification, and this, of course, is preserved by the complex analytic homomorphism $\mathfrak{H} \to SL(N)$).

As a consequence, under the action $\phi(\mathfrak{G})$, the vector ξ_0 describes a model (possibly not faithful) of the space $\mathfrak{G}/\mathfrak{H}$, and in particular the action of $\mathfrak{G}_K$ yields the model of $\mathfrak{G}_K/\mathfrak{H}_K$ considered above, where in agreement with (4.2)

$$\mathfrak{G}_K/\mathfrak{H}_K \subset \mathfrak{G}/\mathfrak{H}.$$

The flow coordinates $\{\xi^1, \cdots, \xi^N\}$ on the model $\mathfrak{G}/\mathfrak{H}$, $\xi^i = \sigma^i(x)$, form a system of spherical functions, which is subject to the representation $\phi(\mathfrak{G})$ and under restriction to $\mathfrak{G}_K/\mathfrak{H}_K$ obviously goes over into the given system of spherical functions (4.3).

Finally, the uniqueness of the choice of the $\sigma^i(x)$ follows from their complex analyticity and the uniqueness of analytic continuation; the initial element of the function $\sigma^i(x)$ is uniquely determined in the neighborhood of any point $x_K \in \mathfrak{G}_K/\mathfrak{H}_K \subset \mathfrak{G}/\mathfrak{H}$ by the real analytic structure of the function $\sigma_K^i(x_K)$ on $\mathfrak{G}_K/\mathfrak{H}_K$.

It follows from the proposition we have just proved that there is a one-to-one correspondence between systems of spherical functions on $\mathfrak{G}/\mathfrak{H}$ and $\mathfrak{G}_K/\mathfrak{H}_K$ by the principle of restriction of the former to the latter. We remark that under this restriction no relations of linear dependence can arise, for otherwise such a relation could be transferred to the original functions by means of initial elements. For the same reason no new algebraic relations can arise, so that the set of all such relations for the spherical functions on $\mathfrak{G}_K/\mathfrak{H}_K$ is exactly the same as that for the corresponding spherical functions on $\mathfrak{G}/\mathfrak{H}$.

In particular, all polynomial expressions of spherical functions in terms of spherical functions on $\mathfrak{G}_K/\mathfrak{H}_K$ are the same as those on $\mathfrak{G}/\mathfrak{H}$.

EXAMPLE. The Grassmannian manifold of p-dimensional planes in $E^n(C)$ can be regarded as a compact homogeneous space $G_n^p = \mathfrak{G}_K/\mathfrak{H}_K$, where $\mathfrak{G}_K = SU(n)$ and $\mathfrak{H}_K = (SU(p) \times SU(q) \cdot T^1)$ with $p+q=n$ and T^1 a "one-dimensional torus" (which is not a direct factor: some factorization is still necessary). By the principle (4.2), the space G_n^p corresponds to the complex analytic space of Example 4, subsection 5, so that all the spherical functions on G_n^p are expressible as polynomials in the coordinates of the tensor $z = \bar{\epsilon}_1\bar{\epsilon}_2$, where $\bar{\epsilon}_1$ and $\bar{\epsilon}_2$ are the direction polyvectors of a p-dimensional plane and the plane of dimension $q = n - p$ orthogonal to it (in the positive definite Hermitian metric).

Received 10 NOV 1967

REFERENCES

[1] P. K. Raševskiĭ, *The structure of tensors which admit a given invariance group,* Dokl. Akad. Nauk SSSR **177** (1967), 275-276 = Soviet Math. Dokl. **8** (1967), 1408-1409. MR **36** #2640.

[2] G. B. Gurevič, *Foundations of the theory of algebraic invariants,* OGIZ, Moscow, 1948; English transl., Noordhoff, Groningen, 1964. MR **11**, 413; **32** #1211.

[3] P. K. Raševskiĭ, *Projective representations of homogeneous spaces,* Mat. Sb. **50 (92)** (1960), 171-202. (Russian) MR **23** #A3803.

Translated by:
D. W. Johnson

Trudy Moskov. Mat. Obsc.
Tom 20 (1969)

Trans. Mos Math. Soc.
Vol. 20 (1969)

SOME PROPERTIES OF SINGULAR SEMISIMPLE ALGEBRAIC GROUPS OVER NONCLOSED FIELDS

B. Ju. VEĬSFEĬLER

Contents

INTRODUCTION 109
§1. Auxiliary propositions 113
§2. Groups of type G_2 116
§3. Groups of type D_4 116
§4. Groups of type F_4 121
§5. Groups of type E_6 123
§6. Groups of type E_7 128
§7. Groups of type E_8 132
REFERENCES 133

Introduction

0. The classification of the classical semisimple algebraic groups over nonclosed fields was achieved by A. Weil in [**26**] (see also [**12**], [**17**] and [**18**]). In particular, his work gives a realization of an arbitrary classical group as a subgroup of the group of regular elements of a suitable associative algebra with an involution. The realizations of groups of type G_2 and F_4 as groups of automorphisms of Cayley algebras and singular Jordan algebras enable us to describe these groups ([**11**], [**14**], [**15**], [**16**], [**24**]). Some of the groups of type D_4, E_6, E_7 and E_8 also turn out to be connected with Jordan algebras ([**6**], [**9**], [**11**], [**16**], [**22**], [**24**]).

In principle these results can be used to get information on the algebraic groups mentioned above, if information about the corresponding algebras (associative, alternative, or Jordan) is already available.

1. In this paper we associate with every algebraic group G defined over a field k (with any parabolic subgroup P such that the factor-space G/P is defined over k) the *invariant* $\operatorname{ind}_k G$ (or $\operatorname{ind}_k(G/P)$), the greatest common divisor of the degrees of all separable extensions K of k that decompose G (such that $(G/P)_k \neq \emptyset$).

The basic hypothesis connected with the concept of index is as follows:

HYPOTHESIS 1. (a) $\operatorname{ind}_k G = 1 \Longleftrightarrow G$ *is decomposable over* k.

(b) $\operatorname{ind}_k G/P = 1 \Longleftrightarrow (G/P)_k \neq \emptyset$.

We prove the hypothesis for the groups D_4, F_4, E_6 and E_7 (over a field k of sufficiently large characteristic). For the classical groups Hypothesis 1 follows from known properties of division algebras and from Springer's theorem on quadratic forms (see §§1.6 and 1.7). Thus the hypothesis remains unproved only for groups of type E_8.

Hypothesis 1 can be generalized in several directions. We indicate some of these.

a. Let G be a connected algebraic group defined over a field k, let K be a normal extension of k, $\Gamma = \Gamma(K/k)$ its Galois group, Γ_p a Sylow p-subgroup of Γ and $K_p = K^{\Gamma_p} = \{x \in K : x^\sigma = x \ \forall \sigma \in \Gamma_p\}$. Further, let $r_p : H^1(\Gamma, G_K) \to H^1(\Gamma_p, G_K)$ be the restriction map.

HYPOTHESIS 2. *If* $\alpha \in H^1(\Gamma, G_K)$ *and* $r_p(\alpha) = 0 \ \forall p$, *then* $\alpha = 0$.

b. Now let $\mathfrak{K}$ be a purely transcendental extension of k (that is, $\exists K \supset k : \mathfrak{K}K$ is purely transcendental over K). Let $\operatorname{ind}_k \mathfrak{K}$ be the greatest common divisor of the numbers $[K : k]$, where K is separable over k and $\mathfrak{K}K$ is purely transcendental over K.

HYPOTHESIS 3 (AND 3′). $\operatorname{ind}_k \mathfrak{K} = 1 \Leftrightarrow \mathfrak{K}$ *is purely transcendental (unirational) over* k.

Note that Hypothesis 2 reduces to 1a if G is a group decomposable over k, and Hypothesis 3 reduces to 1b if $\mathfrak{K}$ is a field of functions on the manifold G/P. We note that Hypothesis 3 can also be stated in terms of Galois cohomology.

2. Suppose that G satisfies Hypothesis 1. We shall show how this can be utilized.

(a) *If* $\operatorname{ind}_k G = n = \prod_{i \in I} p_i^{\kappa_i}$ *(or* $\operatorname{ind}_k G/P = n = \prod_{i \in I} p_i^{\kappa_i}$*)*, $\kappa_i \neq 0 \ \forall i \in I$, *and* K *is a normal splitting field for* G *(or* $(G/P)_K \neq \emptyset$*), then for every* $i \in I$ *there exist subfields* $\mathfrak{K}$ *and* $\mathfrak{K}_1$ *of* K *such that* $\operatorname{ind}_{\mathfrak{K}} G = p_i$ *and* $\operatorname{ind}_{\mathfrak{K}_1} G = 1$ *(or correspondingly for* G/P*) and* $\mathfrak{K}_1$ *is a normal extension of* $\mathfrak{K}$ *with Galois group* Z_{p_i}.

Suppose that $p = p_i$. Set $\Gamma = \Gamma(K/k)$ and write Γ_p for a Sylow p-group of Γ, $K_p = K^{\Gamma_p}$. Then $\operatorname{ind}_{K_p} G = p^\kappa$ (or $\operatorname{ind}_{K_p} G/P = P^\kappa$) and $\Gamma(K/K_p) \cong \Gamma_p$. On performing successive cyclic extensions of K_p within K, we eventually arrive at $\mathfrak{K}_1$. For $\mathfrak{K}$ we take the field obtained at the preceding step, etc.

This remark allows us (modulo Hypothesis 1) to state that k carries over this property to a "not too large" extension $\mathfrak{K}$: G is indecomposable over k (or correspondingly $(G/P)_k = \emptyset$). This is particularly useful if K is a splitting field for some maximal k-torus. For instance:

(b) *If every division algebra over* k *is commutative, then* $\operatorname{ind}_k G/B = 1$ *(*B *is the Borel group of* G*)*.

For let K be a splitting field for a maximal k-torus T. If $p \mid \operatorname{ind}_k G/B$, replace k by the field $\mathfrak{K}$ of (a). We show that under these conditions G contains an isotropic $\mathfrak{K}$-group of type A_{p-1}. This contradicts the assumptions, since it is known that such groups correspond to noncommutative division algebras over $\mathfrak{K}$ (and consequently over k). If $p = 2$, our assertion follows from 1.10. If $p \neq 2$, it can be verified that the image of $\Gamma(K/\mathfrak{K})$ (this is a p-group) in the automorphism group of the root system of G relative to T preserves a subset of the roots of type $A_{p-1} \times A_{p-1} \times \cdots \times A_{p-1}$, whence the existence of anisotropic $\mathfrak{K}$-groups of type A_{p-1} follows.

We give yet another example.

(c) *Suppose that k has the following property: if H is a simply connected k-group of type A_n (perhaps not absolutely simple), then $H^1(k,H)=0$. Suppose further that $\widetilde{G}_q$ and $\overline{G}_q$ are a quasi-decomposable simply connected group and its adjoint group. If G is defined by a cocycle $\alpha\in H^1(k,\overline{G}_q)$ lying in the image of the mapping $H^1(k,\widetilde{G}_q)\to H^1(k,\overline{G}_q)$, then* $\operatorname{ind}_k G/B=1$.

Note that Harder proves this assertion in [**13**]. In the case under consideration his proof can be shortened by using 0.2a.

The examples in 0.2(b), (c) show that the Serre hypothesis [**20**] follows from our Hypothesis 1.

(d) Note the formal similarity between Hypothesis 2 and the "Hasse principle". It might be thought that this resemblance is more than formal, because the method of proof of Hypothesis 1 used here can be applied to a proof of Hasse's principle in the following form:

Let k be the field of algebraic numbers and V the set of all its norms. Then

$$p \mid \operatorname{ind}_k G/P \Longleftrightarrow p \mid \prod_{v\in V} \operatorname{ind}_{k_v} G/P.$$

(Naturally for the proof it is useful to replace k by the $\mathfrak{K}$ of 0.2(a).)

3. Apart from proving Hypothesis 1, we consider in this article groups of type D_4, F_4, E_6 and E_7 (over a field k of sufficiently large characteristic, for instance $\operatorname{Char} k > \operatorname{rg} G$). The common property of the groups considered is that *they contain nontrivial semisimple subgroups.* (More detailed information about which groups are considered will be found at the beginning of the corresponding sections.) For groups of the type indicated we calculate the possible values of the invariants $\operatorname{ind}_k G$ and $\operatorname{ind}_k G/P$ (see 1.13) and indicate sufficient conditions for them to contain proper semisimple subgroups. For groups satisfying the conditions indicated we establish the existence of a separable splitting field K such that $[K:k]=\operatorname{ind}_k G$ (this assertion is stronger than Hypothesis 1a). We also calculate the Galois groups of normal extensions of k containing a field K with the conditions just mentioned.

Besides this we consider certain forms of groups of type E_8 and discover the values of the invariants $\operatorname{ind}_k G$ and $\operatorname{ind}_k G/P$.

The results connected with the calculation of the possible values of $\operatorname{ind}_k G/P$ are close to those of Tits [**25**]. We use parts of the tables of isotropic groups given in [**25**].

The investigation of the groups D_4, F_4, E_6 and E_7 is based on the existence of a natural representation. If this representation is defined over k, the stability groups H of a sufficiently general k-point is a k-subgroup of G and has the type B_3, D_4, F_4 or E_6, respectively. This allows us to apply inductive reasoning to a group G of this type.

4. Generalities. We assume that the reader is familiar with the literature on algebraic groups, in particular, with [**8**], [**10**], [**23**] and [**25**]; we use the results of these papers freely. However, we shall discuss certain facts of a general character in §1.

The classification of algebraic groups is given in the form of tables which are collected in appropriate parts of the text and which contain the information referred to in §0.3. We do not consider here the question of the existence of groups with given invariants. Only in special cases do we formulate hypothesis like: the form X does not exist.

I am deeply grateful to È. B. Vinberg for his constant and invaluable help, for a great deal of useful advice, and also for criticism which resulted in the elimination of deficiencies occurring in the original version. I am also grateful to S. G. Gindikin for his support during the preparation of the paper.

5. Notation and conventions. All extensions of the field k that occur in the paper are separable. For this to be so, we assume that $\mathrm{Char}\, k > \mathrm{rg}\, G$.

k_s is the separable closure of k.

$\mathfrak{S}(k, n)$ (or $\mathfrak{S}(k, \Gamma)$) is the set of separable (or normal) extensions of k of degree n (with Galois group Γ).

$\Gamma(K/k)$ is the Galois group of the field K normal over k.

$K^\Gamma = \{x \in K : x^\sigma = x \forall \sigma \in \Gamma\}$, where K is normal over k and $\Gamma \subset \Gamma(K/k)$.

Γ_p is a Sylow p-subgroup of the finite group Γ.

$k(T)$ is a splitting field for the k-torus T.

For $\Gamma_k(H)$ and $K_k(H)$ see §1.2; $a_k(G) = K_k[(G) : k]$.

A k-group, k-torus etc. is a group or torus etc. defined over k.

$N_G(H)$ and $Z_G(H)$ are the normalizer and centralizer in G of the subgroup H.

$T_d(G, k)$ and $H(G, k)$ are the maximal k-trivial torus in G and its centralizer.

$S(H)$ is the semisimple part of the connected component of the reductive group H.

$R(H)$ and $U(H)$ are the radical and unipotent radical of the group H.

$C(H)$ and $C_0(H)$ are the center and connected center of H.

T^i is an i-dimensional torus.

$\Sigma(H)$ and $\Delta(H)$ are the root system of a group G relative to some maximal torus $T \subset H$, and the system of simple roots.

$P_{i_1 \cdots i_s}(H, k)$ (or $P^{i_1 \cdots i_s}(H, k)$) is the set of parabolic k-subgroups of G corresponding to the simple roots with given number (or the complement in Δ). When $k = k_s$, k is omitted.

$B(k) = P^\emptyset(H, k)$ is the set of Borel groups in H.

$\Delta_k(H)$ is the maximal subset of Δ for which the corresponding parabolic subgroup is defined over k.

$W(H) = \mathrm{Int}\, \Sigma(H)$; $W'(H) = \mathrm{Aut}\, \Sigma(H) = W(H) \cdot \mathrm{Aut}\, \Delta(H)$.

$G(\Sigma_1)$ is the regular subgroup (in the sense of Dynkin) of G corresponding to the subsystem Σ_1 of roots in $\Sigma(G)$; $G(\Sigma_1)$ is generated by a maximal torus T and the root subgroups U_α, $\alpha \in \Sigma_1$.

$G(\mathfrak{A})$ is the group of elements with norm 1 in the algebra $\mathfrak{A}$.

We use the following notation in the tables:

$$i(G) = \mathrm{ind}_k\, G, \quad i(P) = \mathrm{ind}_k\, G/P,$$
$$a = [K_k(G) : k], \quad \alpha(P) = \mathrm{ind}_{K_k(G)} \mathfrak{A}_{K_k(G)}(P).$$

The Galois group of the field which was mentioned at the end of §0.3 is indicated in the graph Γ_0. In the graph H_0 the type of certain singular k-subgroups of G is indicated; in the case where nothing is indicated in this graph, the group G is isotropic in some quadratic extension of k, and the existence of such subgroups follows from 1.11.

6. Numbering of the roots.

A_n 1 — 2 — 3 — ... — n
B_n 1 — 2 — 3 — ... — $(n-1) \Rightarrow n$
C_n 1 — 2 — 3 — ... — $(n-1) \Leftarrow n$
D_n 1 — 2 — 3 — ... — $(n-2)$ — $(n-1)$, with n attached to $(n-2)$

E_n 1 — 2 — 3 — 4 — ... — $(n-1)$, with n attached to 3

G_2 1 $\Rrightarrow$ 2
F_4 1 — 2 $\Rightarrow$ 3 — 4

§1. Auxiliary propositions

1 ([8], [19], [25]). Let T be a maximal k-torus in the group G with separable splitting field $k(T) = K$, let $\Gamma = \Gamma(K/k)$, and let $\Sigma = \Sigma(G)$ be the system of roots of G relative to T. Then the action of Γ on Σ defines a homomorphism $\phi: \Gamma \to \operatorname{Aut}\Sigma$. *If $\phi(\Gamma)$ preserves the subsystem Σ_1 of roots in Σ, then the corresponding reductive group $G(\Sigma_1)$ is clearly defined over k.*

2 ([8], [25]). The Galois group Γ_s of the separable closure k_s of k acts on the classes of parabolic subgroups. This action defines a homomorphism $\phi_0: \Gamma \to \operatorname{Aut}\Delta(G)$ (we remark that ϕ_0 is the composite of ϕ (see 1.1) and the natural projection $\operatorname{Aut}\Sigma \to \operatorname{Aut}\Delta$). Set $\Gamma_0 = \operatorname{Ker}\phi_0$, $\Gamma_k(G) = \Gamma/\Gamma_0$, $K_k(G) = k_s^{\Gamma_0}$. Note that *the field $K_k(G)$ is contained in every extension of k over which all classes of parabolic subgroups of G are defined.*

3 ([8]). Let P be a parabolic subgroup of G such that G/P is defined over k. *If $\Gamma_k(G) = 1$, every linear system of divisors on G/P is defined over k, and is consequently a Severi-Brauer manifold.* In general ($\Gamma_k(G) \neq \{1\}$) the linear system corresponding to the representation with weight $\sum_{j=1}^{s} \lambda_{i_j}$ (here λ_q is the fundamental weight defining $P_q(G)$) is defined over k if $P = P_{i_1\cdots i_s}$. We denote the corresponding Severi-Brauer manifold by $\mathfrak{B}(P)$, and the central division algebra associated with $\mathfrak{B}(P)$ by $\mathfrak{A}_k(P)$.

We remark that *we can also associate a division algebra with a linear system not defined over k, but this time its center is a subfield of a field $K_k(G)$ different from k.*

4 ([19], [20], [25]). Every k-group G is defined by a cocycle of the Galois group Γ_s of the separable closure k_s of k in the group of automorphisms $\operatorname{Aut}\overline{G}_d$ of the adjoint group $\overline{G}_d$ of the same type as G. Denoting the simply connected covering of $\overline{G}_d$ by $\widetilde{G}_d$ and the center of $\widetilde{G}_d$ by $C = C(\widetilde{G}_d)$, we get exact sequences:

(a) $$H^1(k, \overline{G}_d) \to H^1(k, \operatorname{Aut} \overline{G}_d) \xrightarrow{\varphi_0} H^1(k, (\operatorname{Aut} \overline{G}_d)/\overline{G}_d),$$

(b) $$H^1(k, C) \to H^1(k, \overline{G}_d) \to H^1(k, \overline{G}_d) \xrightarrow{\delta} H^2(k, k^*).$$

The mapping ϕ_0 is the homomorphism described in §1.2. If C is cyclic, the elements of $H^2(k, C)$ are naturally identified with division algebras over k.

In general we have that δ associates with every group G some division algebra $\mathfrak{A}_k^{(i)}(G)$. In what follows we omit the superscript i and speak of $\mathfrak{A}_k^{(i)}(G)$ as an algebra over k. This algebra $\mathfrak{A}_k(G)$ is closely connected with the algebras $\mathfrak{A}_k(P)$ (see §1.3). If k is a field of sufficiently large characteristic and $\pi = \pi(\Sigma\lambda_{i_j}) : G \to GL(V)$ is the linear representation with weight $\Sigma\lambda_{i_j}$ (see §1.3), then the diagram

$$\begin{array}{ccccc}
H^1(k, \widetilde{G}_d) & \to & H^1(k, \overline{G}_d) & \xrightarrow{\delta} & H^2(k, C) \\
\pi^* \downarrow & & \pi^* \downarrow & & \pi^* \downarrow \\
H^1(k, GL(V)) & \to & H^1(k, PGL(V)) & \xrightarrow{\delta} & H^2(k, k^*)
\end{array}$$

is commutative and the image of $\mathfrak{A}_k(G)$ in the Brauer group $H^2(k, k^*)$ of k is the algebra $\mathfrak{A}_k(P)$ (see §1.3).

We note that the exact sequence (b) associated with the simply connected cover of $\operatorname{Aut} \overline{G}_d$ leads us to a consideration of noncentral division algebras (see the last paragraph of §1.3). Note further that $\mathfrak{A}_k(G) = k \Leftrightarrow \mathfrak{A}_k(P) = 1 \,\forall P$, and in this case G is defined by a cocycle out of $\psi(H^1(k, \widetilde{G}_d))$.

5. Let G and P be an algebraic group defined over k and a parabolic subgroup of it such that the homogeneous space G/P is defined over k.

a. *If G is isotropic over k and $H = H(G, k)$ is the centralizer of a maximal k-trivial subtorus in G, then* $\operatorname{ind}_k G = \operatorname{ind}_k H$ *and* $\operatorname{ind}_k G/P = \operatorname{ind}_k H/P'$, *where P' is a suitable parabolic subgroup of H.*

Moreover,

b. $\operatorname{ind}_k G = [K_k(G) : k] \cdot \operatorname{ind}_{K_k(G)}(G/B)$ (*here B is the Borel group of G*).

c. $\operatorname{ind}_k \mathfrak{A}_k(P) \mid \operatorname{ind}_k G/P$.

All the assertions follow easily from the definitions.

6. *Let G be a group of type A_n.* It is known (see for instance [**26**]) that G is isogenic over $K = K_k(G)$ to the group $G(\mathfrak{A})$ of elements of norm 1 in the centrally simple associative algebra $\mathfrak{A}$. In this case

$$\operatorname{ind}_K G = \operatorname{ind}_K G/B = \operatorname{ind}_K \mathfrak{A} = m,$$
$$\operatorname{ind}_K G/P_i = (m, i)^{-1} \operatorname{ind}_K \mathfrak{A}.$$

It is well known that $\operatorname{ind}_k \mathfrak{A} = \sqrt{\dim \mathfrak{A}}$ for a central division algebra $\mathfrak{A}$.

We remark further that if $\operatorname{ind}_K G = 2$ (or $= 3$), then G has a normal quadratic (or cubic) splitting field $\widetilde{K}/K$ and contains a maximal K-torus decomposable over $\widetilde{K}$ (see [**1**]).

7. a. If G is a classical group of type B_n, C_n or D_n $(n \neq 4)$ and $\mathfrak{A}$ is a simple central associative algebra with involution J figuring in the Weil construction [**26**], then $\mathfrak{A}$ is equivalent (in the Brauer group) to $\mathfrak{A}_k(P_1)$.

Further, expressing the algebra $\mathfrak{A}$ as the tensor product of a division algebra $\mathfrak{D}$ $(=\mathfrak{A}_k(P_1))$ and a complete matrix algebra $\mathfrak{M}_s$, we see that J induces involutions J_1 and J_2 (see [**1**]) on $\mathfrak{D}$ and on $\mathfrak{M}_s$, respectively. We denote by G_1 and G_2 the groups that are defined according to the Weil construction [**26**] by the pairs $(\mathfrak{D}, J_1)$ and $(\mathfrak{M}_s, J_2)$. The number $\mathrm{ind}_k G/P_i$ is expressed in terms of the indices of suitable parabolic subgroups of G_1 and G_2.

b. If $\mathfrak{A}_k(P_1) = k$, then G is decomposable if G is of type C_n or is isogenic to the orthogonal group of some quadratic form when G is of type B_n or D_n. In this last case Springer's theorem [**21**] asserts:

If $\mathrm{Char}\, k \neq 2$ *and* $(G/P_1)_k = \emptyset$, *then* $\mathrm{ind}_k G/P_1 = 2$.

8. When we need the values of the invariants $\mathrm{ind}_k G$ and $\mathrm{ind}_k G/P$ for the classical groups, we shall use them without calculating them.

9. LEMMA. *If* $\widetilde{K}$ *is a separable extension of* k, *then* $\mathrm{ind}_{\widetilde{K}} G \mid \mathrm{ind}_k G$ (*or* $\mathrm{ind}_{\widetilde{K}} G/P \mid \mathrm{ind}_k G/P$).

(a) Let $\mathfrak{S}(k)$ be the set of all those separable extensions K of k that decompose G (or are such that $(G/P)_K \neq \emptyset$). Clearly, $\mathfrak{S}(k)$ is invariant under $\Gamma(k_s/k)$. Let $\mathfrak{S}$ be any system of separable extensions of k invariant under $\Gamma(k_s/k)$ and closed under extensions. For every separable field K/k we get $\mathfrak{S}_K$, the set of composites of elements of $\mathfrak{S}$ with K; this definition is good in view of the invariance of $\mathfrak{S}$ under $\Gamma(k_s/k)$. Put $\mathrm{ind}_K \mathfrak{S} = \mathrm{g.c.d.}_{K_1 \in \mathfrak{S}_K}[K_1 : K]$. Then our lemma is a special case ($\mathfrak{S} = \mathfrak{S}(k)$) of the following proposition:

If $\mathfrak{S}$ *is as above, then* $\mathrm{ind}_{\widetilde{K}} \mathfrak{S} \mid \mathrm{ind}_k \mathfrak{S}$.

Set $\mathrm{ind}_k \mathfrak{S} = p_1^{\kappa_1} \cdots p_s^{\kappa_s}$ and $\mathrm{ind}_{\widetilde{K}} \mathfrak{S} = p_1^{l_1} \cdots p_s^{l_s}$. We shall show that $l_i \leqq \kappa_i$, $\forall i = 1, 2, \cdots, s$.

(b) If K is separable over k, then

$$\mathrm{ind}_k \mathfrak{S} \mid [K : k] \cdot \mathrm{ind}_K \mathfrak{S}.$$

For $\mathrm{ind}_K \mathfrak{S} = \mathrm{g.c.d.}_{K_1 \in \mathfrak{S}_K}[K_1 : K] = ([K : k])^{-1} \cdot \mathrm{g.c.d.}_{K_1 \in \mathfrak{S}_K}[K_1 : k]$. Since $\mathfrak{S}$ is closed under extensions, we have $\mathfrak{S}_K \subset \mathfrak{S}$, whence our assertion follows.

(c) Suppose that $K \in \mathfrak{S}$, $[K : k] = p^\kappa \cdot n$ and $(p, n) = 1$ (the index i has been omitted). Further, let $\mathfrak{K}$ be the normal closure of K/k, $\Gamma = \Gamma(\mathfrak{K}/k)$ and $\widetilde{\Gamma} = \Gamma(\mathfrak{K}/K)$; let Γ_p and $\widetilde{\Gamma}_p$ be the Sylow p-subgroups of Γ and $\widetilde{\Gamma}$; let $k_p = \mathfrak{K}^{\Gamma_p}$ and $\widetilde{k} = \mathfrak{K}^{\widetilde{\Gamma}_p}$. Then $\mathrm{ind}_{k_p} \mathfrak{S} = p^\kappa$.

For $\mathrm{ind}_{k_p} \mathfrak{S} = p^\alpha$, and since $\widetilde{k} \in \mathfrak{S}_{k_p}$, we see that $\alpha \leqq \kappa$. Applying (b), we get

$$\mathrm{ind}_k \mathfrak{S} = p^\kappa n \mid \mathrm{ind}_{k_p} \mathfrak{S}\, [k_p : k] = p^\kappa [k_p : k],$$

so that $\alpha = \kappa$.

(d) The lemma is true.

Set $K' = \mathfrak{K} \cap \widetilde{K}$, $\mathrm{ind}_{K'} G = p^\beta m$, $(p, m) = 1$, and set $\Gamma' = \Gamma(\mathfrak{K}/K')$. Since Γ_p' is contained in Γ_p (or in a conjugate of Γ_p, which is the same in view of (a)), then $\beta \leqq \kappa$ and we may assume that $k = K'$. Denoting the compositum of $\widetilde{K}$ and $\mathfrak{K}$ by $\mathfrak{K}'$, we get $\Gamma(\mathfrak{K}'/\widetilde{K}) \cong \Gamma$. Since $K'' = \{\lambda \in \mathfrak{K}' : \sigma\lambda = \lambda\ \forall \sigma \in \Gamma_p\} \in \mathfrak{S}_{\widetilde{K}}$ and $[K'' : \widetilde{K}] = p^\kappa n_1$, $(n_1, p) = 1$, it follows that $l \leqq \kappa$, as required.

10. LEMMA ([**13**]). *Let K be a quadratic extension of the field k, and P a minimal parabolic subgroup of G defined over K. Then there exists a Levi decomposition* ([**8**], §0.8) $P = L \cdot U$ *(here L is reductive and U is unipotent) such that L is defined over k.*

Clearly, we may assume that G is k-anisotropic. For any $\sigma \in \Gamma(K/k)$, $\sigma \neq 1$, the subgroup $P \cap P^{\sigma}$ is defined over k and contains a Levi subgroup L ([**8**], §4.18). Since $\mathrm{rg}_k G = 0$, $P \cap P^{\sigma}$ is reductive and thus coincides with L.

11. COROLLARY. *If G is anisotropic over k, then $\Gamma(K/k)$ acts (see* §1.1) *on the system of roots of G relative to a maximal K-trivial subtorus T of L according to the rule $\sigma\alpha = -\alpha$. In particular, every reductive group H spanned on the root subspaces (relative to T) is defined over k.*

The first assertion is clear because $\mathrm{rg}_k G = 0$, $T = C_0(L)$ and $\Gamma(K/k) \cong Z_2$. The second assertion follows from the first and from §1.1.

12. By §1.9, $\mathrm{ind}_k G = \prod_p \mathrm{ind}_{K_p} G$ and $\mathrm{ind}_k G/P = \prod_p \mathrm{ind}_{K_p} G/P$ (in the notation of §1.9). Thus, *to find the "possible" values of the index* (*see* §0.3a), *it is enough to multiply* (*or to multiply termwise*) *the corresponding values over K_p*. With these remarks as a background, we do not write down all possible values of the index, but restrict ourselves to the possible values of the p-component in the cases where we do not succeed in finding the best result.

§2. Groups of type G_2

1. PROPOSITION. *Every group G of type G_2 is decomposable over a quadratic extension of k. If $\mathrm{rg}_k G = 0$, then $\mathrm{ind}_k G = 2$.*

The group $W(G)$ preserves the subsystem Σ_1 of "long" roots of type A_2 in $\Sigma(G)$. Thus, in view of 1.1, G contains a k-subgroup H of type A_2. Let V be the tangent space in G/H at the point $\{H\}$. A representation ρ of H in V defined over k and over $K = K_k(H)$ decomposes as a sum of two three-dimensional representations. Thus H, and therefore G, is decomposable over k. By 1.6, $[K:k] = 1$ or 2, so that the first assertion follows.

Suppose now that $\mathrm{rg}_k G = 0$ and let $\mathfrak{K} \in \mathfrak{S}(k, 2s+1)$. If $\mathrm{rg}_{\mathfrak{K}} G > 0$, then $\mathrm{rg}_{\mathfrak{K}} G = 2$ and $B(\mathfrak{K}) \neq \emptyset$. Take $B \in B(\mathfrak{K})$. We have $\dim B = 8$, $\dim H = 8$, $\dim G = 14$ and $\dim B \cap H \geqq 2$, so that $\mathrm{rg}_{\mathfrak{K}}(H) > 0$. This contradicts 1.6 and 1.8, so that $\mathrm{rg}_{\mathfrak{K}} G = 0$ and $\mathrm{ind}_k G = 2$.

§3. Groups of type D_4

0. Our main interest here centers on the groups D_4 for which $\Gamma_k(G) \supset Z_3$. However, the study of these groups is based on properties of classical groups of type D_4 (i.e. the groups for which $\Gamma_k(G) \subset Z_2$). §§3.1, 3.3 and 3.4 are devoted to classical groups. In §3.5 we indicate sufficient conditions for G to contain proper semisimple subgroups. Hypothesis 1 is proved in §§3.2, 3.6 and 3.7. In the last two subsections we study properties of a group G containing certain special subgroups; the invariants of these groups are laid out in Table 2.

D_4, $\mathrm{rg}\, G > 0$ (§3.1)

	D_40	D_41	D_42	D_43	${}^2D_40$	${}^2D_41$	${}^2D_42$	${}^2D_43$	${}^2D_44$	${}^2D_45$	${}^3D_40$	${}^3D_41$	${}^6D_40$	${}^6D_41$
Δ_k	Δ	$\{1,2\}$	$\{2\}$	$\{1\}$	Δ	$\{1,2\}$	$\{2\}$	$\{1\}$	$\{1\}$	$\{1\}$	Δ	$\{2\}$	Δ	$\{2\}$
a	1	1	1	1	2	2	2	2	2	2	3	3	6	6
$i(G)$	1	2	2 or 4	4	2	4	4	2	4	8	3	6	6	12
$i(P_1)$	1	1	2	1	1	1	2	1	1	1	—	—	—	—
$i(P_2)$	1	1	1	2	1	1	1	2	2	2	1	1	1	1
$i(P_3)$	1	2	2	4	—	—	—	—	—	—	—	—	—	—
$i(P_4)$	1	2	2	4	—	—	—	—	—	—	—	—	—	—
$i(P_{34})$	1	2	2 or 4	4	1	2	2	1	2	4	—	—	—	—
$i(P_{134})$	1	2	2 or 4	4	1	2	2	1	2	4	1	2	1	2
$\alpha(P_1)$	1	1	2	1	1	1	2	1	1	1	1	2	1	2
Γ_0	1	Z_2	Z_2 or Z_2+Z_2	Z_2+Z_2	Z_2	$\Gamma_{(1)}$	$\Gamma_{(1)}$	Z_2	$\Gamma_{(1)}$	$\Gamma_{(2)}$	$\Gamma_{(3)}$	$\Gamma_{(3)}$	$\Gamma_{(3)}$	$\Gamma_{(3)}$

TABLE 1

1. Isotropic groups. A list of these groups and of their invariants appears in Table 1 (naturally we do not distinguish between groups whose invariants differ by a permutation of $\{1,3,4\}$). Expressions for the indices are obtained according to §§1.5—1.8. The possible values of the pair $\{\Gamma_k(G), \Delta_k(G)\}$ can be found in [**25**] (or obtained directly using the results of Weil [**26**]).

2. Proposition. *Suppose that* $\mathrm{rg}_k G = 0$ *and* $K \in \mathfrak{S}(k,2)$. *If* $\mathrm{rg}_{\mathfrak{K}} G > 0$, *then* $\mathrm{rg}_{\mathfrak{K}} G = 0 \ \forall \mathfrak{K} \in \mathfrak{S}(k, 2s+1)$. *In particular,* $2 | \mathrm{ind}_k G/P \,\forall\, P$.

a. We show first that if G contains a regular k-subgroup H of type $A_1 \times A_1 \times A_1 \times A_1$, and $\mathrm{rg}_k H = 0$, then $\mathrm{rg}_k G = 0$. Assume that $\mathrm{rg}_k G > 0$. Take $P \in P_i(G,k)$, $P_i(G,k) \neq 0$. Then $\dim P \geqq 19$, $\dim H = 12$ and $\dim P \cap H \geqq 3$. Since $\mathrm{rg}_k H = 0$ we have $\mathrm{rg}_k H_1 = 0$, where $H_1 = P \cap H$; and thus H_1 is contained in a semisimple part S of P. But then $\mathrm{rg}_k Z(H_1) > 0$, because $Z(H_1) \supset Z(S)$. If H_1 contains a k-subgroup $\widetilde{H}$ of type A_1, then $Z(H_1) \subset H$, because $Z(\widetilde{H}) \subset H$ for every subgroup $\widetilde{H}$ of type A_1 in H. But this gives us the contradiction that $\mathrm{rg}_k H > 0$. Thus $H_1 \simeq T^3$ and $Z(H_1) = T^4$. However, there exists a maximal torus T in H containing H_1. Thus $Z(H_1) = T$ and $\mathrm{rg}_k T = 0$, again a contradiction.

b. If $\Delta_K = \{2\}$ or $\{1,2\}$, then the subgroup (see §1.11)

$$H = G_{\alpha_1} G_{\alpha_1 + 2\alpha_2 + \alpha_3 + \alpha_4} G_{\alpha_3} G_{\alpha_4}$$

is defined over k and $\mathrm{rg}_k H = 0$. Then (1.6) $\mathrm{rg}_{\mathfrak{K}} H = 0$, and by what was proved above, $\mathrm{rg}_{\mathfrak{K}} G = 0$.

c. If $\Delta_K = \{1\}$, then $\mathrm{ind}_K G/P_2 = \mathrm{ind}_K G/P_{34} = 2$ (§§1.6 and 1.8), and therefore, by §1.9, $2 | \mathrm{ind}_k G/P_2$, $\mathrm{ind}_k G/P_{34}$ and $(G/P_2)_{\mathfrak{K}} = \emptyset = (G/P_{34})_{\mathfrak{K}}$. Therefore $P_1(G_1 \mathfrak{K}) \neq \emptyset$ if $\mathrm{rg}_{\mathfrak{K}} G > 0$. Since $\Delta_K = \{1\}$, G contains (see 1.11) a k-subgroup H of type A_3. We have $\mathrm{rg}_{\mathfrak{K}} H \cdot Z(H) = 0$, $\dim H \cdot Z(H) = 16$, $\dim P_1 = 22$ and $\dim (H \cdot Z(H) \cap P_1) \geqq 10$. The group $H_1 = H \cdot Z(H) \cap P_1$ is anisotropic and is therefore contained in a semisimple part S of P_1. Thus $\mathrm{rg}\, H_1 \leqq 3$. These are then the possibilities for H_1: A_3, B_2. If H_1 is of type A_3, then $H_1 = H$ so that $\mathrm{rg}_{\mathfrak{K}} Z(H) > 0$, which is false. If H_1 is of type B_2, then $\mathfrak{A}_{\mathfrak{K}}(H, P_2) = \mathfrak{K}$, and since $[\mathfrak{K} : k] = 2s+1$, it follows that $\mathfrak{A}_k(H, P_2) = k$; that is, H contains a k-group of type B_2. Then H is isotropic over $K_1 \in \mathfrak{S}(k,2)$, and $\Delta_{K_1} \neq \{1\}$; that is, this case reduces to the preceding one.

3. Lemma. *If* G/P_1 *is defined over* k *and* $\mathrm{ind}_k \mathfrak{A}_k(P_1) = 2^m$, $m \leqq 2$, *then* G *contains a* k*-subgroup* H *of type* A_3.

If $m = 0$, then G is isogenic with the orthogonal group and contains a subgroup B_3 (and therefore also a subgroup A_3). If $m = 1$ or 2, we let $\mathfrak{A}$ stand for the algebra occurring in the Weil construction [**26**] defining the (classical) group G. Because of our conditions (see §1.7), $\mathfrak{A} = \mathfrak{D} \otimes \mathfrak{M}_q$, where $q = 2^{3-m}$, $\mathfrak{M}$ is a matrix algebra and $\mathfrak{D} = \mathfrak{A}_k(P_1)$. The involution J respects $\mathfrak{D}$ and $\mathfrak{M}_q$ and induces involutions J_1 and J_2 on them. Let G_1 and G_2 be the groups defined for $m = 1$ (or for $m = 2$) by the pairs $(\mathfrak{D}, J_1)$ and $(\mathfrak{M}_q, J_2)$

(or by $(\mathfrak{M}_q, J_2)$ and $(\mathfrak{D}, J_1)$). These involutions J_1 and J_2 are either both symmetric or both skew-symmetric. If they are symmetric, then $G_1 = T^1$ and $G_2 = A_1 \times A_1$. It can be checked (and here we may assume that $k = k_s$) that $Z(G_1)$ is a group of type A_3; that is, the assertion is true in this case.

If J_1 and J_2 are skew-symmetric, then $G_1 = A_1$ and $G_2 = B_2$ $(= C_2)$. Over k_s we may assume (up to isogeny) that $G = SO(8)$, $G_1 = SO(3)$ and $G_2 = SO(5)$. Let T be a maximal torus in G_1. Then $Z(T) = SO(6)$; that is, $Z(T) = A_3$, as required.

4. **Proposition.** *Let G be a group of type D_4, and let $\Gamma_k(G) \subset Z_2$; then there exists an $i = 1, 3$ or 4 such that G/P_i is defined over k and $\mathrm{ind}_k \mathfrak{A}_k(G/P_i) \leqq 2$. Then G contains a subgroup H of type A_3 and $\exists\, K \in \mathfrak{S}(K, 2) : \Delta_K \supset \{1\}$.*

The case $\mathrm{rg}_k G > 0$ is discussed using 3.1. So we assume that $\mathrm{rg}_k G = 0$. Since none of the properties we are considering here alter when k is replaced by a K such that $[K:k] = 2s+1$ and $\mathrm{rg}_K G = 0$, we may assume that G contains a k-torus T such that $\Gamma = \Gamma(k(T)/k)$ is a 2-group. Looking at the Sylow 2-subgroup of $\mathrm{Aut}\,\Sigma$, we see that it preserves two orthogonal root subsystems of type $A_1 \times A_1$. By 1.1, the corresponding subgroups $G_{\alpha_1} G_\beta$ $(\beta = \alpha_1 + 2\alpha_2 + \alpha_3 + \alpha_4)$ and $G_{\alpha_3} G_{\alpha_4}$ are defined over k. If at least one of the groups $G_{\alpha_1}, G_\beta, G_{\alpha_3}, G_{\alpha_4}$ were defined over k, then decomposing it over $K_1 \in \mathfrak{S}(k, 2)$ we would get $\mathrm{rg}_{K_1} G > 0$, from which it follows at once that there is an $s = 1, 3$ or 4 such that G/P_s is defined over k and $i(s) = \mathrm{ind}_k \mathfrak{A}_k(P_s) \leqq 4$. A similar assertion obtains in the case where (let us say) $G_{\alpha_1} G_\beta$ does not contains a subgroup of type A_1. Then $G_{\alpha_1} G_\beta \simeq G(\mathfrak{A})$, where $\mathfrak{A}$ is an algebra with center $K_2 \in \mathfrak{S}(k, 2)$. In this case $\exists\, K_3 \in \mathfrak{S}(K_2, 2)$ such that $\mathrm{rg}_{K_3} G \geqq 2$ (K_3 decomposes $\mathfrak{A}$), and again it follows that $i(s) \leqq 4$. In this situation we apply Proposition 3.4 and find a k-subgroup $\widetilde{H}$ of type A_3. Over $K = k(Z(\widetilde{H}))$, G is isomorphic to one of $D_4 0, D_4 1, D_4 3$, ${}^2D_4 0$ or ${}^2D_4 1 - {}^2D_4 5$, and we shall assume that $(G/P_1)_K \neq \emptyset$. We can also assume that G/P_1 is defined over k. It follows that $i(1) \leqq 2$, and this completes the proof of the proposition.

5. **Proposition.** *Suppose that $\mathrm{rg}_k G = 0$, $\Gamma_k(G) \supset Z_3$, $K \subset K_k(G)$ and $K \in \mathfrak{S}(k, 3)$. In the following cases G contains proper semisimple k-subgroups:*

a. *If G is isogenic over K with an orthogonal group.*

b. *If $\mathrm{rg}_K G > 0$.*

Let $\{1, \sigma, \tau\}$ be different embeddings of K in $K_k(G)$.

a. In the first case G contains a K-group H of type B_3, $\dim H = 21$, $\dim(H \cap H^\sigma \cap H^\tau) \geqq 7$, $H_1 = H \cap H^\sigma \cap H^\tau$ is defined over k, and the assertion follows.

b. Suppose that G does not contain proper semisimple k-subgroups. Let P be a parabolic K-group in G, $H = P \cap P^\sigma \cap P^\tau$ and $H_1 = Z(H)$. The tori H and H_1 are defined over k, and $\mathrm{rg}_K H_1 > 0$. Let T be a one-dimensional K-decomposable torus in H_1. Then the torus $T_1 = T T^\sigma T^\tau$ is two dimensional (because if $\dim T_1 = 3$, then $T_1 = R_{K/k} T$ and $\mathrm{rg}_k T_1 > 0$) and defined over k.

By assumption $Z(T_1)$ is a maximal torus. Set $\mathfrak{K} = k(T_1)$. As $Z(T_1)$ is a maximal torus, G is quasi-decomposable over $\mathfrak{K}$. On the other hand, $\mathfrak{K} \supset K_k(G)$, so that $\Gamma_{\mathfrak{K}}(G) = 1$; that is, G is decomposable over $\mathfrak{K}$. It is easy to see that, under the action of $\Gamma(\mathfrak{K}/k)$ $(\subset S_3 \times Z_2)$ on the system of roots of G relative to $Z(T_1)$, some subsystems of type A_2 are left invariant.

Application of Hypothesis 1.1 gives our assertion.

6. Lemma. *If k has no quadratic extension and* $\mathrm{rg}_k G = 0$, *then* $\mathrm{ind}_k G/P_2 = 3$.

The conditions give $\Gamma_k(G) = Z_3$. If H is a proper semisimple subgroup of G, then $H = A_2$ (the remaining cases are impossible, because $\mathrm{rg}_k G = 0$ and $\mathfrak{S}(k,2) = \emptyset$).

Suppose that $\mathfrak{K} \in \mathfrak{S}(k,m)$, $3 \nmid m$ and $\mathrm{rg}_{\mathfrak{K}} G > 0$. Since G is decomposable over the join $\mathfrak{K}'$ of $\mathfrak{K}$ and $K = K_k(G)$, and $[\mathfrak{K}' : \mathfrak{K}] = 3$, it follows that G is quasi-decomposable over $\mathfrak{K}$. Let $P \in P_2(G, \mathfrak{K})$, $H_1 = Z(H)$ and $H_2 = HH_1$. Then $\dim P = 19$, $\dim H_2 = 10$ and $\dim P \cap H_2 \geqq 1$. Since $\mathrm{rg}_{\mathfrak{K}} H_2 = 0$, $H_3 = P \cap H_2$ is contained in some semisimple part S of P. In order not to arrive at a contradiction, we must assume that H_3 is a two-dimensional anisotropic torus in $H(G, \mathfrak{K}) = Z(T_d(G, \mathfrak{K}))$. Then $\mathfrak{K}(H_3) = \mathfrak{K}'$ and H_3 is conjugate under some element of H to some subtorus H_4 of H_2 defined over k. Then $Z(H_4)$ is a k-subgroup of type A_2. Thus $\mathrm{rg}_k Z(H_4) = 0$; that is, $\mathrm{rg}_{\mathfrak{K}} Z(H_4) = 0$, so that $\mathrm{rg}_{\mathfrak{K}} Z(H_3) = 0$; and this contradicts the supposition that $\mathrm{rg}_{\mathfrak{K}} G > 0$.

7. Theorem. $\mathrm{ind}_k G/P = 1 \Leftrightarrow (G/P)_k \neq \emptyset$.

We can restrict ourselves to the case $\mathrm{rg}_k G = 0$. If $K \in \mathfrak{S}(k,2)$ and $\mathrm{rg}_K G > 0$, we apply the arguments of §1.2. So we may assume that k is closed with respect to quadratic extensions. Under these conditions we can apply Lemma 3.6.

8. Proposition. *If G contains a k-subgroup H of type A_2, then* $\mathrm{ind}_k G \mid 12$ *and* $\mathrm{ind}_k G/P_2 \mid 6$. *If* $\Gamma_k(G) = Z_3$, *then* $\mathrm{ind}_k G \mid 6$.

Let K be a cubic subfield of $K_k(G)$ and set $K_2 = \{K, K_k(H)\}$ for the join of K and $K_k(H)$; then K_2 decomposes H, and if $T = T_d(H, K_2)$, then $Z(T) \simeq T^4$; that is, G is quasi-decomposable over K_2. Thus G is decomposable over $K_3 = \{K_k(H), K_k(G)\}$. Our assertions follow from what has been said.

9. Proposition. *If* $3 \mid \mathrm{ind}_k G/P_2$, *then every semisimple k-subgroup of G is of type A_2. Conversely, if G contains a semisimple k-subgroup H such that either H is of type A_2 and* $\mathrm{ind}_k H = 2$, *or else H is not of type A_2, then G is isotropic over* $K \in \mathfrak{S}(k,2)$, *and* $3 \mid \mathrm{ind}_k G/P_2$. *If* $\Gamma_k(G) \supset Z_3$, *then* $\mathrm{ind}_k G/P_2 = 2$.

In view of Proposition 3.5 it is enough to prove the second assertion. If G contains subgroups A_1, B_2, B_3 and G_2, then these groups—and therefore G itself—are isotropic over a suitable $K \in \mathfrak{S}(k,2)$. If H is of type A_3, then $Z(H)$ decomposes over $K \in \mathfrak{S}(k,2)$. If H is of type A_2 and $\mathrm{ind}_k H = 2$, then H is decomposable over $K_k(H)$ (see §1.6). If $H = A_1 \times A_1 \times A_1$ and $H =$

$R_{\mathfrak{K}/k}H_1$, where $\mathfrak{K} \in \mathfrak{S}(k,3)$, then $Z(H) = A_1$ and this case reduces to the preceding one. If $H = A_1 \times A_1$, then H is decomposable over two quadratic extensions, so that G is a classical group and we can apply 3.4.

So we have shown that $\exists K \in \mathfrak{S}(k,2) : \mathrm{rg}_K G > 0$. If $3 \mid \mathrm{ind}_k G/P_2$, then $\Gamma_k(G) \supset Z_3$. If now $\Gamma_k(G) \supset Z_3$ and $K \in \mathfrak{S}(k,2) : \mathrm{rg}_K G > 0$, then $\Gamma_K(G) \supset Z_3$; so that $\Delta_K \supset \{2\}$; that is, $\mathrm{ind}_k G/P_2 = 2$. This proves the proposition.

D_4, $\Gamma_k(G) \supset Z_3$, $\Delta_k = \varnothing$ (§§3.5, 3.6, 3.8, 3.9)

Type	${}^3D_4 2$	${}^3D_4 3$	${}^6D_4 2$	${}^6D_4 3$	${}^3D_4 4$	${}^3D_4 5$	${}^6D_4 4$	${}^6D_4 5$
a	3	3	6	6	3	3	6	6
$i(G)$	6	6	12	12	6	6	12	12
$i(P_{134})$	6	6	6	6	2	4	2	4
$i(P_2)$	3	6	3	6	2	2	2	2
$\alpha(P_1)$	2	1	2	1	1	2	1	2
Γ_0'	Z_6	Z_6	$S_3 \times Z_2$	$S_3 \times Z_2$	Z_6	$\Gamma_{(4)}$	$S_3 \times Z_2$	$\Gamma_{(4)}$
H_0	A_2	A_2	A_2	A_2	A_1	A_1	A_1	A_1

TABLE 2

10. Explanation of the tables. We do not include the classical anisotropic groups in the table, because on the one hand this would enlarge them significantly, and on the other hand it falls under §1.9 anyway.

In the tables we use the following notation:

$\Gamma_{(1)}$ is the Galois group of the normal closure of a sequence of two quadratic extensions.

$\Gamma_{(2)}$ is the Galois group of the normal closure of the field $\mathfrak{S}(K, Z_2 + Z_2)$, where $K \in \mathfrak{S}(k,2)$.

$\Gamma_{(3)}$ is the Galois group of the normal closure of a field $K \in \mathfrak{S}(K_k(G), 2)$.

$\Gamma_{(4)}$ is the direct product of some group $\Gamma_{(3)}$ with a Z_2.

§4. Groups of type F_4

0. In this section we determine the possible values of the invariants of groups of type F_4. These reductions can be extracted from [**4**], [**5**] and other papers. We recall that, in view of [**25**], only the values Δ, $\{4\}$ and $\emptyset$ are possible for Δ_k.

1. The Weil group of a group G of type F_4 preserves a subsystem of the roots of type D_4 (long roots) and consequently *the corresponding subgroup H is defined over k* (see §1.1). As is easy to see (or to verify), H is simply connected. Let C be its center. Since Char $\neq 2$, $C_{\bar{k}} = Z_2 + Z_2$. For any $c \in C$, $Z(c) = H_c$ is a group of type B_4. The group $\Gamma_k(H)$ acts on C, and $C_k = C^{\Gamma_k(H)}$.

2. $\Delta_k = \{4\} \Rightarrow F_41$.

If L is a Levi subgroup in $P \in P_4(G,k)$, $U = U(P)$, $U_{2\alpha} = [U,U]$ and $S = S(L)$, then S acts on $U_{2\alpha}$ as a linear group; and since $\dim U_{2\alpha} = 7$ and S is of type B_3, it follows that S is isogenic with $SO(7)$ over k. Thus $\exists K \in \mathfrak{S}(k,2) : \mathrm{rg}_K S > 0$, i.e. $\mathrm{rg}_K G \geqq 2$, so $\mathrm{rg}_K G = 4$, and G is decomposable over K. Because of §1.5, $\mathrm{ind}_k G = \mathrm{ind}_k P_i = \mathrm{ind}_k S = 2$ $(i = 1,2,3)$. We note that in this case H is of type ${}^2D_41$, $\Gamma_k(H) = Z_2$ and $C_k(H) = Z_2$.

3. $\mathrm{rg}_k G = 0$ and $\exists K \in \mathfrak{S}(k,2) : \mathrm{rg}_K G = 4 \Rightarrow F_42$.

Under the conditions $\Gamma_k(H) = 1$ and $C_k = Z_2 + Z_2$, it follows that $P_4(G,\mathfrak{K}) \neq \emptyset$ if $\mathfrak{K} \in \mathfrak{S}(k, 2s+1)$ and $\mathrm{rg}_{\mathfrak{K}} G > 0$; and that $\dim P = 37$ if $P \in P_4(G,\mathfrak{K})$. Then for $c \in C_k$ we have

$$\mathrm{rg}_k H_c = \mathrm{rg}_{\mathfrak{K}} H_c = 0,\ \dim H_c = 36,\ \dim P \cap H_c \geqslant 73 - 52 = 21.$$

Since $H_1 = P \cap H_c \subset H_c$, $\mathrm{rg}_{\mathfrak{K}} H_1 = 0$. Thus H_1 is contained in the semisimple part of P, and, by dimension arguments, they coincide. Then $Z(H_1) = T$, $\mathrm{rg}_{\mathfrak{K}} T = 1$ and $T \subset H_c$, which is impossible. Thus $\mathrm{ind}_k G = \mathrm{ind}_k G/P = 2\ \forall P$.

4. LEMMA. *If $3 \nmid \mathrm{ind}_k H/P_2(H)$, then G is decomposable over $K \in \mathfrak{S}(k,2)$.*

Since a cocycle $\alpha \in H^1(k, G_d)$ defining G is the image of a cocycle $\beta \in H^1(k, N_{G_d}(H_d))$ defining H and H is simply connected, it follows that $\exists K \in \mathfrak{S}(k,2) : \mathrm{rg}_K H > 0$ (by Proposition 3.9); that is, that G is of type F_41 or F_40 over K. If necessary we replace H by some other group and get, by Corollary 1.11; that H contains a k-subgroup H of type B_3, and therefore a k-subgroup H_1 of type A_3. This subgroup is simply connected, and a cocycle defining it is a cocycle out of $H^1(k, N_{\tilde{H}_d}(H_1)_d)$. Thus H_1 is decomposable over $K = K_k(H_1)$. Therefore $\mathrm{rg}_K G \geqq 2$; that is, $\mathrm{rg}_K G = 4$.

5. F_43, G_44, F_45.

$3 | \mathrm{ind}_k H/P_2(H)$; if $K \in \mathfrak{S}(k,3)$, $K \subset K_k(H)$, then G has types F_40, F_41, F_42, respectively, over K. If G is of type F_43 or F_45, then $\exists K_1 \in \mathfrak{S}(k, Z_3)$: G is of type F_40 or F_42 over K_1.

We show that

$$3\,|\,\mathrm{ind}_k G,\quad 3\,|\,\mathrm{ind}_k G/P\ \forall P.$$

Taking a quadratic extension K_1 of k we see that G is of type F_43 over K_1. If $\mathfrak{K}/K_1$ decomposes G, then $\dim B \cap H > 0$, where $B \in P^{\emptyset}(G,\mathfrak{K}) \neq \emptyset$, $\dim B = 28$ and $\dim H = 28$. Thus $3 | [\mathfrak{K} : K_1]$, so that $3 | \mathrm{ind}_k G$, $\mathrm{ind}\, G/P\ \forall P$.

6. THEOREM. $\mathrm{ind}_k G/P = 1 \Leftrightarrow (G/P)_k \neq \emptyset$.

This follows from §§4.1, 4.2 and 4.4.

F_4

Type	F_40	F_41	F_42	F_43	F_44	F_45
Δ_k	Δ	$\{4\}$	$\varnothing$	$\varnothing$	$\varnothing$	$\varnothing$
$i(G)$	1	2	2	3	6	6
$i(P_s)$ $s=1,2,3$	1	2	2	3	6	6
$i(P_4)$	1	1	2	3	3	6
Γ_0	1	Z_2	Z_2	Z_3	S_3	Z_6

TABLE 3

§5. Groups of type E_6

0. In this section we consider groups G of type E_6 satisfying any of the conditions:

$\Gamma_k(G)=1$, $\mathfrak{A}_k(P_1)=k$ (§§5.1 and 5.2);

$\Gamma_k(G)=Z_2$, $\mathfrak{A}_k(P_1)=K$, where $K=K_k(G)$ (§5.3);

$\Gamma_k(G)=Z_2$, $\mathrm{rg}_K G>0$, where $K=K_k(G)$ (§§5.4 and 5.6);

$\Gamma_k(G)=1$, $\mathrm{ind}_k\mathfrak{A}(P_1)=3$ (§5.5).

It can be proved that $27\nmid \mathrm{ind}_k G$ (§5.7), and that Hypothesis 1 is satisfied by groups of type E_6 (§5.8).

By [**25**], Δ_k can assume only the following values: Δ, $\{1,5\}$, $\{3,6\}$, $\emptyset$ when $\Gamma_k(G)=1$; and Δ, $\{1,5,6\}$, $\{3,6\}$, $\{1,5\}$, $\{6\}$, $\emptyset$ when $\Gamma_k(G)=Z_2$.

1. *If* $\Gamma_k(G)=1$ *and* $\mathfrak{A}_k(P_1)=k$, *then*:

a. *The representation* π_{λ_1} (*the fundamental representation corresponding to* P_1) *is defined over* k.

b. *The stability group* H *of a sufficiently general* k-*point in this representation is a* k-*group of type* F_4.

c. *The cocycle* $\alpha\in H^1(k,\overline{G}_d)$ *defining* G *lies in the image of the map* $H^1(k,H_d)\to H^1(k,\overline{G}_d)$.

d. $\delta(\alpha)=1$, *where* $\delta: H^1(k,\overline{G}_d)\to H^2(k,C(\widetilde{G}_d))$.

2. $\Gamma_k(G)=1$, $\mathfrak{A}_k(P_1)=k\Longrightarrow E_60-E_63$.

By the above discussion, G is completely defined by H, and we write $E_6i=\mathfrak{q}(F_4m)$ if H is of type F_4m and is contained in a group of type E_6i.

We have $\mathfrak{q}(F_40)=\mathfrak{q}(F_41)=E_60$, $\mathfrak{q}(F_42)=E_61$, $\mathfrak{q}(F_43)=\mathfrak{q}(F_44)=E_62$, $\mathfrak{q}(F_54)=E_63$. The splitting fields of these groups turn up as splitting fields for H. In our calculation of the invariants of G we use assertions 1.5 and 1.9 and the following proposition:

If G is of type $E_6 2$, then $\operatorname{ind}_k G = \operatorname{ind}_k G/P = 3\ \forall\ P$.

For suppose that $K \in \mathfrak{S}(k, m)$, $3 \nmid m$ and $\operatorname{rg}_K G > 0$. Then G is of type $E_6 0$ over K. Let B be the Borel K-group in G. Then $\dim B = 42$, $\dim H = 52$ and $\dim B \cap H \geqq 16$, so that $\operatorname{rg}_K H > 0$; and this is impossible because $3 \nmid m$ and $3 \mid \operatorname{ind}_k H$.

3. $\Gamma_k(G) = Z_2$, *and if* $K = K_k(G)$, *then* $\mathfrak{A}_K(P_1) = K \Rightarrow {}^2E_6 0 - {}^2E_6 11$ (*G is of type $E_6 0 - E_6 3$ over K*).

a. $\exists\, \mathfrak{K} \in \mathfrak{S}(k, 2) : \operatorname{rg}_{\mathfrak{K}} G = 6 \Rightarrow {}^2E_6 0 - {}^2E_6 3$.

The case $\Delta_k = \{1, 5\}$ is impossible by results of Araki ([**17**], §4). We show that $\operatorname{ind}_k G/P = 2$ for groups of type $E_6 1$ (whenever G/P is defined over k). Take $K_1 \in \mathfrak{S}(k, 2l+1)$ and $\operatorname{rg}_{K_1} G > 0$. Then G has type ${}^2E_6 0$, ${}^2E_6 2$ or ${}^2E_6 3$ over K_1; in particular, $\Delta_{K_1} \supset \{6\}$. Suppose that $P \in P_6(G, K_1)$ and that H is a k-subgroup of G of type $A_5 \times A_1$, $H_1 = H \cap P$. We have $\dim P = 57$, $\dim H = 38$ and $\dim H_1 \geqq 17$. It can be checked (we do not stop to do this here, since it belongs to the theory of the "classical" group H) that H_1 is conjugate over K_1 or over an extension of K_1 to an odd-degree subgroup defined over k. Since $\operatorname{rg}_{K_1} Z(H_1) > 0$, we have obtained a contradiction.

b. G is of type $E_6 1$ over $K \Rightarrow {}^2E_6 4$, ${}^2E_6 5$.

If $\operatorname{rg}_k G > 0$, then $\Delta_k = \{1, 5\}$.

In all cases G contains a simply-connected k-subgroup H of type D_4. Using Proposition 3.4 and studying the specific features of the case under consideration, we find $K_1 \in \mathfrak{S}(k, 2) : H$ is of type ${}^2D_4 0$ over K_1. Then G is of type ${}^2E_6 0$ over K_1 (the case $E_6 0$ disappears, because then we return to case 5.3a). From this we get expressions for the indices and the existence of a K_2 such that $\operatorname{rg}_{K_2} G = 6$.

c. G is of type $E_6 2$ or $E_6 3$ over $K \Rightarrow {}^2E_6 6 - {}^2E_6 11$.

G contains a K-subgroup H of type F_4, and $3 \mid \operatorname{ind}_K H$.

Suppose that $\sigma \in \Gamma(K/k)$, $\sigma \neq 1$ and $H_1 = H \cap H^\sigma$. Then $\dim H_1 \geqq 26$, and since $3 \mid \operatorname{ind}_K H_1$, it follows that H_1 is of type D_4. A splitting field appears as a splitting field of H.

4. $\Delta_k = \{6\} \Rightarrow {}^2E_6 12$, ${}^2E_6 13$, ${}^2E_6 2$.

If G is not of type ${}^2E_6 2$, then $\mathfrak{A}_K(P_1) \neq K$, where $K = K_k(G) \neq k$. In this case $H(G, k) = H \cong A_5$, and H is isogenic over $K_k(H) = K_k(G) = K$ to $G(\mathfrak{A})$, where $\mathfrak{A}$ is a central simple algebra. We show that $2 \nmid \operatorname{ind}_K \mathfrak{A}$.

If $2 \mid \operatorname{ind}_K \mathfrak{A}$, then $\exists\, K_1 \in \mathfrak{S}(k, 3) : \operatorname{ind}_{K_1} \mathfrak{A} = 2$. However, it then follows that $\operatorname{rg}_{K_1} G = 6$, and since H is the centralizer of some decomposable torus, $\operatorname{rg}_{K_1} H = 5$ so that $\mathfrak{A} \otimes K_1 \cong \mathfrak{M}_6$, as required.

5. $\Gamma_k(G) = 1$, $\operatorname{ind}_k \mathfrak{A}_k(P_1) = 3 \Rightarrow E_6 4 - E_6 9$.

Under these conditions, $\exists\, K \in \mathfrak{S}(k, Z_3) : \mathfrak{A}_K(P_1) = K$ (see §1.6). Let $\Gamma(K/k) = \{1, \sigma, \sigma^2\}$.

a. $\operatorname{rg}_k G > 0 \Rightarrow E_6 4$ $(\Delta_k = \{3, 6\})$.

b. $\operatorname{rg}_k G = 0$, and G is of type $E_6 0$ over K or $E_6 1 \Rightarrow E_6 5 - E_6 7$.

Take $P \in P_{15}(G, K)$, $H = P \cap P^\sigma \cap P^{\sigma^2}$ and $H_1 = Z(H)$.

Then $\dim P = 54$, $\dim H \geqq 6$ and $\operatorname{rg}_K H_1 \geqq 2$.

It follows from this that G contains a k-torus $T \cong T^2 : k(T) = K$. If $\mathrm{rg}_K G = 6$, then G contains a maximal k-torus T_1 decomposable over K.

We consider the case $\mathrm{rg}_K G = 6$ and show that $3 \mid \mathrm{ind}_k G/P \; \forall P$.

Take $\mathfrak{K} \in \mathfrak{S}(k, m)$, $3 \nmid m$, $\mathrm{rg}_{\mathfrak{K}} G > 0$. Since $\mathfrak{A}_k(P_1) \neq k$, we have that $\mathfrak{A}_{\mathfrak{K}}(P_1) \neq \mathfrak{K}$ and thus $(G/P_i)_{\mathfrak{K}} = \emptyset \; \forall i = 1, 2, 4, 5$. If $(G/P_3)_{\mathfrak{K}} \neq \emptyset$, then $(G/P_6)_{\mathfrak{K}} \neq \emptyset$; and take $P \in P_6(G, \mathfrak{K})$. Suppose further that $H_2 \cong A_2 \times A_2 \times A_2$ is a regular subgroup of G relative to T_1. We have that $\dim H_2 = 24$, $\dim P = 57$ and $\dim P \cap H_2 \geqq 3$. Since $\mathrm{rg}_{\mathfrak{K}} H_2 = 0$, it follows that $\mathrm{rg}_{\mathfrak{K}} H_3 = 0$, where $H_3 = P \cap H_2$. The group $H_4 = Z(H_3)$ is isotropic over $\mathfrak{K}$. We shall show that $H_4 \subset H_2$, and thereby achieve a contradiction. For if H_3 is non-abelian, then $H_3 \supset A_2$ and $Z(H_3) \subset H_2$. If H_3 is a torus, then it follows from the properties of H_2 that $\dim H_3 \geqq 4$. If H_3 is a regular torus, then H_4 is a maximal torus in H_2. If H_3 is a singular torus, then H_4 contains one of the groups A_2 in H_2; that is, again $H_4 \subset H_2$.

b. G is of type $E_6 2$, $E_6 3 \Rightarrow E_6 8$, $E_6 9$.

G contains a K-subgroup H of type F_4, and $3 \mid \mathrm{ind}_K H$.

We have $H_1 = H \cap H^\sigma$ and $\dim H_1 \geqq 26$. Since $3 \mid \mathrm{ind}_K H$, H_1 is of type D_4 and $\dim H_1 = 28$. Thus, if $H_2 = H \cap H^\sigma \cap H^{\sigma^2}$, then $H_2 = H_1 \cap H_1^\sigma$. Since H_1 and H_1^σ are contained in a group H^σ of type F_4, we have $\dim H_2 \geqq 28 + 28 - 52 = 4$. Since $H_2 \subset H$, it follows that H contains a k-torus $T = T^2$. For if H_2 is non-abelian, then H_2 contains A_2. If H_2 is a k-torus, the torus we look for is a complement of H_2 in a maximal torus.

Set $\mathfrak{K} = k(T)$. Then $\Gamma(\mathfrak{K}/k) \subset S_3 \times Z_2$, $\mathrm{rg}_{\mathfrak{K}} G \geqq 2$, G is of type $E_6 0$ or $E_6 4$ over $\mathfrak{K}$. If $\mathrm{rg}_{\mathfrak{K}} G = 6$, then G is of type $E_6 6$ or $E_6 7$. So we may assume that $\mathrm{rg}_{\mathfrak{K}} G < 5$; that is, $\mathrm{rg}_{\mathfrak{K}} G = 2$, G has type $E_6 4$ over $\mathfrak{K}$, and G has a splitting field with group $\Gamma \subset Z_3 \times S_3 \times Z_2$, and $9 \nmid \mathrm{ind}_k G/P_6 = \mathrm{ind}_k G/P_3$. It is also clear from what we said above that $9 \mid \mathrm{ind}_k G/P_i$, $i = 1, 2, 4, 5$ (otherwise we get groups of type $E_6 6$ and $E_6 7$).

6. $K = K_k(G)$, $\mathrm{rg}_K G > 0$, $\mathfrak{A}_K(P_1) = K \Rightarrow {}^2E_6 12 - {}^2E_6 15$.

If $\Delta_k = \{6\}$, see 5.4.

If $\Delta_k = \{3, 6\}$, this group is ${}^2E_6 14$.

Therefore we may assume that $\mathrm{rg}_k G = 0$. By our conditions G is of type $E_6 4$ over K. It follows from this (§1.1) that G contains a k-subgroup $H \cong A_5$; $Z(H) = H_1 \cong A_1$.

Let us assume that $\exists K_1 \in \mathfrak{S}(k, 3) : \mathrm{ind}_{K_1} H = 2$. Since $\mathfrak{A}_K(P_1) \neq K$, it follows that $3 \mid \mathrm{ind}_K G/P_i$, $i = 1, 2, 4, 5$, and therefore that $\Delta_K \cap \{1, 2, 4, 5\} \neq \emptyset$. If $\Delta_K \supset \{3\}$, then $\Delta_K \supset \{6\}$. Hence G is of one of the types ${}^2E_6 0$—${}^2E_6 3$ over K_1. We want to show that $\mathrm{ind}_k G/P_3 = \mathrm{ind}_k G/P_6 = 2$. If G is of type ${}^2E_6 1$ over K_1, we apply 1.10. If $\mathrm{rg}_{K_1} H > 0$, then considering $H(H, K_1) = Z(T_d(H, K_1))$ we would get $\mathrm{rg}_{K_1} H > 0$ (here we use the fact that $\Delta_{K_1} \supset \{6\}$). Thus $\mathrm{rg}_{K_1} H = 0$. Suppose that $P \in P_6(G, K_1)$. Then $\dim P = 57$, $\dim HH_1 = 38$ and $\dim P \cap HH_1 \geqq 17$.

The group $H_2 = P \cap HH_1$ is anisotropic over K_1, and these are the possibilities for $S(H_2) : A_2 \times A_2$, A_3, $A_3 \times A_1$ etc. In all cases it follows that

E_6, $\mathfrak{A}_{K_k(G)}(P_1) = K_k G$ (§§5.1—5.3)

Type	E_60	E_61	E_62	E_63	${}^2E_60$	${}^2E_61$	${}^2E_62$	${}^2E_63$	${}^2E_34$	${}^2E_65$	${}^2E_66$	${}^2E_67$	${}^2E_68$	${}^2E_69$	${}^2E_610$	${}^2E_611$
a	1	1	1	1	2	2	2	2	2	2	2	2	2	2	2	2
Δ_k	Δ	$\{1,5\}$	$\varnothing$	$\varnothing$	Δ	$\varnothing$	$\{6\}$	$\{1,56\}$	$\{1,5\}$	$\varnothing$	$\varnothing$	$\varnothing$	$\varnothing$	$\varnothing$	$\varnothing$	$\varnothing$
$i(G)$	1	2	3	6	2	2	2	2	4	4	6	6	6	6	12	12
$i(P_s)$ $s=1,5$	1	1	3	3	—	—	—	—	—	—	—	—	—	—	—	—
$i(P_s)$ $s=2,4$	1	2	3	6	—	—	—	—	—	—	—	—	—	—	—	—
$i(P_3)$	1	2	3	6	1	2	2	2	2	2	3	6	6	6	6	6
$i(P_6)$	1	2	3	6	1	2	1	1	2	2	3	6	3	3	6	6
$i(P_{15})$	1	1	3	3	1	2	2	1	1	2	3	6	6	3	3	6
$i(P_{24})$	1	2	3	6	1	2	2	2	2	2	3	6	6	6	6	6
Γ_0	1	Z_2	Z_3	Z_6	Z_2	Z_2	Z_2	Z_2	Z_2+Z_2		$\Gamma_{(1)}$	$\Gamma_{(1)}$	$\Gamma_{(1)}$	$\Gamma_{(1)}$	$\Gamma_{(2)}$	$\Gamma_{(2)}$
H_0			F_4	F_4		D_4				D_4	D_4	D_4	D_4	D_4	D_4	D_4

TABLE 4

E_6, $\alpha(P_1) = 3$ (§§5.4—5.6)

Type	$E_6 4$	$E_6 5$	$E_6 6$	$E_6 7$	$E_6 8$	$E_6 9$	${}^2E_6 12$	${}^2E_6 13$	${}^2E_6 14$	${}^2E_6 15$
a	1	1	1	1	1	1	2	2	2	2
Δ_k	$\{3,6\}$	$\varnothing$	$\varnothing$	$\varnothing$	$\varnothing$	$\varnothing$	$\{6\}$	$\{6\}$	$\{3,6\}$	$\varnothing$
$i(G)$	3	6	3	6	9	18	6	6	6	6
$i(P_s)$ $s=1,5$	3	3	3	3	9	9	—	—	—	—
$i(P_s)$ $s=2,4$	3	6	3	6	9	18	—	—	—	—
$i(P_3)$	1	2	3	6	3	6	2	2	1	2
$i(P_6)$	1	2	3	6	3	6	1	1	1	2
$i(P_{15})$	3	3	3	3	9	9	3	6	3	6
$i(P_{24})$	3	6	3	6	9	18	6	6	3	6
Γ_0	Z_3	Z_6	Z_3	Z_6	$\Gamma_{(3)}$	$\Gamma_{(3)}$	$\Gamma_{(1)}$	$\Gamma_{(1)}$	$\Gamma_{(1)}$	$\Gamma_{(1)}$
H_0		D_4	D_4	D_4	A_2	A_2	A_2			D_4

TABLE 5

$\mathrm{rg}_{K_1} H > 0$ or $\mathrm{rg}_{K_1} H_1 > 0$, which is impossible. Therefore G has type ${}^2E_6 1$ over K_1, and the assertion is established modulo the existence of K_1.

To find a K_1 we use the construction of Weil [25] for H. If $\mathfrak{A} = \mathfrak{D} \otimes \mathfrak{M}_2$ is the algebra over K that occurs in this construction, then the involution defining H induces an involution J on $\mathfrak{D}$ ([3], X—14). The group H' corresponding to the pair $(\mathfrak{D}, J)$ has type A_2. Here $3 \mid \mathrm{ind}_k H \Leftrightarrow 3 \mid \mathrm{ind}_k H'$. For K_1 we can take a subfield of a splitting field of some maximal torus in H.

7. PROPOSITION. $27 \nmid \mathrm{ind}_k G$.

Let T be any maximal torus in G, $K = k(T)$, $\Gamma = \Gamma(K/k)$ and $K' = K^{\Gamma_3}$. We may assume (1.9) that $k = K'$. Then $\Gamma_k(G) = 1$, and $\Gamma(K/k) = \Gamma$ is a 3-group. Γ acts on $\Sigma = \Sigma(G)$ (see §1.1) and preserves a subsystem Σ_1 of Σ of type $A_2 \times A_2 \times A_2$. Write $H = G(\Sigma_1)$ for the corresponding group. If H does not contain regular k-subgroups A_2, then $H = G(\mathfrak{A})$, where $\mathfrak{A}$ is a division algebra over some $K_1 \in \mathfrak{S}(k, Z_3)$. Then H (and therefore also G)

is decomposable over some $K_2 \in \mathfrak{S}(K_1, Z_3)$ (as in §1.6), so that $\mathrm{ind}_k G \leqq 9$. If H does contain a k-subgroup H_1 of type A_2, then H_1 is decomposable (§1.6) again) over $K_1 \in \mathfrak{S}(k, Z_3)$ and G is of the type $E_6 4$ or $E_6 0$ over K_1; that is, G is decomposable over $K_2 \in \mathfrak{S}(K_1, Z_3)$; and so finally $\mathrm{ind}_k G \leqq 9$, as required.

8. THEOREM. *Hypothesis* 1 *is true for groups of type* E_6.

By 5.1—5.7 we may assume that $\Gamma_k(G) = 1$ and that $\mathrm{ind}_k \mathfrak{A}_k(P_1) = 9$. Suppose that $\exists K \in \mathfrak{S}(k, m) : \mathrm{rg}_K G > 0$. Then $\mathrm{ind}_K \mathfrak{A}_K(P_1) \leqq 3$; that is, K diminishes the index of the algebra $\mathfrak{A}_k(P_1)$, so that $3 | m$.

9. The following notation is used in Tables 4 and 5:

$\Gamma_{(1)}$ is a subgroup of some $S_3 \times Z_2$;

$\Gamma_{(2)}$ is a subgroup of some $S_3 \times Z_2 \times Z_2$;

$\Gamma_{(3)}$ is a subgroup of some $Z_3 \times S_3 \times Z_2$.

§6. Groups of type E_7

0. In this section we consider groups of type E_7 having these properties:

a. $3 \nmid \mathrm{ind}_k G$ (§§6.1, 6.3, 6.4, 6.5 and 6.7).

b. $3 | \mathrm{ind}_k G$ and $\mathrm{ind}_k \mathfrak{A}_k(P_6) \leqq 2$ (§§6.2, 6.6 and 6.8).

c. $3 | \mathrm{ind}_k G$, $\mathrm{ind}_k \mathfrak{A}_k(P_6) = 4$ and $\exists K \in \mathfrak{S}(k, 2) : G$ contains a K-group of type E_6 (§6.9).

Further we prove that $8 \nmid \mathrm{ind}_k \mathfrak{A}_k(G)$ (in §6.1) and $9 \nmid \mathrm{ind}_k G$ (in §6.2); and also that groups of type E_7 satisfy Hypothesis 1.

We note at once that *the stability group of a sufficiently general point in a* 56-*dimensional representation of a group* E_7 *is a group of type* E_6 [**1**].

1. THEOREM. (a) *If* $\mathrm{ind}_k \mathfrak{A}_k(P_6) = 2^t$, *then* $t \leqq 2$ *and* $2^{t+2} \nmid \mathrm{ind}_k G$.

(b) *If* $\mathrm{ind}_k \mathfrak{A}_k(P_6) = 4$, *then* $\mathfrak{A}_k(P_6) \sim \mathfrak{D}_1 \otimes \mathfrak{D}_2$, *where* $\mathfrak{D}_1$ *and* $\mathfrak{D}_2$ *are quaternion algebras over* k.

(c) $4 \nmid \mathrm{ind}_k G/P_1$.

(d) *If* $\mathrm{ind}_k \mathfrak{A}_k(P_6) = 4$, *then* $4 | \mathrm{ind}_k G/P_i$ $(i = 4, 6, 7)$, $(i = 4, 7)$ *and* $4 | \mathrm{ind}_k G/P_i$ $(i = 2, 3, 6)$.

(e) *If* $8 | \mathrm{ind}_k G$ *and* $\mathrm{ind}_k \mathfrak{A}_k(P_6) = 4$, *then* $8 | \mathrm{ind}_k G/P_i$ $(i = 4, 7)$ *and* $4 | \mathrm{ind}_k G/P_i$ $(i = 2, 3, 6)$.

Let T be a maximal torus in G, let $\Gamma = \Gamma(k(T)/k)$, let Γ_2 be a Sylow 2-group of Γ, and let $k' = (k(T))^{\Gamma_2}$; as usual (see §1.9), we replace k by k'. Since Γ_2 preserves a subsystem of roots of type $D_6 \times A_1$, it follows (see §1.1) that G contains a k-subgroup $H = H_1 \cdot H_2$ of type $D_6 \times A_1$. Let $K \subset \mathfrak{S}(k, 2)$ be a splitting field for H_2. Over this field, $\{1\} \subset \Delta_k$ (because G contains a root K-subgroup). Thus $H_3 = H(G, K) \subset D_6$, and because of Corollary 1.11 we may assume that $H_3 \subset H_1$. Then $H_1 \subset P$ for some $P \in P_1(G, K)$, and the representation of H_1 on $U/[U, U]$ (where $U = U(P)$) is equivalent to a spinor representation of H_1. The stability group H_4 of a sufficiently general K-point is a group of type A_5 [**1**]. We have (see §1.6) $\mathrm{ind}_K H_4 = 2^s$, $s \leqq 2$

for H_4, and $H_4 = G(\mathfrak{A})$ over $\mathfrak{K} = K_K(H_4)$, where now $\text{ind}_{\mathfrak{K}}\mathfrak{A} = 2^{s-1}$. The first assertion follows from this. Assertion (b) comes from a theorem of Albert [3] and the fact that $(\mathfrak{A}_k(P_6))^2 \sim 1$; (c) was, in fact, proved in the very beginning of this section.

To prove (d) we assume that $\text{ind}_k G/P_i = 2$, where $i \in \{4, 6, 7\}$. Take a $\mathfrak{K} \in \mathfrak{S}(k, 2(2s+1)) : (G/P_i)_{\mathfrak{K}} \neq \emptyset$. Then G is decomposable over $\mathfrak{K}$, so that $\text{ind}_k \mathfrak{A}_k(P_1) \leqq 2$. Assertion (e) is verified in the same way as (d) was.

2. PROPOSITION. *$3^2 \nmid \text{ind}_k G$. If $3 | \text{ind}_k G$, then $3 | \text{ind}_k G/P_i$ $(i = 2, 3, 4, 5, 7)$ and $3 \nmid \text{ind}_k G/P_6$.*

Making the usual extension K corresponding to a Sylow 3-subgroup, we see that G contains a subgroup H of type E_6 and that $\Delta_K \supset \{6\}$. It follows that $3 \nmid \text{ind}_k G/P_6$. If $3 | \text{ind}_k H$, then $3 | \text{ind}_k G$. The representation of H on $U(P_6)$ is equivalent to a linear 27-dimensional representation. Therefore H is of type $E_6 3$, and the assertion follows.

3. $E_7 0$—$E_7 4$ (*G is decomposable over some $K \in \mathfrak{S}(k, 2)$*).

The cases $\Delta_k = \{6\}$ and $\Delta_k = \{5\}$ are impossible in view of §4 of [7]. We show that $\mathfrak{A}_k(P_1) = k$ if G is of type $E_7 4$. Indeed, $H = H(G, k)$ contains a k-subgroup H_1 of type A_5, with H_1 decomposable over $K_k(H_1)$. Thus H_1 is defined by a cocycle from a simply-connected group. It follows that G is defined by a cocycle from a simply-connected group, and our assertion follows from this.

Let us now consider a group $E_7 1$. This type is defined by the embedding of a subgroup $H = H_1 \cdot H_2$ of type $D_6 \times A_1$ in E_7. In this case (since H_2 is anisotropic) H cannot be defined by a cocycle from a simply-connected group. Thus $\text{ind}_k \mathfrak{A}_k(P_6) = 2$. In other words, $\text{ind}_k G/P_i = 2$, $i = 1, 4, 6, 7$.

Assume that $\text{ind}_k G/P_i = 1$, $i \in \{2, 3, 5\}$, and take $\mathfrak{K} \in \mathfrak{S}(k, 2s+1)$ such that $\text{rg}_{\mathfrak{K}} G > 0$. Over $\mathfrak{K}$ we have $(G/P_i)_{\mathfrak{K}} \neq \emptyset$, $i \in \{2, 3, 5\}$, G is decomposable over some $K' \in \mathfrak{S}(\mathfrak{K}, 2)$, and $\text{ind}_{\mathfrak{K}} G/P_m = 2$, $m = 1, 4, 6, 7$. The only case that does not contradict this last condition is $\Delta_{\mathfrak{K}} = \{5\}$. But this is impossible by the result of Araki ([7], §4) already quoted.

4. $E_7 5$, $E_7 6$ $(\Delta_k \supset \{5\}$, $\exists K \in \mathfrak{S}(k, 2) : \text{rg}_K G = 7)$.

Take $\mathfrak{K} \in \mathfrak{S}(k, 2)$ such that $\text{rg}_{\mathfrak{K}} G_{\alpha_6} = 1$. Then G is of type $E_7 2$ over $\mathfrak{K}$. A group H of type D_4 is defined over k, so that $\exists K_1 \in \mathfrak{S}(k, 2)$ for which G has type $E_7 3$ over K_1. Thus $\text{ind}_k G = \text{ind}_k G/P_7 = \text{ind}_k G/P_4 = 4$ and $\text{ind}_k G/P_i \leqq 2$ $\forall i \neq 4, 6, 7$. Moreover, G is decomposable over $\{K_1, \mathfrak{K}\}$.

5. $E_7 7$ $(\Delta_k \supset \{1\}$, $\exists K \in \mathfrak{S}(k, 2) : \text{rg}_K G = 7)$.

As we saw in §6.1, $H = H(G, k)$ contains a k-subgroup H_1 of type A_5 in this case. Setting $K = K_k(H_1)$ we see that G is type $E_7 3$ over K. On the other hand, $H_1 = G(\mathfrak{A})$ over K, where $\mathfrak{A} = \mathfrak{D} \otimes \mathfrak{M}_3$, $\mathfrak{D}$ being a quaternion division algebra over K. A result of Weil [26] tells us that $\mathfrak{D}$ has an involution over k, so that, by Albert [3], $\mathfrak{D} = K \otimes \mathfrak{D}_1$ in this case, where $\mathfrak{D}_1$ is a quaternion division algebra over k. Let $K_1 \in \mathfrak{S}(k, 2)$ be a splitting field for $\mathfrak{D}_1$. Then the field $\{K_1, K\}$ splits G. The indices are determined in an obvious fashion (or by 1.5).

6. E_78, E_79 $(\Delta_k \supset \{6\},\ \exists K \in \mathfrak{S}(k,2) : \mathrm{rg}_K G = 7)$.

Then $H(G,k)$ is of type E_62 or E_63.

7. E_710—E_712 $(\mathrm{rg}_k G = 0,\ 3 \nmid \mathrm{ind}_k G)$.

a. If $\mathrm{ind}_k \mathfrak{A}_k(P_6) = 1$, then G is of type E_72 or E_74 and $\mathrm{rg}_k G > 0$.

b. E_71 and E_711 $(\mathrm{ind}_k \mathfrak{A}_k(P_6) = 2$ and $\mathrm{rg}_k G = 0)$.

Suppose that $K \in \mathfrak{S}(k,2)$ splits $\mathfrak{A}_k(P_6)$. Then G is of type E_72 or E_74 over K. By Theorem 6.1 a splitting field is the join of two successive quadratic extensions.

c. E_710 and E_712 $(\mathrm{ind}_k \mathfrak{A}_k(P_6) = 4$ and $\mathrm{rg}_k G = 0)$.

In this case (see Theorem 6.1) $\mathfrak{A}_k(P_6) = \mathfrak{D}_1 \otimes \mathfrak{D}_2$. If $K \in \mathfrak{S}(k,2)$ splits $\mathfrak{D}_1$, then G has one of the types E_71, E_75, E_76, E_77 or E_711 over K. In all cases G contains a K-subgroup H of type $D_6 \times A_1$. Considering the k-subgroup $H_1 = H \cap H^\sigma$, where $\sigma \in \Gamma(K/k)$, and its centralizer, we find that G is isotropic over a suitable $K_1 \in \mathfrak{S}(k,2)$. The values of the indices come from Theorem 6.1d and e. A splitting field can be obtained as the join of a splitting field (with group $Z_2 + Z_2$) for $\mathfrak{A}_k(P_1)$ (see Theorem 6.1b) and (if necessary) a quadratic splitting field of a group of type E_72 or E_74. We can also employ the device used in §6.5; then we obtain first a quadratic field (splitting field for the centralizer of a D_6), and next one with the group $Z_2 + Z_2$ (splitting field of a group of type E_77).

8. E_713—E_717 $(rg_k G = 0,\ 3 \mid \mathrm{ind}_k G,\ \mathrm{ind}_k \mathfrak{A}_k(P_6) \leqq 2)$.

a. E_713 $(\mathrm{rg}_k G = 0,\ \mathfrak{A}_k(P_6) = k)$.

The values of the indices, and a splitting field, are easy to determine.

b. E_714—E_717 $(\mathrm{rg}_k G = 0,\ \mathrm{ind}_k \mathfrak{A}_k(P_6) = 2)$.

Suppose that $K \in \mathfrak{S}(k,2)$ splits $\mathfrak{S}_k(P_6)$. These are the possibilities: G is of type E_78, E_79 or E_713 over K. In the first two cases G contains a k-subgroup H of one of the types ${}^2E_66$—${}^2E_611$.

If G is of type E_713 over K, then G contains a K-subgroup H of type $E_6 \times T'$, and we have $\sigma \in \Gamma(K/k)$ and $\dim H \cap H^\sigma \geqq 79 + 79 - 133 = 25$, so that (in view of the properties of groups of type E_63) $H_1 = H \cap H^\sigma$ always contains a regular k-subgroup H_2 of type A_2, and $Z(H_2) = H_3$ is a subgroup of type A_5. Using the device of §6.5 we see that H_3 (and therefore G) has a splitting field whose Galois group is a subgroup of $Z_2 \times Z_2 \times S_3$, where the first Z_2 is $\Gamma_k(H_3)$, the second is the Galois group of the splitting field of a quaternion algebra over k (see §6.5), and finally $\Gamma_k(H_3) \times S_3$ contains the Galois group of a splitting field of some group of type A_2 (see §5.6).

9. E_718 $(\mathrm{rg}_k G = 0,\ 3 \mid \mathrm{ind}_k G,\ \mathrm{ind}_k \mathfrak{A}_k(P_6) = 4,$ *and* $\exists K \in \mathfrak{S}(k,2)$ *such that* G *contains a* K*-subgroup of type* $E_6)$.

Just as in §6.8b, G contains a subgroup of type $A_2 \times A_5$, from which we get our splitting field. Since $8 \nmid \mathrm{ind}_k H_3$ (H_3 comes from §6.8b), it follows that $8 \nmid \mathrm{ind}_k G$, so that G is of type E_712 over a cubic extension; and this gives the values of the indices.

10. THEOREM. *Hypothesis* 1 *is valid for groups of type* E_7.

E_7

Type	E_70	E_71	E_72	E_73	E_74	E_75	E_76	E_77	E_78	E_79	E_710	E_711	E_712	E_713	E_714	E_715	E_716	E_717	E_718
Δ_k	Δ	$\varnothing$	$\{1,5,6\}$	$\{1,2,3,5\}$	$\{1\}$	$\{1,5\}$	$\{5\}$	$\{1\}$	$\{6\}$	$\{6\}$	$\varnothing$	$\varnothing$	$\varnothing$	$\varnothing$	$\varnothing$	$\varnothing$	$\varnothing$	$\varnothing$	$\varnothing$
$i(G)$	1	2	2	2	2	4	4	4	3	6	8	4	4	6	6	6	12	12	12
$i(P_1)$	1	2	1	1	1	1	2	1	3	3	2	2	2	3	6	3	3 or 6	6	6
$i(P_2)$	1	2	2	1	2	2	2	2	3	6	4	4	2 or 4	6	6	3	6	12	6 or 12
$i(P_3)$	1	2	2	1	2	2	2	2	3	6	4	4	2 or 4	6	6	3	6	12	6 or 12
$i(P_4)$	1	2	2	2	2	4	4	4	3	6	8	4	4	6	6	6	12	12	12
$i(P_5)$	1	2	1	1	2	1	1	2	3	3	2 or 4	2 or 4	2 or 4	6	6	3	3 or 6	6 or 12	6 or 12
$i(P_6)$	1	2	1	2	2	2	2	2	1	1	4	2 or 4	4	2	2	2	2	2 or 4	4
$i(P_7)$	1	2	2	2	2	4	4	4	3	6	8	4	4	6	6	6	12	12	12
$\alpha(P_6)$	1	2	1	2	1	2	2	2	1	1	4	2	4	1	2	2	2	2	2
Γ_0	1	Z_2	Z_2	Z_2	Z_2	$Z_2 + Z_2$			Z_3	Z_6	§6.7c	§6.7b	$Z_2 + Z_2$	Z_6	Z_6	Z_6	(§6.8b) $Z_2 \times Z_2 \times S_3$		
H_0											$D_6 \times A_1$			E_6	E_6	E_6	$A_2 \times A_5$		

TABLE 6

This was proved in §§6.2, 6.3 and 6.7.

11. Generalities. a. HYPOTHESIS. $\operatorname{ind}_k \mathfrak{A}(P_6) < 4$.

b. We consider all groups of type E_7, as well as the groups for which $3 \mid \operatorname{ind}_k G$ and $\operatorname{ind}_k \mathfrak{A}_k(P_6) = 4$, and which do not contain groups of type E_6 over any quadratic extension.

c. In the cases where the table indicates that $\operatorname{ind}_k G/P_i = 4$ or 2, etc., this is not necessarily because both cases are possible. For instance, we have included under the label "group of type $E_7 16$" those groups that reduce to one of the groups $E_7 5$, $E_7 6$, $E_7 7$ after a cubic extension, and which consequently are well known to be distinct. In other cases the nonuniqueness is connected with the fact that our methods have not allowed us to get a more exact result.

§7. Groups of type E_8

0. Here we consider groups of type E_8 satisfying these conditions:
$\Delta_k \supset \{7\}$ (in §7.1);
G is decomposable over some $K \in \mathfrak{S}(k, 2)$ (in §7.2);
G is decomposable over some $K \in \mathfrak{S}(k, Z_3)$ (in §7.4b).
In addition, we prove that $2^5 \nmid \operatorname{ind}_k G$ (§7.3) and $3^3 \nmid \operatorname{ind}_k G$ (§7.4a).
By Tits [**25**], the following values are possible for Δ_k:

$$\emptyset, \{1\}, \{7\}, \{1,7\}, \{6,7\}, \{1,5,6,7\}, \Delta.$$

1. $E_8 0 - E_8 5$ $(\Delta_k \supset \{7\})$.
In this case $P_7(G,k) \neq \emptyset$, and on taking $P \in P_7(G,k)$, $U = U(P)$, $P = U \cdot H$ and $S = S(H)$, we see that a 56-dimensional representation of the group S of type E_7 is defined over k; that is, $\mathfrak{A}_k(S, P_6) = k$. Thus S is of one of these types: $E_7 0$, $E_7 2$, $E_7 4$, $E_7 8$, $E_7 9$ or $E_7 13$, and G has a corresponding type $E_8 0$—$E_8 5$.

2. $E_8 6$ (*G is decomposable over $K \in \mathfrak{S}(k,2)$ and* $\Delta_k \not\supset \{7\}$).
By [**7**], §4, the case $\Delta_k = \{1\}$ is impossible. If $\Delta_k = \emptyset$ and $\mathfrak{K} \in \mathfrak{S}(k, 2s+1)$ is such that $\operatorname{rg}_{\mathfrak{K}} G > 0$, then $\Delta_{\mathfrak{K}} \supset \{7\}$, since $\Delta_{\mathfrak{K}} = \{1\}$ is impossible. Suppose that $P \in P_7(G, \mathfrak{K})$, H is a k-subgroup of G of type $E_7 \times A_7$. Then $\dim H = 136$, $\dim P = 191$ and $\dim P \cap H \geqq 79$. Since $H_1 = P \cap H$ is anisotropic over $\mathfrak{K}$, it contains P in its semisimple part, so that $S(H_1)$ is of type E_6 or E_7 and $\operatorname{rg} H_1 \geqq 7$. Therefore $Z(H_1) = H_2$ is a decomposable $\mathfrak{K}$-group of type A_1. This contradicts the fact that $\operatorname{rg}_{\mathfrak{K}} H_1 = 0$. Thus $2 \mid \operatorname{ind}_k G/P \ \forall P$.

3. THEOREM. $2^5 \nmid \operatorname{ind}_k G$.

As usual, we make an extension over which $\Gamma = \Gamma(k(T)/k)$ is a 2-group. Then Γ preserves a subsystem of type D_8 and G contains (see §1.1) a k-group H of type D_8. We suppose that H is isotropic over an extension K of degree 4, and show that $\operatorname{ind}_K G \leqq 4$. For if $\Delta_K \supset \{7\}$, then G is of one of the types $E_8 0$—$E_8 2$ over K, so that it is decomposable over some $\mathfrak{K} \in \mathfrak{S}(k, 2)$. If $\Delta_K \not\supset \{7\}$, then $\Delta_K = \{1\}$, and we consider a representation of $H_1 = H(G, K)$

on the corresponding unipotent radical U. The representation of H_1 on $[U, U]$ is equivalent to the natural representation of a group of type D_7, and to a spinor representation over $U/[U, U]$. Thus H_1 is isogenic over K to an orthogonal group (and in addition is defined by a cocycle from a simply-connected group). Over a suitable $\mathfrak{K} \in \mathfrak{S}(k, 2)$ we have $\mathrm{rg}_{\mathfrak{K}} H_1 > 0$, so that $\mathrm{ind}_k G \leqq 2^4$, as above.

Thus it remains to show that $\exists K \in \mathfrak{S}(k, 4)$ such that $\mathrm{rg}_K H > 0$. For this purpose we employ the Weil construction [**26**]. Let $\mathfrak{A}$ be an associative algebra over k with involution J of the first type defining H. Then $\mathfrak{A} = \mathfrak{M}_s \otimes \mathfrak{D}$, where $\mathfrak{D}$ is a division algebra. We assume first that $s \neq 1$. Then J induces an involution in $\mathfrak{M}_s$, and we take for K a quadratic field over which the pair $(\mathfrak{M}_s, J/\mathfrak{M}_s)$ is isotropic. Then H is also isotropic over K. If $s = 1$, we first construct an auxiliary quadratic extension $\mathfrak{K}$ such that $\mathrm{ind}_{\mathfrak{K}} \mathfrak{A} < 2^4$ and return to the situation discussed above.

REMARK. The representation of H on the tangent space to the factor space G/H is equivalent to a spinor representation and is defined over k. It is clear from a comparison of the 2-components of the Weyl groups of G and H that H is an inner form of type D_8. Possibly it could be proved by using these two properties that $2^4 \nmid \mathrm{ind}_k G$.

4. Cubic extensions.

a. $27 \nmid \mathrm{ind}_k G$.

Let T be a maximal k-torus in G. As usual, we may assume that $\Gamma = \Gamma(k(T)/k)$ is a 3-group. Then Γ preserves a subsystem of the roots of type $E_6 \times A_2$ in $\Sigma(G)$, so that G contains commuting k-groups H_1 and H_2 of types E_6 and A_2. Suppose that $K \in \mathfrak{S}(k, Z_3)$ splits H_2 (see §1.6). Then $\Delta_K \supset \{6, 7\}$; that is, G is of type $E_8 3$ over K; hence $\mathrm{ind}_K G = 3$, and our assertion follows.

b. G is decomposable over $K \in \mathfrak{S}(k, Z_3)$ $(E_8 3, E_8 7)$.

Looking at $P \cap P^{\sigma} \cap P^{\sigma^2}$, where $P \in P_{67}(G, K)$ and $\sigma \in \Gamma(K/k)$, and applying an inductive argument, we see that G contains a k-torus T decomposable over K. If $\mathrm{rg}_k G = 0$, we show that $\mathrm{ind}_k G/P = 3 \, \forall P$. For if $\mathfrak{K} \in \mathfrak{S}(k, m)$ and $\mathrm{rg}_{\mathfrak{K}} G > 0$, then $P_7(G, \mathfrak{K}) \neq \emptyset$. Further, G contains a k-subgroup H of type $E_6 \times A_2$ (E_6 is of type $E_6 6$, $A_2 = G(\mathfrak{A})$, $\dim H = 86$ and $\dim H \cap \widetilde{P} \geqq 29$, where $\widetilde{P} \in P_7(G, \mathfrak{K})$). It follows that $H \cap \overline{P}$ contains a (proper) subgroup of type D_4. Then $Z(H \cap P)$ is isotropic over $\mathfrak{K}$, and the usual methods give rise to a contradiction to the fact that H is $\mathfrak{K}$-anisotropic.

c. G contains a torus decomposable over a 3-extension $(E_8 8)$.

In this case G contains k-subgroups H_1 and H_2 of types $E_6 8$ and A_2 (see §7.4a), $\mathrm{ind}_k G/P_i = 3$ $(i = 6, 7)$ and $\mathrm{ind}_k G/P_i = 9$ $(i = 1, 2, 3, 4, 5, 8)$; G has a splitting field with group $Z_3 + Z_3$.

Received 22 NOV 1967

REFERENCES

[1] È. B. Vinberg, *Invariant linear connections in a homogeneous space,* Trudy Moskov. Mat. Obšč. **9** (1960), 191-210. (Russian) MR **31** #690.

[2] E. B. Dynkin, *Semisimple subalgebras of semisimple Lie algebras,* Mat. Sb. **30 (72)** (1952), 349-462; English transl., Amer. Math. Soc. Transl. (2) **6** (1957), 111-244. MR **13**, 904.

[3] A. A. Albert, *Structure of algebras*, Amer. Math. Soc. Colloq. Publ., vol. 24, Amer. Math. Soc., Providence, R. I., 1939. MR **1**, 99.
[4] ———, *A construction of exceptional Jordan division algebras*, Ann. of Math. (2) **67** (1958), 1-28. MR **19**, 1036.
[5] ———, *On exceptional Jordan division algebras*, Pacific J. Math. **15** (1965), 377-404. MR **32** # 130.
[6] H. P. Allen, *Jordan algebras and Lie algebras of type D_4*, J. Algebra **5** (1967), 250-265. MR **34** # 7593.
[7] S. Araki, *On root systems and an infinitesimal classification of irreducible symmetric spaces*, J. Math. Osaka City Univ. **13** (1962), 1-34; Russian transl., Matematika **10** (1966), no. 1, 90-126. MR **27** # 3743.
[8] A. Borel and J. Tits, *Groupes réductifs*, Inst. Hautes Études Sci. Publ. Math. No. 27 (1965), 55-150; Russian transl., Matematika **11** (1967), no. 1, 43-111. MR **34** # 7527.
[9] R. B. Brown, *A new type of nonassociative algebras*, Proc. Nat. Acad. Sci. U.S.A. **50** (1963), 947-949. MR **28** # 2135.
[10] Séminaire C. Chevalley, 1956-1958, *Classification des groupes de Lie algébriques*, Secrétariat mathématique, Paris, 1958. MR **21** # 5696.
[11] C. Chevalley and R. D. Schafer, *The exceptional simple Lie algebras F_4 and E_6*, Proc. Nat. Acad. Sci. U.S.A. **36** (1950), 137-141. MR **11**, 577.
[12] J. Dieudonné, *La géométrie des groupes classiques*, 2nd rev. ed., Springer-Verlag, Berlin, 1963. MR **28** # 1239.
[13] G. Harder, *Über die Galoiskohomologie halbeinfacher Matrizengruppen*. I, Math. Z. **90** (1965), 404-428. MR **32** # 7677.
[14] H. Hijikata, *A remark on the groups of type G_2 and F_4*, J. Math. Soc. Japan **15** (1963), 159-164. MR **27** # 1442.
[15] N. Jacobson, *Composition algebras and their automorphisms*, Rend. Circ. Mat. Palermo (2) **7** (1958), 55-80. MR **21** # 66.
[16] ———, *Some groups of transformations defined by Jordan algebras*. I, II, III, J. Reine Angew. Math. **201** (1959), 178-195; **204** (1960), 74-98; **207** (1961), 61-85. MR **21** # 5666; **28** # 3065; **28** # 3066.
[17] ———, *Clifford algebras for algebras with involution of type D*, J. Algebra **1** (1964), 288-300. MR **29** # 5849.
[18] ———, *Triality and Lie algebras of type D_4*, Rend. Circ. Mat. Palermo (2) **13** (1964), 129-153. MR **31** # 5932.
[19] I. Satake, *On the theory of reductive algebraic groups over a perfect field*, J. Math. Soc. Japan **15** (1963), 210-235; Russian transl., Matematika **9** (1965), no. 2, 19-44. MR **27** # 1438.
[20] J.-P. Serre, *Cohomologie galoisienne*, Lecture Notes in Math., no. 5, Springer-Verlag, Berlin and New York, 1964. MR **31** # 4785.
[21] T. A. Springer, *Sur les formes quadratiques d'indice zéro*, C. R. Acad. Sci. Paris **234** (1952), 1517-1519. MR **13**, 815.
[22] ———, *Quelques résultats sur la cohomologie galoisienne*, Colloq. Théorie des Groupes Algébriques (Bruxelles, 1962), Librairie Universitaire, Louvain et Gauthier-Villars, Paris, 1962, pp. 129-135. MR **32** # 2417.
[23] J. Tits, *Groupes semi-simples isotropes*, Colloq. Théorie des Groupes Algébriques (Bruxelles, 1962), Librairie Universitaire, Louvain et Gauthier-Villars, Paris, 1962, pp. 137-147. MR **26** # 6174.
[24] ———, *Algèbres alternatives, algèbres de Jordan et algèbres de Lie exceptionnelles*. I: *Construction*, Nederl. Akad. Wetensch. Proc. Ser. A. **69** = Indag. Math. **28** (1966), 223-237. MR **36** # 2658.
[25] ———, *Classification of algebraic semisimple groups*, Proc. Sympos. Pure Math., vol. 9, Amer. Math. Soc., Providence, R. I., 1966, pp. 33-62. MR **37** # 309.
[26] A. Weil, *Algebras with involutions and the classical groups*, J. Indian Math. Soc. **24** (1960), 589-623; Russian transl., Matematika **7** (1963), no. 4, 31-55. MR **25** # 147.

Translated by J. Wiegold

Trudy Moskov. Mat. Obšč.
Tom 20 (1969)

Trans. Moscow Math. Soc.
Vol. 20 (1969)

ON WEAK SOLUTIONS OF THE DIRICHLET AND NEUMANN PROBLEMS

V. G. MAZ'JA

To S. G. Mihlin on his sixtieth birthday

Contents

INTRODUCTION 135

CHAPTER I. Conditions for the solvability in $\overset{\circ}{L}{}_q^{(1)}(D)$ of the Dirichlet problem 137

§1. Basic notation and auxiliary assertions 137

§2. Estimates for the classical solution of the Dirichlet problem 141

§3. Unique solvability of the Dirichlet problem in $\overset{\circ}{L}{}_q^{(1)}$ $(q \leqq 2)$ 147

CHAPTER II. Solvability in $L_q^{(1)}$ $(q \leqq 2)$ of the Neumann problem in a domain with nonsmooth boundary 150

§4. Estimates for the first derivatives of a solution of the Neumann problem 150

§5. Theorems on the unique solvability of the Neumann problem. Estimates in L_∞ and L_q. 157

§6. Subareal mappings and a test of solvability conditions for the Neumann problem 161

§7. On the exactness of the solvability conditions for the Neumann problem 167

BIBLIOGRAPHY 171

Introduction

During the past decade the properties of the solutions of linear elliptic equations with principal part in divergence form have been intensively investigated under very weak assumptions concerning the coefficients. In this direction are papers devoted to estimates for the Hölder norm [**1-6**], for the maximum modulus or the norm in $L_q(D)$ [**7-12**] and for the norm of the gradient in $L_q(D)$ [**10, 13, 14**], and to a study of the behavior of a solution in a neighborhood of a singular point at infinity and near an irregular boundary [**15-20**] and to a proof of Harnack's inequality [**21**].

In the present paper we prove existence and uniqueness theorems for weak solutions of the Dirichlet and Neumann problems, and obtain estimates for such solutions.

Some of the results were given without proof in [**10**], and some are being published for the first time.

---

*The present translation incorporates suggestions made by the author.

In Chapter I we consider the Dirichlet problem with zero boundary condition for the operator

$$P = -\nabla (A\nabla) + \mathbf{a}\nabla + a,$$

where A is a matrix such that for any real vector ξ

$$\xi A\xi \geqslant \varkappa |\xi|^2, \quad \varkappa = \text{const} > 0. \tag{0.1}$$

We assume that $\mathbf{a} \in L_r(D)$ $(r > n)$, $a \in L_{n/2}(D)$ and the elements of the matrix $A(x)$ are measurable and bounded in D. Here and below D is an open subset of R^n.

If a function f is pth power summable in D, $1 < p < 2n/(n+2)$, then the equation $Pu = f$ cannot in general have a solution with finite energy integral. In this case one is able to prove the existence and uniqueness of a generalized solution in $\overset{\circ}{L}{}_q^{(1)}(D)$ $(q = pn/(n-p) < 2)$ and to estimate the norm of the gradient of this solution in $L_q(D)$ in terms of the norm of f in $L_p(D)$.

The condition that the elements of A be bounded is used only in proving unique solvability, and can probably be weakened. In any case, in not one of the estimates given below do the constants depend on the least upper bound of the eigenvalues of A. In this connection we note a paper of N. Meyers [**13**], in which an estimate is given for the gradient of a solution of the equation $\nabla(A\nabla u) = \operatorname{div}\phi$ in $L_p(D)$ in terms of the norm of ϕ in $L_p(D)$ for indices p sufficiently close to two. This closeness is determined by the ratio of the least upper and greatest lower bounds of the eigenvalues of A.

The theorems proved in Chapter I given new conditions for the uniqueness of a solution of the Dirichlet problem for the operator P. One of these conditions is the inequality

$$\sup\operatorname{vrai} a^- < \lambda \exp\left(-\frac{k}{\varkappa} v_D^{n^{-1}-r^{-1}} \|\mathbf{a}\|_r\right),$$

where λ is the first eigenvalue of the Dirichlet problem for the operator $-\nabla(A\nabla)$, $\|\cdot\|_r$ is the norm in the space $L_r(D)$ $(r > n)$, v_E is n-dimensional measure of a set E and k is a constant depending only on n and r. It would perhaps be interesting to compare this test with the results obtained with the use of geometric methods by A. D. Aleksandrov [**22**] on conditions for the uniqueness of a solution of a second order elliptic equation that is not in divergence form.

In Chapter II we define a generalized solution of the Neumann problem with homogeneous boundary condition in a space $L_q^{(1)}(D)$ $(0 < q \leqq 2)$ and prove the unique solvability of this problem. In this connection the index q depends on the properties of the boundary of the domain D.

For an arbitrary domain D the Neumann problem can turn out to be unsolvable in $L_q^{(1)}(D)$ for arbitrarily small $q > 0$; but it is possible to prove the existence and uniqueness of a generalized solution with finite integral

$$\int_D M(|\nabla u|)\,dx,$$

where $M(x)$ is a convex upwards function that depends only on the domain D and tends to ∞ for $x \to +\infty$.

The boundary ∂D of a domain is characterized by the behavior for $t \to +0$ of the function

$$\lambda(t) = \inf_{\{E\}} s(D \cap \partial E),$$

where s is $(n-1)$-dimensional area and E is any subset of D such that the part of its boundary which is interior to D is locally Lipschitz and $v_D/2 \geqq v_E \geqq t$. We obtain estimates depending on $\lambda(t)$ for the maximum modulus and the norm in $L_q(D)$ of a solution of the Neumann problem.

We introduce the notion and study the properties of a "subareal" mapping of a domain, i.e. a mapping under which an $(n-1)$-dimensional area of subsets of D is not essentially increased. The use of such mappings permits one to test solvability conditions for the Neumann problem for specific domains.

We show that a condition for the solvability of the Neumann problem in $L_q^{(1)}$ $(q \geqq 1)$ as well as conditions for the solution being bounded and belonging to L_q are best possible in a well defined sense.

We note that the method used here can be applied to other boundary value problems, including nonhomogeneous ones. In particular, similar results can be obtained under the assumption that a linear combination of the solution and its normal derivative is given on the boundary (see [**10**]).

Another possible generalization, which was also left out of this article, is a sharpening of the conditions concerning the coefficients and boundary by making use of a property such as capacity (cf. [**22**], [**24**]).

Chapter I

CONDITIONS FOR THE SOLVABILITY IN $\overset{\circ}{L}{}_q^{(1)}(D)$ OF THE DIRICHLET PROBLEM

§1. Basic notation and auxiliary assertions

Let D be an open subset of n-dimensional Euclidean space R^n $(n \geqq 3)$,[1)] and suppose the n-dimensional measure of D is finite.

We introduce the following notation: ∂E and $\overline{E}$ are the boundary and closure of a set $E \subset D$, $v(E) = v_E$ is the n-dimensional measure of E and $s(E) = s_E$ is the Hausdorff $(n-1)$-dimensional measure of E. If $u(x)$ is a function defined in D, then

$$M(t) = \{x : |u(x)| \geqslant t\}, \quad E(t) = \{x : |u(x)| = t\}, \quad K(T, t) = M(t) \setminus M(T),$$

[1)] We exclude the case $n = 2$ in this chapter only to avoid complicating the discussion.

where $T \geqq t$. In this connection $s_t \equiv S_{E(t)}$, $v_t \equiv v_{M(t)}$ and u^+ and u^- are the positive and negative parts of u. Also, let v_n be the n-dimensional volume of the unit ball in R^n and let $\alpha_n = nv_n^{1/n}$ be the exact constant in the isoperimetric inequality

$$s_{\partial E} \geqslant \alpha_n v_E^{\frac{n-1}{n}},$$

which we will apply only for subsets E that are bounded by infinitely differentiable manifolds.

Let $\|\cdot\|_p$ be the norm in the space $L_p = L_p(D)$ and let $p' = p(p-1)^{-1}$. In the sequel $\mathscr{K}$ is the space of infinitely differentiable functions with compact supports in D. By the space $\overset{\circ}{L}{}_q^{(1)}$ we mean the completion of $\mathscr{K}$ in the metric of the integral

$$\int |\nabla u|^q \, dx.$$

Here and below an integral without indicated limits means an integral over D. All functions considered in the paper are real.

Let $A(x)$ denote a matrix satisfying condition (0.1). Let $\mathbf{a}(x) \in L_r$, where $r > n$, and $a(x) \in L_{n/2}$. We also assume that the elements of $A(x)$ are measurable and bounded in D.

Let $u, v \in \overset{\circ}{L}{}_2^{(1)}$. We define the bilinear forms

$$\mathfrak{A}(u, v) = \int \nabla v A \nabla u dx, \quad \mathfrak{B}(u, v) = \int v \mathbf{a} \nabla u dx,$$

$$\mathfrak{C}(u, v) = \int a v u dx, \quad \mathfrak{P}(u, v) = \mathfrak{A}(u, v) + \mathfrak{B}(u, v) + \mathfrak{C}(u, v).$$

The notation $\mathfrak{P}_D(u, v)$ will sometimes be used in place of $\mathfrak{P}(u, v)$.

A function $U(x) \in \overset{\circ}{L}{}_2^{(1)}$ is called a *generalized solution in* $\overset{\circ}{L}{}_2^{(1)}$ *of the Dirichlet problem for the equation*

$$PU \equiv -\nabla(A \nabla U) + \mathbf{a} \nabla U + aU = f, \tag{1.1}$$

where $f \in L_{2n/(n+2)}$, if

$$\mathfrak{P}(U, v) = \int f v dx \tag{1.2}$$

for all functions $v \in \overset{\circ}{L}{}_2^{(1)}$. A solution in $\overset{\circ}{L}{}_2^{(1)}$ of the Dirichlet problem for the more general equation

$$PU = f + \operatorname{div} \varphi, \tag{1.3}$$

where $\phi \in \overset{\circ}{L}_2$, is analogously defined by means of the integral identity

$$\mathfrak{P}(U, v) = \int f v dx - \int \varphi \nabla v dx. \tag{1.4}$$

Suppose the function f belongs to a space L_p $(2n/(n+2) > p \geqq 1)$. We will say that a function $u \in \overset{\circ}{L}{}_q^{(1)}$ $(q = pn/(n-p))$ is a *generalized solution in* $\overset{\circ}{L}{}_q^{(1)}$ of the Dirichlet problem for the equation $Pu = f$ if the following condition is satisfied: for some sequence f_m, $f_m \in L_{2n/(n+2)}$, converging to f in L_p a sequence of solutions in $\overset{\circ}{L}{}_2^{(1)}$ of the Dirichlet problems for the equations $Pu_m = f_m$ converges in $\overset{\circ}{L}{}_q^{(1)}$ to u.

We note further that $c, c_1, c_2, \cdots$ below denote constants depending only on the "dimensionless" parameters n, p, q and r.

We mention the following well-known inequality of Hardy (see [**25**], Russian page 296):

$$\int_0^\infty x^{-r} F^p(x)\,dx \leqslant \left(\frac{p}{|r-1|}\right)^p \int_0^\infty x^{-r}(xf)^p\,dx, \tag{1.5}$$

where $p > 1$, $r \neq 1$ and

$$F(x) = \int_0^x f(t)\,dt \quad \text{for } r > 1,$$

$$F(x) = \int_x^\infty f(t)\,dt \quad \text{for } r < 1.$$

Lemma 1 formulated below is a corollary of a formula of A. S. Kronrod [**26, 27**] for the n-dimensional variation of a function (see also [**28**]).

LEMMA 1. *Suppose $u(x)$ is a function in* $\mathrm{Lip}(D)$, *i.e. satisfying a local Lipschitz condition in D, and $\Phi(x)$ is a Borel measurable function in D. Then*

$$\int_D \Phi(x)\,|\nabla u|\,dx = \int_0^\infty d\tau \int_{E(\tau)} \Phi(x)\,ds. \tag{1.6}$$

LEMMA 2. *For any function $u(x) \in \mathrm{Lip}(D) \cap L_p^{(1)}$ $(p > 1)$ and almost all $t \geqq 0$*

$$s_t^{p'} \leqslant -v_t' \left(\int_{E(t)} |\nabla u|^{p-1}\,ds\right)^{\frac{1}{p-1}}. \tag{1.7}$$

PROOF. Hölder's inequality implies that for $T > t > 0$

$$\left(\int_{K(T,t)} |u|^{p-1} |\nabla u|\,dx\right)^{p'} \leqslant \int_{K(T,t)} |u|^p dx \left(\int_{K(T,t)} |\nabla u|^p\,dx\right)^{\frac{1}{p-1}}.$$

Using Lemma 1, we get

$$\left(\int_t^T \tau^{p-1} s_\tau d\tau\right)^{p'} \leqslant \int_{K(T,t)} |u|^p dx \left(\int_t^T d\tau \int_{E(\tau)} |\nabla u|^{p-1} ds\right)^{\frac{1}{p-1}}.$$

We divide both sides of this inequality by $(T-t)^{p'}$ and estimate the first factor in the right side:

$$\left(\frac{1}{T-t}\int_t^T \tau^{p-1} s_\tau d\tau\right)^{p'} \leqslant T^p \frac{v_t - v_T}{T-t} \left(\frac{1}{T-t}\int_t^T d\tau \int_{E(\tau)} |\nabla u|^{p-1} ds\right)^{\frac{1}{p-1}}.$$

Passing to the limit for $T \to t$, we obtain inequality (1.7) for almost all $t > 0$.

The following inequality (to within a constant and the method of proof) is due to S. L. Sobolev [**29**].

LEMMA 3. *For any function* $u \in \overset{\circ}{L}{}_p^{(1)}(D)$

$$\| u \|_q \leqslant p^{\frac{1}{p}} q^{\frac{1}{p'}} \alpha_n^{-1} \| \nabla u \|_p, \tag{1.8}$$

where $q = pn/(n-p)$ *and* $n > p$.

PROOF. It is clearly sufficient to prove (1.8) for a function $u \in \mathscr{K}(D)$. With each such function we associate a function $\psi(t)$ defined by the equality

$$\psi(t) = \int_0^t \frac{d\tau}{\left(\int_{E(\tau)} |\nabla u|^{p-1} ds \right)^{\frac{1}{p-1}}}. \tag{1.9}$$

By virtue of Lemma 2 and the isoperimetric inequality $s_t \geqq \alpha_n v_t^{(n-1)/n}$ we have

$$\psi(t) \leqslant -\int_0^t \frac{v_\tau' d\tau}{s_\tau^{p'}} \leqslant -\alpha_n^{-p'} \int_0^t \frac{v_\tau' d\tau}{v_\tau^{\frac{p'}{n'}}} \leqslant q (p')^{-1} \alpha_n v_t^{-\frac{p'}{q}}. \tag{1.10}$$

Further, by repeating with obvious changes the arguments presented in [**24**] on pages 1151-1152, we get that the inverse $t(\psi)$ of the function $\psi(t)$ is absolutely continuous and that the following identity is valid:

$$\int |\nabla u|^p dx = \int_0^{\psi(\max|u|)} [t'(\psi)]^p d\psi. \tag{1.11}$$

Using the identity

$$\int |u|^q dx = q \int_0^\infty v_t t^{q-1} dt$$

and the monotonicity of the function v_t, we get

$$\int |u|^q dx = q \int_0^\infty v_t^{1-\frac{p}{q}} t^{q-p} v_t^{\frac{p}{q}} t^{p-1} dt$$

$$\leqslant \frac{q}{p} \int_0^\infty \left(\int_0^t v_\tau^{\frac{p}{q}} d\tau^p \right)^{\frac{q-p}{p}} v_t^{\frac{p}{q}} dt^p = \left(\int_0^\infty v_\tau^{\frac{p}{q}} d\tau^p \right)^{\frac{q}{p}}.$$

In the last integral we make the change of variable $\tau = t(\psi)$. Then

$$\| u \|_q^p \leqslant p \int_0^\infty v_{t(\psi)}^{\frac{p}{q}} [t(\psi)]^{p-1} t'(\psi) d\psi \leqq$$

$$\leqslant p \sup_t v_t^{\frac{p}{q}} [\psi(t)]^{p-1} \Bigl(\int_0^{\psi(\max|u|)} [t'(\psi)]^p d\psi \Bigr)^{\frac{1}{p}} \Bigl(\int_0^{\psi(\max|u|)} [t(\psi)]^p \frac{d\psi}{\psi^p} \Bigr)^{\frac{1}{p'}},$$

which by virtue of Hardy's inequality (1.5) does not exceed

$$\frac{p^p}{(p-1)^{p-1}} \sup_t v_t^{\frac{p}{q}} [\psi(t)]^{p-1} \int_0^{\psi(\max|u|)} [t'(\psi)]^p d\psi.$$

Taking (1.11) into consideration, we obtain

$$\| u \|_q^p \leqslant \frac{p^p}{(p-1)^{p-1}} \sup v_t^{\frac{p}{q}} [\psi(t)]^{p-1} \| \nabla u \|_p^p.$$

This result and (1.10) imply estimate (1.8).

§2. Estimates for the classical solution of the Dirichlet problem

In the lemmas proved in this section we assume that the boundary ∂D and the functions A, $\mathbf{a}$, a, f and ϕ are infinitely differentiable in $\overline{D}$, and that there exists a solution $u \in \overset{\circ}{L}{}_2^{(1)}$ of equation (1.3). As is known, the function u is infinitely differentiable in $\overline{D}$ and vanishes on ∂D. Hence for almost all $t > 0$ the contour sets $E(t)$ are infinitely differentiable compact manifolds without boundary. In the sequel we will have in mind only such values of t.

LEMMA 4. *Suppose g is a nonnegative Borel measurable function in D. Then*

$$\int_0^\infty \int_{E(\tau)} |g|\, ds \Bigl(\int_{E(\tau)} |\nabla u|\, ds \Bigr)^{-1} d\tau \leqslant k v_D^{\frac{r-n}{rn}} \| g \|_r, \tag{2.1}$$

where $r > n$ and

$$k = \Bigl(n \frac{r-1}{r-n} \Bigr)^{\frac{1}{r'}} \alpha_n^{-1}.$$

PROOF. Using the Hölder inequality three times and Lemma 2, we get

$$\int_{E(\tau)} |g|\, ds \leqslant \Bigl(\int_{E(\tau)} |g|^r \frac{ds}{|\nabla u|} \Bigr)^{\frac{1}{r}} \Bigl(\int_{E(\tau)} |\nabla u|\, ds \Bigr)^{\frac{1}{r}} s_\tau^{\frac{r-2}{r}}$$

$$\leqslant s_\tau^{-1} (-v'_\tau)^{\frac{1}{r'}} \Bigl(\int_{E(\tau)} |g|^r \frac{ds}{|\nabla u|} \Bigr)^{\frac{1}{r}} \int_{E(\tau)} |\nabla u|\, ds.$$

From this result and the isoperimetric inequality

$$s_\tau \geqslant \alpha_n v_\tau^{\frac{n-1}{n}} \tag{2.2}$$

we immediately obtain the following estimate for the integral in the left side of (2.1):

$$\int_0^\infty \int_{E(\tau)} |g|\,ds\Big(\int_{E(\tau)} |\nabla u|\,ds\Big)^{-1} d\tau$$

$$\leqslant \alpha_n^{-1}\Big(-\int_0^\infty v_\tau^{\frac{(1-n)r}{n(r-1)}} v'_\tau d\tau\Big)^{\frac{1}{r'}} \Big(\int_0^\infty \int_{E(\tau)} |g|^r \frac{ds}{|\nabla u|} dt\Big)^{\frac{1}{r}}.$$

In order to complete the proof it suffices to note that the first integral in the right side does not exceed

$$\frac{n(r-1)}{r-n} v_D^{\frac{r-n}{n(r-1)}},$$

LEMMA 5. *For almost all* $\tau > 0$

$$\begin{aligned}\int_{E(\tau)} \frac{\nabla u A \nabla u}{|\nabla u|} ds \leqslant \int_{M(\tau)} (|f| + a^- |u|)\,dx \exp\Big(\int_\tau^\infty \Phi(\sigma)\,d\sigma\Big) \\ + \int_{E(\tau)} |\varphi|\,ds + \int_\tau^\infty \int_{E(t)} |\varphi|\,ds\,\Phi(t) \exp\Big(\int_t^\tau \Phi(\sigma)\,d\sigma\Big) dt,\end{aligned} \tag{2.3}$$

where

$$\Phi(t) = \Big(\int_{E(t)} \frac{\nabla u A \nabla u}{|\nabla u|} ds\Big)^{-1} \int_{E(t)} |\mathbf{a}|\,ds. \tag{2.4}$$

PROOF. Integrating equation (1.3) over the set M_τ and applying the Gauss-Green formula, we get

$$-\int_{E(\tau)} \nabla u A \mathbf{n} ds + \int_{M(\tau)} \mathbf{a}\nabla u dx + \int_{M(\tau)} a u dx = \int_{M(\tau)} f dx + \int_{E(\tau)} \varphi \mathbf{n}\, ds,$$

where $\mathbf{n}$ is the outward (with respect to $M(\tau)$) normal on $E(\tau)$. Hence

$$\int_{E(\tau)} \frac{\nabla u A \nabla u}{|\nabla u|} ds - \int_{M(\tau)} |\mathbf{a}|\,|\nabla u|\,dx \leqslant \xi(\tau), \tag{2.5}$$

where

$$\xi(\tau) = \xi_1(\tau) + \xi_2(\tau),$$

$$\xi_1(\tau) = \int_{M(\tau)} (|f| + a^- |u|)\,dx, \quad \xi_2(\tau) = \int_{E(\tau)} |\varphi|\,ds.$$

Inequality (2.5) can be rewritten in the form

$$-\frac{d}{d\tau}\Big\{\exp\Big(-\int_\tau^\infty \Phi(\sigma)\,d\sigma\Big)\int_\tau^\infty \int_{E(t)} |\mathbf{a}|\,ds dt\Big\} \leqslant \Phi(\tau)\,\xi(\tau) \exp\Big(-\int_\tau^\infty \Phi(\sigma)\,d\sigma\Big). \tag{2.6}$$

As was shown in Lemma 4, the function $\Phi(\sigma)$ is summable. We note further that

$$\int_{\tau}^{\infty}\int_{E(\tau)} |\mathbf{a}|\,ds\,dt \leqslant \int_D |\mathbf{a}||\nabla u|\,dx \leqslant \|\mathbf{a}\|_2\,\|\nabla u\|_2 < \infty.$$

Thus the function in braces in inequality (2.6) is absolutely continuous, and hence

$$\int_{\tau}^{\infty}\int_{E(\tau)} |\mathbf{a}|\,ds\,dt \leqslant \int_{\tau}^{\infty} \Phi(\tau)\,\xi(t)\exp\Big(\int_t^{\tau}\Phi(\sigma)\,d\sigma\Big).$$

Since $\xi_1(t)$ does not increase, we consequently obtain

$$\int_{\tau}^{\infty}\int_{E(t)} |\mathbf{a}|\,ds\,dt \leqslant \xi_1(t)\Big[\exp\Big(\int_{\tau}^{\infty}\Phi(\sigma)\,d\sigma\Big)-1\Big]+\int_{\tau}^{\infty}\Phi(t)\,\xi_2(t)\exp\Big(\int_t^{\tau}\Phi(\sigma)\,d\sigma\Big)\,dt,$$

which in combination with (2.5) implies estimate (2.3).

Lemmas 4 and 5 directly imply

LEMMA 6. *For almost all* $\tau > 0$

$$K\int_{E(\tau)} \frac{\nabla u A \nabla u}{|\nabla u|}\,ds \leqslant \int_{M(\tau)} (|f|+a^-|u|)\,dx+\int_{\tau}^{\infty}\int_{E(t)} |\varphi|\,ds\Phi(t)\,dt + K\int_{E(\tau)}|\varphi|\,ds, \quad (2.7)$$

where

$$K = \exp\Big(-\frac{k}{\varkappa}v_D^{\frac{r-n}{rn}}\,\|\mathbf{a}\|_r\Big). \quad (2.8)$$

LEMMA 7. *Suppose*

$$\|a^-\|_{n/2} < \frac{n-2}{4n}\,\alpha_n^2\varkappa K, \quad (2.9)$$

where K is the constant in (2.8). *Then*

$$\|\nabla u\|_2 \leqslant C\,(\|f\|_{\frac{2n}{n+2}} + \|\varphi\|_2) \quad (2.10)$$

with the constant C depending on $\varkappa$, $\|a^-\|_{n/2}$, $\|\mathbf{a}\|_r$, v_D, n *and* r.

PROOF. Integrating inequality (2.7) from 0 to ∞ and using Lemma 1, we get

$$\varkappa K\int_D (\nabla u)^2\,dx \leqslant \int_D (|u||f|+a^-u^2)\,dx + \int_0^{\infty} d\tau \int_{\tau}^{\infty}\int_{E(t)} |\varphi|\,ds\Phi(t)\,dt$$
$$+K\int_D |\varphi||\nabla u|\,dx.$$

As was shown in [**24**] (Corollary 3), condition (2.9) guarantees the estimate

$$\int a^- u^2\, dx \leqslant (\varkappa K - \delta) \int (\nabla u)^2\, dx, \tag{2.11}$$

where $\delta = \text{const} > 0$. Therefore

$$\int (\nabla u)^2\, dx \leqslant C_1 \Big(\|f\|_{\frac{n}{2}} \|u\|_{\frac{2n}{n-2}} + \int |\varphi|^2 dx + \int_0^\infty \int_{E(\tau)} |\varphi|\, ds \Phi(\tau)\, \tau\, d\tau \Big).$$

Applying Lemma 3 and recalling the definition of the function $\Phi(\tau)$, we get

$$\int (\nabla u)^2\, dx \leqslant C_2 (\|f\|^2_{\frac{n}{2}} + \|\varphi\|_2^2)$$

$$+ C_3 \Big\{ \varkappa^{-1} \int_0^\infty \Big(\int_{E(\tau)} |\varphi|\, ds \Big)^2 \Big(\int_{E(\tau)} |\nabla u|\, ds \Big)^{-1} d\tau \Big\}^{\frac{1}{2}} \Big\{ \int_0^\infty \Big(\int_{E(\tau)} |\mathrm{a}|\, ds \Big)^2 \Big(\int_{E(\tau)} |\nabla u|\, ds \Big)^{-1} \tau\, d\tau \Big\}^{\frac{1}{2}}.$$

Further, using Lemma 1, we find that

$$\int_0^\infty \Big(\int_{E(\tau)} |\varphi|\, ds \Big)^2 \Big(\int_{E(\tau)} |\nabla u|\, ds \Big)^{-1} d\tau \leqslant \int_0^\infty \int_{E(\tau)} |\varphi|^2 \frac{ds}{|\nabla u|}\, d\tau = \int_D |\varphi|^2\, dx. \tag{2.12}$$

Thus, for any positive ϵ,

$$\int (\nabla u)^2\, dx \leqslant C_4(\varepsilon) (\|f\|^2_{\frac{n}{2}} + \|\varphi\|_2^2) + 2\varepsilon \int_0^\infty \Big(\int_{E(\tau)} |\mathrm{a}|\, ds \Big)^2 \Big(\int_{E(\tau)} |\nabla u|\, ds \Big)^{-1} \tau\, d\tau.$$

The last summand in the right side does not exceed

$$2\varepsilon \int_0^\infty \int_{E(\tau)} \frac{|\mathrm{a}|^2}{|\nabla u|}\, ds\, \tau\, d\tau = \varepsilon \int_D |\mathrm{a}|^2 u^2\, dx \leqslant \varepsilon \|\mathrm{a}\|_n^2 \|u\|^2_{\frac{2n}{n-2}}.$$

Again using Lemma 3, we obtain (2.10).

REMARK 1. From the proof of Lemma 7 it readily follows that estimate (2.10) remains valid if condition (2.9) is replaced by the inequality

$$\sup_D \operatorname{vrai} [a^-(x)] < \lambda K, \tag{2.13}$$

where λ is the first eigenvalue of the Dirichlet problem for the operator $-\nabla A \nabla$ in $\overset{\circ}{L}{}_2^{(1)}$.

LEMMA 8. *Suppose* $\phi \equiv 0$ *and*

$$\|a^-\|_{\frac{n}{2}} < c\varkappa K, \tag{2.14}$$

where

$$c = \Big(1 - 2\frac{p}{n}\Big) \frac{\alpha_n^2}{pp'}, \quad q = \frac{pn}{n-p}, \quad 1 < p \leqslant \frac{2n}{n+2}.$$

Then the following estimate holds for the function u:

$$\| \nabla u \|_q \leqslant \frac{cc_1}{c\varkappa K - \| a^- \|_{\frac{n}{2}}} \| f \|_p, \tag{2.15}$$

where

$$c_1 = p' \left(\frac{n-p}{n} \right)^{\frac{1}{q}} \alpha_n^{-1}.$$

PROOF. From inequality (2.7) we get

$$\varkappa^q K^q \int_0^\infty \int_{E(\tau)} |\nabla u|^{q-1} ds\, d\tau \leqslant \int_0^\infty \xi^q(\tau) \int_{E(\tau)} |\nabla u|^{q-1} ds \left(\int_{E(\tau)} |\nabla u|\, ds \right)^{-q} d\tau, \tag{2.16}$$

where

$$\xi(\tau) = \int_{M(\tau)} (|f| + a^- |u|)\, dx. \tag{2.17}$$

On the left side we have $(\kappa K \| \nabla u \|_q)^q$ by Lemma 1. Since $q < 2$,

$$\int_{E(\tau)} |\nabla u|^{q-1} ds \leqslant \left(\int_{E(\tau)} |\nabla u|\, ds \right)^{q-1} s_\tau^{2-q}.$$

Applying the inequality

$$\int_{E(\tau)} |\nabla u|\, ds \geqslant s_\tau^2 (-v_\tau')^{-1},$$

proved in Lemma 2, we get

$$\int_{E(\tau)} |\nabla u|^{q-1} ds \leqslant -\left(\int_{E(\tau)} |\nabla u|\, ds \right)^q v_\tau' s_\tau^{-q}.$$

Consequently

$$\varkappa^q K^q \| \nabla u \|_q^q \leqslant -\int_0^\infty \xi^q(\tau)\, s_\tau^{-q} v_\tau'\, d\tau.$$

It follows upon applying the isoperimetric inequality (2.2) that

$$\varkappa^q K^q \| \nabla u \|_q^q \leqslant -\alpha_n^{-q} \int_0^\infty \xi^q(\tau)\, v_\tau^{q\frac{1-n}{n}} v_\tau'\, d\tau. \tag{2.18}$$

Let $\zeta(t)$ denote a nonincreasing function of one variable that is equimeasurable with the function $|f| + a^- |u|$. Clearly

$$\xi(\tau) \leqslant \int_0^{v_\tau} \zeta(t)\, dt.$$

This together with (2.17) implies

$$\varkappa^q K^q \| \nabla u \|_q^q \leqslant -\alpha_n^{-q} \int_0^\infty \left[\int_0^{v_\tau} \zeta(t)\, dt \right]^q v_\tau^{q\frac{1-n}{n}} v_\tau'\, d\tau \leqslant \alpha_n^{-q} \int_0^{v_D} \left[\int_0^\mu \zeta(t)\, dt \right]^q \mu^{q\frac{1-n}{n}}\, d\mu.$$

We denote the latter integral by J. Integrating by parts, we obtain

$$J = \frac{qn}{q(n-1)-n} \int_0^{v_D} \left[\int_0^{\mu} \zeta(t)\, dt \right]^{q-1} \zeta(\mu)\, \mu^{1+q\frac{1-n}{n}}\, d\mu$$
$$\leqslant p' J^{\frac{1}{q'}} \left(\int_0^{\infty} \zeta^q(\mu)\, \mu^{\frac{q}{n}}\, d\mu \right)^{\frac{1}{q}} \tag{2.20}$$

Since the function $\zeta(u)$ does not increase,

$$\int_0^{v_D} \zeta^q(\mu)\, \mu^{\frac{q}{n}+1}\, d\mu \leqslant \int_0^{v_D} \left[\int_0^{\mu} \zeta^{\frac{qn}{n+q}}(t)\, dt \right]^{\frac{q}{n}} \zeta^{\frac{qn}{n+q}}(\mu)\, d\mu$$
$$= \frac{n-p}{n} \left(\int_0^{v_D} \zeta^p(t)\, dt \right)^{\frac{q+n}{n}}.$$

Thus

$$\varkappa K \| \nabla u \|_q \leqslant p' \left(\frac{n-p}{n} \right)^{\frac{1}{p}} \alpha_n^{-1} \left(\int_0^{v_D} \zeta^p(t)\, dt \right)^{\frac{1}{p}}.$$

We now note that

$$\left(\int_0^{v_D} \zeta^p(t)\, dt \right)^{\frac{1}{p}} = \| |f| + a^- |u| \|_p \leqslant \| f \|_p + \| a^- u \|_p.$$

Consequently

$$\varkappa K \| \nabla u \|_q \leqslant p' \left(\frac{n-p}{n} \right)^{\frac{1}{q}} \alpha_n^{-1} \left(\| f \|_p + \| a^- \|_{\frac{n}{2}} \| u \|_{\frac{pn}{n-2p}} \right). \tag{2.21}$$

By virtue of Lemma 3

$$\| u \|_{\frac{pn}{n-2p}} \leqslant q \left(\frac{n-q}{n} \right)^{-\frac{1}{q'}} \alpha_n^{-1} \| \nabla u \|_q,$$

which together with (2.21) proves the lemma.

REMARK 2. Suppose the coefficient a in equation (1.1) is nonnegative. Let $N(t)$ denote a nondecreasing function defined for $t \geqq 0$ and such that the function $t^{-1}N(t)$ is convex upwards. By slightly changing the beginning of the proof of Lemma 8 (using in this connection Jensen's inequality in place of Hölder's inequality), we obtain the following estimate:

$$\int N(|\nabla u|)\, dx \leqslant \int_0^{v_D} N \left[C v^{\frac{1-n}{n}} \int_0^{v} f^*(\tau)\, d\tau \right] dv, \tag{2.22}$$

where $C=(\kappa\alpha_n K)^{-1}$ and $f^*(\tau)$ is a nonincreasing function that is equimeasurable with $|f(x)|$. We note that if $D=\{x:|x|\leqq 1\}$ and $u(x)$ is a solution of the Dirichlet problem for the equation $-\Delta u=f(|x|)$, where $f(|x|)$ is a nonincreasing function, then inequality (2.22) becomes an equality.

An analogous estimate can be proved under the assumption that the free term of equation (1.1) is a measure. In particular, we get that the Green function satisfies the following inequality for all $s\in D$:

$$\int N(|\nabla_x G(x,s)|)\,dx \leqslant \int_0^{v_D} N\left[Cv^{\frac{1-n}{n}}\right]dv. \tag{2.23}$$

§3. Unique solvability of the Dirichlet problem in $\overset{\circ}{L}{}_q^{(1)}$ $(q\leqq 2)$

We first prove a theorem on the existence and uniqueness of a generalized solution in $\overset{\circ}{L}{}_2^{(1)}$.

THEOREM 1. *Suppose D is an open set of finite measure and suppose the elements of the matrix A are bounded measurable functions,* $\mathbf{a}\in L_r$ $(r>n)$, $a\in L_{n/2}$, $f\in L_{2n/(n+2)}$, $\phi\in L_2$. *Suppose further*

$$\|a^-\|_{\frac{n}{2}} < \frac{n-2}{4n}\,\alpha_n^2\varkappa\exp\left(-\frac{k}{\varkappa}v_D^{\frac{r-n}{rn}}\|\mathbf{a}\|_r\right), \tag{3.1}$$

where κ is the constant of ellipticity and k is a "dimensionless" constant defined by the equality

$$k=\left(n\frac{r-1}{r-n}\right)^{\frac{r-1}{r}}\alpha_n^{-1}.$$

Then there exists a unique solution $U\in\overset{\circ}{L}{}_2^{(1)}$ of equation (1.3), *and estimate* (2.10) *holds for this solution.*

PROOF. *Existence.* We first assume in addition that the functions A, $\mathbf{a}$ and a are infinitely differentiable in D and that $f,\phi\in\mathscr{K}(D)$. Let $\{D_m\}$ be an increasing sequence of open sets with boundaries of class C^∞ which converges to D. Let V_m denote an infinitely differentiable solution of the Dirichlet problem for equation (1.3) in D_m. The existence of such solutions guarantees the following estimate, implied by Lemma 7:

$$\|\nabla V_m\|_2 \leqslant C\left(\|f\|_{\frac{2n}{n+2}}+\|\varphi\|_2\right), \tag{3.2}$$

where C depends on $\|\mathbf{a}\|_2$, $\|a^-\|_{n/2}$, κ, v_D, n and r. (The functions V_m are assumed to be extended by zero outside the D_m.) From $\{V_m\}$ we select a subsequence $\{V_{m_k}\}$ that is weakly convergent in $\overset{\circ}{L}{}_2^{(1)}$. Let w denote the weak limit of this subsequence. Clearly,

$$\mathfrak{P}_D(w,v)=\lim_{m\to\infty}\mathfrak{P}_{D_m}(w_m,v)=\lim_{m\to\infty}\int_{D_m}(fv-\varphi\nabla v)\,dx=\int_D(fv-\varphi\nabla v)\,dx,$$

where v is any function in $\mathscr{K}(D)$. Estimate (2.10) holds for w by virtue of (3.2).

We now assume that the coefficients A, $\mathbf{a}$ and a of the operator P satisfy the conditions of the theorem while the functions f and $\boldsymbol{\phi}$ belong to $\mathscr{K}(D)$. Approximating A, $\mathbf{a}$ and a by mean functions, we construct a sequence of Dirichlet problems on D for equations with infinitely differentiable coefficients. Since estimate (3.2) with a constant C not depending on ν holds for the solutions w_ν of these problems, it is possible to select a subsequence from the sequence $\{w_\nu\}$ that weakly converges in $\overset{\circ}{L}{}_2^{(1)}$ to a solution $W \in \overset{\circ}{L}{}_2^{(1)}$ of the equation $PW = f + \operatorname{div} \boldsymbol{\phi}$.

We get rid of the assumption $f, \boldsymbol{\phi} \in \mathscr{K}$. Suppose $f \in L_{2n/(n+2)}$ and $\boldsymbol{\phi} \in L_2$, and suppose $\{f_l\}$ and $\{\boldsymbol{\phi}_l\}$ are sequences of functions in $\mathscr{K}$ which converge to f in $L_{2n/(n+2)}$ and to $\boldsymbol{\phi}$ in L_2 respectively. As was just shown, for each l there exists a solution $U_l \in \overset{\circ}{L}{}_2^{(1)}$ of the equation $PU_l = f_l + \operatorname{div} \boldsymbol{\phi}_l$, and estimate (3.2) holds for it. From $\{U_l\}$ we select a weakly convergent (in $\overset{\circ}{L}{}_2^{(1)}$) subsequence, for which we retain the notation $\{U_l\}$. The limit of $\{U_l\}$ is denoted by U. Passing to the limit in the equality

$$\mathfrak{P}_D(U_l, v) = \int_D (f_l v - \varphi_l \nabla v)\, dx,$$

we get that

$$\mathfrak{P}(U, v) = \int_D (fv - \varphi \nabla v)\, dx,$$

i.e. that U is the desired solution.

Uniqueness. We define in $\overset{\circ}{L}{}_2^{(1)}$ the scalar product

$$[u, v] = \mathfrak{A}(u, v)$$

and introduce operators T_1 and T_2 in $\overset{\circ}{L}{}_2^{(1)}$ by means of the equalities

$$[T_1 u, v] = \mathfrak{B}(u, v), \quad [T_2 u, v] = \mathfrak{C}(u, v).$$

By virtue of the conditions $\mathbf{a} \in L_r$ $(r > n)$ and $a \in L_{n/2}$ these operators are completely continuous in $\overset{\circ}{L}{}_2^{(1)}$. Equality (1.4) is equivalent to the equation

$$u + (T_1 + T_2)u = \Psi, \tag{3.3}$$

where $\Psi \in \overset{\circ}{L}{}_2^{(1)}$. Therefore the just-proved existence of the solution of (1.4) (and hence also of (3.3)) for any right side implies the uniqueness of this solution.

LEMMA 9. *Suppose all of the requirements of Theorem* 1 *with respect to* $\mathbf{a}$, a *and* f *are satisfied, and suppose* $\boldsymbol{\phi} = 0$ *while the function* a^- *satisfies the more stringent* (*than* (3.1)) *condition* (2.14).

Then estimate (2.15) *holds for the solution* $U \in \overset{\circ}{L}{}_2^{(1)}$ *of equation* (1.1) (*the existence and uniqueness of this solution are proved in Theorem* 1).

PROOF. We will make use of the notation introduced in the proof of Theorem 1.

From the subsequence $\{V_{m_k}\}$ we select a subsequence, for which we retain the notation $\{V_{m_k}\}$, such that the arithmetic means

$$\sigma_k = \frac{1}{k}(V_{m_1} + V_{m_2} + \ldots + V_{m_k}\}$$

strongly converge in $\overset{\circ}{L}{}_2^{(1)}$. This is always possible by virtue of the Banach-Saks theorem. Since the sequence $\{\sigma_k\}$ strongly converges in $\overset{\circ}{L}{}_q^{(1)}$, by using estimate (2.5), which is valid for each of the functions V_{m_j}, we obtain

$$\|\nabla w\|_q \leqslant \overline{\lim_{k\to\infty}} \frac{1}{k} \sum_{j=1}^{k} \|\nabla V_{m_j}\|_q \leqslant \frac{cc_1}{c\varkappa K - \|a^-\|_{n/2}} \|f\|_p.$$

Applying an analogous argument first to the sequence $\{w_\nu\}$ and then to $\{U_l\}$, we obtain estimate (2.15) for the function U. The lemma is proved.

We now prove a theorem on the existence and uniqueness of a solution in $\overset{\circ}{L}{}_q^{(1)}$.

THEOREM 2. *Suppose D is an open set of finite measure and suppose the elements of the matrix A are bounded measurable functions, $\mathbf{a} \in L_r$ $(r > n)$, $a \in L_{n/2}$, $f \in L_p$ $(1 < p < 2n/(n+2))$. Suppose further*

$$\|a^-\|_{n/2} < c\varkappa \exp\left(-\frac{k}{\varkappa} v_D^{\frac{r-n}{rn}} \|\mathbf{a}\|_r\right), \tag{3.4}$$

where κ is the constant of ellipticity while c and k are "dimensionless" constants defined by the equalities

$$c = \left(1 - \frac{2p}{n}\right)\frac{\alpha_n^2}{pp'}, \qquad k = \left(n\,\frac{r-1}{r-n}\right)^{\frac{r-1}{r}} \alpha_n^{-1}.$$

Then there exists one and only one solution $U \in \overset{\circ}{L}{}_q^{(1)}$ $(q = pn/(n-p))$ of equation (1.1), *and for this solution*

$$\|\nabla U\|_q \leqslant \frac{cc_1}{c\varkappa K - \|a^-\|_{n/2}} \|f\|_p, \tag{3.5}$$

where

$$c_1 = p'\left(\frac{n-p}{n}\right)^{\frac{1}{q}} \alpha_n^{-1}.$$

PROOF. Let $\{f_m\}$ be any sequence of functions in $L_{2n/(n+2)}$ which converge to f in L_p. By virtue of Lemma 9 a corresponding sequence of solutions $\{U_m\}$ in $\overset{\circ}{L}{}_2^{(1)}$ converge in $\overset{\circ}{L}{}_q^{(1)}$ to a function U. The latter is the desired solution. The uniqueness of this solution also follows from Lemma 9. The theorem is proved.

CHAPTER II

SOLVABILITY IN $L_q^{(1)}$ ($q \leqq 2$) OF THE NEUMANN PROBLEM IN A DOMAIN WITH NONSMOOTH BOUNDARY

§4. Estimates for the first derivatives of a solution of the Neumann problem

Let D denote a domain in R^n ($n \geqq 2$) of finite n-dimensional measure, and let $A(x)$ denote a matrix whose elements are bounded and measurable functions. Suppose, as before, $A(x)$ satisfies condition (0.1). Outside D we put $A(x) = \kappa I$, where I is the identity matrix.

Let $L_q^{(1)}$ ($q > 0$) be the completion of the set of functions in $C^{(\infty)}(D)$ whose first derivatives belong to L_q, in the metric of the integral

$$\int |\nabla u|^q dx.$$

As in the first chapter, an integral without indicated limits means an integral over D. Let $L_q^{(1)}/Z$ denote the factor space of $L_q^{(1)}$ modulo the subspace Z of constants. For $q \geqq 1$ the space $L_q^{(1)}/Z$ is a Banach space with norm $\|\nabla u\|_q$.

We also introduce a space of which $L_q^{(1)}$ is a special case for $q < 1$. Let $M(x)$ be a convex upwards function on $[0, \infty)$ such that $M(0) = 0$ and $M(x) \to \infty$ for $x \to \infty$. Let $L_M^{(1)}(D)$ denote the completion of the set of functions in $C^{(\infty)}(D)$ such that

$$\int M(|\nabla u|)\, dx < \infty$$

in the metric of this integral. The factor space of $L_M^{(1)}$ modulo Z (denoted by $L_M^{(1)}/Z$) is a metric space. An element $\{u + c\}$ of either $L_q^{(1)}/Z$ or $L_M^{(1)}/Z$ is denoted by u .

We confine ourselves to a consideration of the particularly simple operator

$$Q = -\nabla (A(x) \nabla).$$

Let

$$\mathfrak{Q}(u, v) = \int_D \nabla u A \nabla v dx, \tag{4.1}$$

where $u, v \in L_2^{(1)}$. Also, let f be a summable function that is orthogonal to the unit function in D.

An element $u^{\cdot} \in L_2^{(1)}/Z$ is called a *generalized solution in* $L_2^{(1)}/Z$ *of the Neumann problem* if

$$\mathfrak{Q}(u, v) = \int_D f v dx \tag{4.2}$$

for all functions $v \in L_2^{(1)} \cap L_\infty$ (the set of such functions is dense in $L_2^{(1)}$). By a solution in $L_2^{(1)}$ we will mean any function in the class $u^{\cdot}$.

If $f \in \mathscr{K}$ then the Neumann problem is uniquely solvable in $L_2^{(1)}/Z$ for any domain D. In fact,

$$\left|\int_D fvdx\right| \leqslant \|f\|_2 \inf_{\gamma\in Z} \|v-\gamma\|_{L_2(G)}, \tag{4.3}$$

where G is any domain that has a smooth boundary, is situated in D together with its closure and contains the support of f. Therefore, by virtue of Poincaré's inequality for the domain G, the functional

$$\int_D fvdx$$

is bounded in $L_2^{(1)}/Z$, which implies the unique solvability of the Neumann problem in $L_2^{(1)}/Z$.

From (4.2) and (4.3) there directly follows the estimate

$$\|\nabla u\|_2 \leqslant K(f), \tag{4.4}$$

where $K(f)$ is a constant depending on f and the constant of ellipticity κ. We note further that the function $u(x)$ can be approximated in $L_2^{(1)}$ by a sequence of solutions $u_m \in L_2^{(1)} \cap C^{(\infty)}(D)$ of the Neumann problems for the equations $\nabla(A_m \nabla u_m) = f$, where $\{A_m\}$ is a sequence of averages of the matrix A. In fact, the identity

$$\int_D \nabla(u_m - u) A_m \nabla v dx = \int_D \nabla u (A - A_m) \nabla v dx$$

implies the estimate

$$\varkappa \|\nabla(u_m - u)\|_2 \leqslant \|(A - A_m)\nabla u\|_2,$$

which together with estimate (4.4) and the convergence in n-dimensional measure of the A_m to A leads to the required result.

REMARK 3. What has just been stated permits one to make the following deduction, which will be useful below. If an estimate

$$\|\nabla u\|_q \leqslant C\|f\|_p, \tag{4.5}$$

where $q \in (0, 2]$, $f \in \mathscr{K}(D)$ and C is a constant depending only on D, κ, p, q and n, has been proved under the assumption of infinite differentiability in D of a solution $u \in L_2^{(1)}$ of problem (4.2), then it also holds without this assumption.

Suppose the function f is summable and orthogonal to the unit function in D. We will say that an element $u^{\cdot} \in L_q^{(1)}/Z$ $(0 < q < 2)$ is a *generalized solution in* $L_q^{(1)}/Z$ *of the Neumann problem for the equation* $Qu = f$ if the following condition is satisfied. For some sequence f_m, $f_m \in \mathscr{K}$, $f_m \perp 1$, which converges to f in L, a sequence of solutions in $L_2^{(1)}/Z$ of the Neumann problems for the equations $Qu_m = f_m$ converges in $L_q^{(1)}/Z$ to u . By a solution in $L_q^{(1)}$ is meant any function in the class u . A solution in $L_M^{(1)}/Z$ or in $L_M^{(1)}$ is defines analogously.

(We note that when $1 \leqq q < 2$ the definition of a solution $u \in L_q^{(1)}/Z$ implies the validity of the integral identity (4.2) for all $v \in L_{q'}^{(1)} \cap L_\infty$. As an example of J. Serrin shows [**14**], the validity of the integral identity for a function $u \in L_q^{(1)}$ when $q < n/(n-1)$ does not imply that u is a generalized solution in $L_q^{(1)}$ of the Neumann problem in the sense defined above.)

Let

$$P_\tau = \{x : u(x) \geqslant \tau\}, \quad F_\tau = \{x : u(x) = \tau\}.$$

Below, without loss of generality, we will prove estimates for the solutions of the Neumann problem for that function u in a class $u^\cdot \in L_q^{(1)}/Z$ or a class $u^\cdot \in L_M^{(1)}/Z$ which satisfies the condition

$$\inf\left\{t : v(P_t) \leqslant \frac{1}{2} v(D)\right\} = 0. \tag{4.6}$$

When speaking of a solution in $L_q^{(1)}$ or in $L_M^{(1)}$, we will always assume that equality (4.6) holds.

Let $\lambda(t)$ denote

$$\inf s(D \cap \partial E)$$

on the set of all $E \subset D$ such that $D \cap \partial E$ is a locally Lipschitz manifold and $t \leqq v_E \leqq v_D/2$.

LEMMA 10. *Suppose $f \in \mathscr{K}(D)$, $f \perp 1$ and u is a solution in $C^{(\infty)}(D) \cap L_2^{(1)}(D)$ of the Neumann problem for the equation $Qu = f$. Then*

$$\int M(|\nabla u|)\,dx \leqslant 2 \int_0^{\frac{1}{2}v_D} M\left[\frac{\varkappa^{-1}}{\lambda(\mu)} \int_0^\mu f^*(\tau)\,d\tau\right] d\mu, \tag{4.7}$$

where $f^*(\tau)$ is a nonincreasing function that is equi-measurable with $|f(x)|$ while the nonnegative function $M(t)$ tends to infinity for $t \to \infty$ and is either convex upwards or such that the function $t^{-1}M(t)$ is convex upwards on the semiaxis $t \geqq 0$.

PROOF. Suppose $t > \tau \geqq 0$ and let $v(x) = t - \tau$ if $x \in P_t$, and $v(x) = (u(x) - \tau)^+$ if $x \notin P_t$. Substituting v into equality (4.2), we get

$$\int_{P_\tau \setminus P_t} \nabla u A \nabla u dx = \int_{P_\tau} f v dx.$$

We divide both sides of this inequality by $t - \tau$ and pass to the limit for $t \to \tau$. Then, by Lemma 1, for almost all $\tau \geqq 0$

$$\varkappa \int_{F_\tau} |\nabla u|\,ds = \int_{P_\tau} f dx. \tag{4.8}$$

(An analogous inequality is obtained for almost all $\tau \leqq 0$.) By Lemma 2

$$-v'(P_\tau) \int_{F_\tau} |\nabla u|\,ds \geqslant s^2(F_\tau). \tag{4.9}$$

From this result and (4.8) we get

$$\varkappa s^2(F_\tau) \leqslant -v'(P_\tau)\Big|\int_{P_\tau} f\,dx\Big|. \tag{4.10}$$

It follows from Lemma 1 that

$$J_+ = \int_{P_0} M(|\nabla u|)\,dx = \int_0^\infty dt \int_{F_t} M(|\nabla u|)\frac{ds}{|\nabla u|}. \tag{4.11}$$

If $M(t)$ is a convex upwards function, then by virtue of Jensen's inequality

$$\Big(\int_{F_t}\frac{ds}{|\nabla u|}\Big)^{-1}\int_{F_t} M(|\nabla u|)\frac{ds}{|\nabla u|} \leqslant M\Big[\Big(\int_{F_t}\frac{ds}{|\nabla u|}\Big)^{-1} s(F_t)\Big],$$

which on the basis of Lemma 1 is equivalent to the inequality

$$\int_{F_t} M(|\nabla u|)\frac{ds}{|\nabla u|} \leqslant -v'(P_t)\,M\Big[-\frac{s(F_t)}{v'(P_t)}\Big].$$

Applying (4.10), we get from this result and (4.11) that

$$J_+ \leqslant -\int_0^\infty M\Big[\frac{1}{\varkappa s(F_t)}\Big|\int_{P_t} f\,dx\Big|\Big]\,v'(P_t)\,dt. \tag{4.12}$$

Suppose now $t^{-1}M(t)$ is a convex upwards function. By virtue of Jensen's inequality

$$\frac{1}{s(F_t)}\int_{F_t}\frac{M(|\nabla u|)}{|\nabla u|}\,ds \leqslant \Big[\frac{1}{s(F_t)}\int_{F_t}|\nabla u|\,ds\Big]^{-1} M\Big[\frac{1}{s(F_t)}\int_{F_t}|\nabla u|\,ds\Big],$$

which together with (4.9) implies

$$\int_{F_t}\frac{M(|\nabla u|)}{|\nabla u|}\,ds \leqslant -v'(P_t)\,M\Big[\frac{1}{s(F_t)}\int_{F_t}|\nabla u|\,ds\Big].$$

Taking equality (4.8) into consideration, we again arrive at (4.12).

By noting that the sets F_τ are locally infinitely differentiable manifolds in D for almost all $\tau \geqq 0$ and recalling the definition of $\lambda(t)$, we get

$$s(E_t) \geqslant \lambda[v(P_t)]. \tag{4.13}$$

Let $f_+^*(\tau)$ denote a nondecreasing function that is equi-measurable on P_0 with $|f|$. It now follows from (4.12) and (4.13) that

$$J_+ \leqslant -\int_0^\infty M\Big[\frac{1}{\varkappa\lambda[v(P_t)]}\int_0^{v(P_t)} f_+^*(\xi)\,d\xi\Big]\,v'(P_t)\,dt.$$

Thus

$$J_+ \leqslant \int_0^{v(P_0)} M\Big[\frac{1}{\varkappa\lambda(v)}\int_0^{v} f_+^*(\xi)\,d\xi\Big]\,dv. \tag{4.14}$$

The estimate

$$J_- \equiv \int\limits_{D\setminus P_0} M(|\nabla u|)dx \leqslant \int\limits_0^{v(D\setminus P_0)} M\left[\frac{1}{\varkappa\lambda(v)}\int\limits_0^v f_-^*(\xi)\,d\xi\right]dv,$$

where $f_-^*(\xi)$ is a nonincreasing function that is equi-measurable on $D\setminus P_0$ with $|f|$, is proved analogously. Noting that $f_\pm^*(\xi) \leqq f^*(\xi)$, we obtain the assertion of the lemma.

REMARK 4. Suppose the conditions of Lemma 10 are satisfied. Then the following estimates can be readily obtained from equality (4.8), Lemma 2 and the definition of $\lambda(t)$:

$$\int M(u_+)\,dx \leqslant \int\limits_0^{v_{P_0}} M\left(\varkappa^{-1}\int\limits_\mu^{v_{P_0}} \frac{dt}{\lambda^2(t)}\int\limits_0^t f_+^*(\tau)\,d\tau\right)d\mu, \tag{4.15}$$

$$\int M(u_-)\,dx \leqslant \int\limits_0^{v_{D\setminus P_0}} M\left(\varkappa^{-1}\int\limits_\mu^{v_{D\setminus P_0}} \frac{dt}{\lambda^2(t)}\int\limits_0^t f_-^*(\tau)\,d\tau\right)d\mu, \tag{4.16}$$

where $M(t)$ is a nondecreasing function on the semiaxis $t>0$. From this result we obtain the inequality

$$\inf_{\gamma\in Z}\int M(|u-\gamma|)\,dx \leqslant 2\int\limits_0^{{}^1\!/_2 v_D} M\left(\varkappa^{-1}\int\limits_\mu^{{}^1\!/_2 v_D} \frac{dt}{\lambda^2(t)}\int\limits_0^t f^*(\tau)\,d\tau\right)d\mu. \tag{4.17}$$

LEMMA 11. *Suppose $p>1$ and*

$$\varliminf_{t\to 0} t^{-\alpha}\lambda(t) > 0, \tag{4.18}$$

where $1 \geqq \alpha \geqq (n-1)/n$ for $n>2$ and $1 \geqq \alpha > 1/2$ for $n=2$.[2] *Suppose further $f\in \mathscr{K}(D)$, $f \perp 1$ and u is a solution in $C^{(\infty)}(D)\cap L_2^{(1)}(D)$ of the Neumann problem for the equation $Qu=f$. Then the following estimate holds for the function u:*

$$\|\nabla u\|_q \leqslant p'\,(p/q)^{1/q}\varkappa^{-1}\mathfrak{A}\,\|f\|_p, \tag{4.19}$$

where $1<q\leqq 2$, $q=p[1-p(1-\alpha)]^{-1}$ and

$$\mathfrak{A} = \sup\left\{\frac{t^\alpha}{\lambda(t)} : 0<t<\frac{1}{2}v_D\right\}.$$

PROOF. In this as well as in the following two lemmas we will make use of the notation introduced in the proof of Lemma 10. From (4.14) and (4.18) it follows that

---

[2] Clearly, inequality (4.18) is impossible for $\alpha < (n-1)/n$. The case $n=2$, $\alpha = 1/2$ is not considered only in order to avoid complicating the discussion.

$$J_+ \leqslant \varkappa^{-q} \int_0^{v(P_0)} \left(\frac{1}{\lambda(\mu)} \int_0^v f_+^*(\xi)\, d\xi \right)^q dv \leqslant (\mathfrak{A}\varkappa^{-1})^q \int_0^{v(P_0)} \left(\int_0^v f_+^*(\xi)\, d\xi \right)^q v^{-\alpha q} dv.$$

Hardy's inequality can be applied to the latter integral:

$$J_+ \leqslant \left(\frac{q}{\alpha q - 1} \right)^q (\mathfrak{A}\varkappa^{-1})^q \int_0^{v(P_0)} [t^{1-\alpha} f_+^*(t)]^q\, dt.$$

Further, using the monotonicity of f_+^*, we get

$$J_+ \leqslant p'^q (\mathfrak{A}\varkappa^{-1})^q \int_0^{v(P_0)} \left(\int_0^t [f_+^*(\tau)]^p\, d\tau \right)^{\frac{q-p}{p}} [f_+^*(t)]^p\, dt$$

$$= p'^q p q^{-1} (\mathfrak{A}\varkappa^{-1})^q \left(\int_0^{v(P_0)} [f^*(t)]^p\, dt \right)^{q/p}.$$

Consequently

$$J_+ \leqslant p'^q p q^{-1} (\mathfrak{A}\varkappa^{-1})^q \left(\int_{P_0} |f|^p\, dx \right)^{q/p}.$$

The estimate

$$J_- \leqslant p'^q p q^{-1} (\mathfrak{A}\varkappa^{-1})^q \left(\int_{D \setminus P_0} |f|^p\, dx \right)^{q/p}$$

is proved analogously. Using the inequalities $a^\gamma + b^\gamma \leqq (a+b)^\gamma$, where $a, b > 0$ and $\gamma \geqq 1$, we obtain (4.19).

The lemma just proved characterizes the domains in which higher powers of the first derivatives of the solution than those of the free term are summable. The opposite case is considered in the following lemma.

LEMMA 12. *Suppose* $\infty \geqq p > 1$ *and*

$$\int_0 \left[\frac{t^\alpha}{\lambda(t)} \right]^{\frac{1}{\alpha - 1}} \frac{dt}{t} < \infty, \tag{4.20}$$

where $2 - 1/p \geqq \alpha > 1$. *Suppose further* $f \in \mathscr{K}(D)$, $f \perp 1$ *and* u *is a solution in* $L_2^{(1)} \cap C^{(\infty)}(D)$ *of the Neumann problem for the equation* $Qu = f$. *Then the following estimate holds for the function* u:

$$\| \nabla u \|_q \leqslant 2^{\alpha-1} p' \varkappa^{-1} \mathfrak{B} \| f \|_p, \tag{4.21}$$

where $q = p[1 + p(\alpha - 1)]^{-1}$ *and*

$$\mathfrak{B} = \left\{ \int_0^{1/2 v_D} \left[\frac{t}{\lambda(t)} \right]^{\frac{1}{\alpha - 1}} dt \right\}^{\alpha - 1}.$$

PROOF. We first note that inequality (1.5) implies the estimate

$$\mathfrak{B}_1 \equiv \left\{ \int\limits_0^{^1/_2 v_D} \left(\mu^{q-1} \int\limits_{\mu}^{^1/_2 v_D} \frac{d\tau}{\lambda^q(\tau)} \right)^{\frac{p}{p-q}} d\mu \right\}^{\frac{p-q}{pq}}$$

$$\leqslant \left(\frac{p'}{q} \right)^{\frac{1}{q}} \left\{ \int\limits_0^{^1/_2 v_D} \left[\frac{\mu}{\lambda(\mu)} \right]^{\frac{pq}{p-q}} d\mu \right\}^{\frac{p-q}{pq}} = \left(\frac{p'}{q} \right)^{\frac{1}{q}} \mathfrak{B}. \tag{4.22}$$

Putting $M(\tau) = \tau^q$ in (4.14), we get

$$J_+ \leqslant q\varkappa^{-q} \int\limits_0^{v(P_0)} \int\limits_{\mu}^{v(P_0)} \frac{d\tau}{\lambda^q(\tau)} \left(\int\limits_0^{\mu} f_+^*(\tau)\, d\tau \right)^{q-1} f_+^*(\mu)\, d\mu .$$

Hence, applying Hölder's inequality, we deduce

$$J_+ \leqslant q\varkappa^{-q} \left\{ \int\limits_0^{v(P_0)} [f_+^*(\mu)]^p\, d\mu \right\}^{\frac{1}{p}} \left\{ \int\limits_0^{v(P_0)} \left[\int\limits_0^{\mu} f_+^*(t)\, dt \right]^p \frac{d\mu}{\mu^p} \right\}^{\frac{q-1}{p}} \mathfrak{B}_1^q .$$

Inequalities (1.5) and (4.22) now imply

$$J_+ \leqslant q\,(p')^{q-1} \mathfrak{B}_1^q \varkappa^{-q} \left(\int\limits_0^{v(P_0)} [f_+^*(\mu)]^p\, d\mu \right)^{\frac{q}{p}} \leqslant p'^q \mathfrak{B}^q \varkappa^{-q} \left(\int\limits_{P_0} |f|^p dx \right)^{q/p} .$$

Analogously,

$$J_- \leqslant (p' \mathfrak{B} \varkappa^{-1})^q \left(\int\limits_{D \setminus P_0} |f|^p dx \right)^{q/p} .$$

Combining the latter inequalities and noting that the estimate $a^\gamma + b^\gamma \leqq 2^{1-\gamma}(a+b)^\gamma$ holds for $a, b > 0$ and $\gamma < 1$, we obtain (4.21).

LEMMA 13. *Suppose* $\infty \geqq p \geqq 1$ *and*

$$\int\limits_0 \left[\frac{t^\alpha}{\lambda(t)} \right]^q \frac{dt}{t} < \infty, \tag{4.23}$$

where $q = p[1 + p(\alpha - 1)]^{-1}$, $\alpha > 2 - p^{-1}$ *for* $p > 1$ *and* $\alpha = 1$ *for* $p = 1$. *Suppose further* $f \in \mathscr{K}(D)$, $f \perp 1$ *and* u *is a solution in* $L_2^{(1)} \cap C^{(\infty)}(D)$ *of the Neumann problem for the equation* $Qu = f$. *Then the following estimate holds for the function* u:

$$\| \nabla u \|_q \leqslant 2^{\alpha-1} \varkappa^{-1} \mathfrak{C} \| f \|_p, \tag{4.24}$$

where $q = p[1 + p(\alpha - 1)]^{-1}$ *and*

$$\mathfrak{C} = \left\{ \int\limits_0^{^1/_2 v_D} \left[\frac{t^{p-1}}{\lambda^p(t)} \right]^{\frac{1}{1+p(\alpha-1)}} dt \right\}^{\frac{1}{p} + \alpha - 1} .$$

PROOF. Putting $M(\tau) = \tau^q$ and applying Hölder's inequality, we get

$$J_+ \leqslant \varkappa^{-q} \int_0^{v(P_0)} \frac{v^{q/p'}}{\lambda^q(v)} \left(\int_0^v [f_+^*(t)]^p \, dt \right)^{q/p} dv$$

$$\leqslant \varkappa^{-q} \left\{ \int_0^{v(P_0)} [f_+^*(t)]^p \, dt \right\}^{q/p} \int_0^{v(P_0)} \frac{v^{q/p'}}{\lambda^q(v)} dv \leqslant (\varkappa^{-1}\mathfrak{C})^q \left(\int_{P_0} |f|^p dx \right)^{q/p}.$$

Analogously,

$$J_- \leqslant (\varkappa^{-1}\mathfrak{C})^q \left(\int_{D \setminus P_0} |f|^p dx \right)^{q/p}.$$

Consequently estimate (4.24) is valid.

§5. Theorems on the unique solvability of the Neumann problem. Estimates in L_∞ and L_q

In the following theorem we prove the existence and uniqueness of a solution in $L_M^{(1)}/Z$ of the Neumann problem for an arbitrary domain D.

THEOREM 3. *Suppose D is an arbitrary domain of finite measure in R^n, and suppose the elements of the matrix A are bounded measurable functions while f is a given function in L that is orthogonal to the unit function in D.*

Suppose further M is a function satisfying one of the conditions formulated on page 152 and such that[3]

$$\int_0 M\left(\frac{1}{\lambda(\mu)} \right) d\mu < \infty. \tag{5.1}$$

Then in $L_M^{(1)}/Z$ there exists one and only one solution of the Neumann problem for the equation $Qu = f$. The following estimate holds for this solution:

$$\int_D M\left(\varkappa \frac{|\nabla u|}{\|f\|_1} \right) dx \leqslant 2 \int_0^{^1/_2 v_D} M\left(\frac{1}{\lambda(\mu)} \right) d\mu. \tag{5.2}$$

PROOF. We first show that $\lambda(t) > 0$ for $t > 0$. Let D' denote a strictly interior subdomain of D with an infinitely differentiable boundary for which

$$v(D \setminus D') < \frac{1}{2} t.$$

Let E be any subset of D with a locally Lipschitz boundary such that

$$\frac{1}{2} v_D \geqslant v_E \geqslant t > 0.$$

It is obvious that there exists a constant $k > 0$ not depending on E such that

[3] Such functions $M(t)$ exist.

$$s(D' \cap \partial E) \geqslant k[v(D' \cap E)]^{\frac{n-1}{n}} > k2^{\frac{1-n}{n}} t^{\frac{n-1}{n}}$$

It is easily seen that, for any nonincreasing function $\zeta(\mu)$ such that $\zeta(\mu) \to \infty$ for $\mu \to +0$, a function M can be found which satisfies one of the conditions formulated on page 152 as well as the condition

$$\int_0 M[\zeta(\mu)]\, d\mu < \infty.$$

Here we put $\zeta(\mu) = [\lambda(\mu)]^{-1}$.

It can be assumed without loss of generality that $\|f\|_1 < \kappa$. Let f_m be an arbitrary sequence of functions in K which converges to $f \not\equiv 0$ in L and let $\{u_m\}$ be a corresponding sequence of solutions of Neumann problems in $L_2^{(1)}$.

From Lemma 10 it follows that

$$\int M(|\nabla u_m|)\, dx \leqslant 2 \int_0^{1/2 v D} M\left(\frac{\varkappa^{-1}}{\lambda(\mu)} \|f_m\|_1\right) d\mu. \tag{5.3}$$

The fact that $\{f_m\}$ is a Cauchy sequence in L implies that $\{u_m\}$ is a Cauchy sequence in $L_M^{(1)}/Z$ and hence converges to an element $u \in L_M^{(1)}/Z$. This element is the desired solution.

We prove the uniqueness of a solution in $L_M^{(1)}/Z$. Let $u^{(i)}$ $(i = 1, 2)$ be two solutions in $L_M^{(1)}/Z$ of the Neumann problem for the equation $Qu = f$, and let $u_m^{(i)}$ be a sequence of solutions in $L_2^{(1)}/Z$ of the Neumann problems for the equations $Qv = f_m^{(i)}$, where $f_m^{(i)} \to f$ in L, which converges to $u^{(i)}$. Lemma 10 implies the estimate

$$\int M[|\nabla(u_m^{(1)} - u_m^{(2)})|]\, dx \leqslant 2 \int_0^{1/2 v D} M\left(\frac{\varkappa^{-1}}{\lambda(\mu)} \|f_m^{(1)} - f_m^{(2)}\|_1\right) d\mu.$$

Consequently $u^{(1)} = u^{(2)}$.

From (5.3) we obtain the estimate

$$\int M(|\nabla u|)\, dx \leqslant 2 \int_0^{1/2 v D} M\left(\frac{\varkappa^{-1}}{\lambda(\mu)} \|f\|_1\right) d\mu,$$

first under the condition $\kappa^{-1}\|f\|_1 < 1$ and then, by a limit passage, in the case $\kappa^{-1}\|f\|_1 = 1$. A renormalization of the solution now gives inequality (5.2).

We prove a theorem on the unique solvability of the Neumann problem in $L_q^{(1)}/Z$.

Theorem 4. *Suppose D is a domain of finite measure, and suppose the elements of the matrix A are bounded measurable functions while f is a given function in $L_p(D)$ $(\infty \geqq p \geqq 1)$ that is orthogonal to the unit function in D. Suppose further that the domain D satisfies one of the following conditions:*

$$\lim_{t \to 0} t^{-\alpha} \lambda(t) > 0, \tag{5.4}$$

if $1 \geqq \alpha \geqq (n-1)/n$ *for* $n > 2$, $1 \geqq \alpha > 1/2$ *for* $n = 2$ *and* $2(3-2\alpha)^{-1} \geqq p > 1$;

$$\int_0 \left[\frac{t}{\lambda(t)} \right]^{\frac{1}{\alpha-1}} dt < \infty, \tag{5.5}$$

if $1 + 1/p' \geqq \alpha > 1$, *where* $\infty \geqq p > 1$;

$$\int \left[\frac{t^{p-1}}{\lambda^p(t)} \right]^{\frac{1}{1+p(\alpha-1)}} dt < \infty, \tag{5.6}$$

if $\alpha > 1 + 1/p'$ *for* $\infty \geqq p > 1$ *and* $\alpha = 1$ *for* $p = 1$.

Then there exists one and only one generalized solution in $L_q^{(1)}/Z$ *of the Neumann problem for the equation* $Qu = f$, *where* $q = p[1 - p(1-\alpha)]^{-1}$. *The following estimate holds for this solution*:

$$\| \nabla u \|_q \leqslant C_i \| f \|_p, \tag{5.7}$$

where $i = 1, 2, 3$ *and the* C_i *are the constants in inequalities* (4.19), (4.21) *and* (4.24) *respectively*.

Proof. Suppose $\{f_m\}$ is an arbitrary sequence of functions in $\mathscr{K}$ that are orthogonal to the unit function in D, which converges to f in L_p, and suppose $\{u_m\}$ is a corresponding sequence of solutions of Neumann problems in $L_2^{(1)}/Z$. Recalling Remark 3 on page 151, we get from estimates (4.19), (4.21) and (4.24) that $\{u_m^{\cdot}\}$ is a Cauchy sequence in $L_q^{(1)}/Z$ and hence converges to an element $u^{\cdot} \in L_q^{(1)}/Z$, which is the desired solution.

The uniqueness of a solution in $L_q^{(1)}/Z$ follows from Theorem 3.

Let

$$\operatorname{osc} u = \sup_D \operatorname{vrai} u - \inf_D \operatorname{vrai} u.$$

In the following two theorems we prove estimates for a solution of the Neumann problem in L_∞ and L_q.

Theorem 5. *Suppose the conditions of Theorem* 4 *are satisfied with respect to A and f, and suppose*

$$\int_0 \left[\frac{\mu^{1-\frac{1}{2p}}}{\lambda(\mu)} \right]^{2p'} \frac{d\mu}{\mu} < \infty. \tag{5.8}$$

Then there exists one and only one solution in $L_2^{(1)}/Z$ *of the Neumann problem for the equation* $Qu = f$, *and the following estimate holds for this solution*:

$$\operatorname{osc}_D u \leqslant \varkappa p' \left(2 \int_0^{1/2 v_D} \left[\frac{\mu}{\lambda^2(\mu)} \right]^{p'} d\mu \right)^{\frac{1}{p'}} \| f \|_p. \tag{5.9}$$

Proof. We first note that (5.8) implies the relation

$$\lim_{\mu \to 0} \mu^{\frac{1}{2p}-1} \lambda(\mu) = \infty,$$

which by virtue of Theorem 4 guarantees the existence and uniqueness of a solution in $L_2^{(1)}/Z$.

It suffices to prove estimate (5.9) under the assumption that A is infinitely differentiable in D and f belongs to $\mathcal{K}(D)$. By virtue of equality (4.8), for $\tau \geqq 0$

$$\varkappa \int_{F_\tau} |\nabla u|\, ds \leqslant \int_{P_\tau} |f|\, dx.$$

Consequently

$$\tau \leqslant \varkappa^{-1} \int_0^\tau \int_{P_c} |f|\, dx \frac{dc}{\int_{F_c} |\nabla u|\, ds}.$$

Let $f_+^*(\mu)$ denote a nonincreasing function that is equi-measurable on P_0 with $|f|$. Then, using Lemma 2, we get

$$\tau \leqslant -\varkappa^{-1} \int_0^\tau \int_0^{v(P_c)} f_+^*(\xi)\, d\xi \frac{v'(P_c)}{s^2(F_c)}\, dc.$$

We now note that $s(F_c) \geqq \lambda[v(P_c)]$. Therefore

$$\tau \leqslant \int_0^{v(P_\tau)} \int_0^\mu f_+^*(\xi)\, d\xi \frac{d\mu}{\lambda^2(\mu)} = \varkappa \int_0^{v(P_\tau)} f_+^*(\mu) \int_\mu^{v(P_\tau)} \frac{d\xi}{\lambda^2(\xi)}\, d\mu. \tag{5.10}$$

Hence

$$\sup \operatorname{vrai} u^+ \leqslant \varkappa \left(\int_0^{1/2 v_D} [f_+^*(\mu)]^p\, d\mu \right)^{\frac{1}{p}} \left(\int_0^{1/2 v_D} \left[\int_\mu^{1/2 v_D} \frac{d\xi}{\lambda^2(\xi)} \right]^{p'} d\mu \right)^{\frac{1}{p'}}.$$

Applying Hardy's inequality (1.5), we obtain

$$\sup \operatorname{vrai} u^+ \leqslant \varkappa p' \|f\|_{L_p(P_0)} \left(\int_0^{1/2 v_D} \left[\frac{\mu}{\lambda^2(\mu)} \right]^{p'} d\mu \right)^{\frac{1}{p'}}. \tag{5.11}$$

The assertion of the theorem now follows from (5.11) and an analogous estimate for u^-.

We conclude this section by giving another estimate for the norm in L_q of a solution of the Neumann problem.

THEOREM 6. *Suppose that condition* (5.4), *where* $\alpha \geqq 1$, *and the conditions of Theorem* 4 *with respect to* A *and* f *are satisfied. Suppose further* $1 < p < \frac{1}{2}(1-\alpha)^{-1}$.

Let u *denote a solution of the Neumann problem for the equation* $Qu = f$ *which is in* $L_{p/[1+p(1-\alpha)]^{-1}}$ *if* $p < 2(3-2\alpha)^{-1}$ *and in* $L_2^{(1)}$ *if* $p \geqq 2(3-2\alpha)^{-1}$.

Then the following estimate holds:

$$\inf_{\gamma \in Z} \|u - \gamma\|_q \leqslant p' p^{\frac{1}{p}} q^{\frac{1}{p'}} \mathfrak{A}^2 \varkappa^{-1} \|f\|_p, \tag{5.12}$$

where $q = p[1 - 2p(1-\alpha)]^{-1}$.

PROOF. It suffices to prove estimate (5.12) for a solution u of the equation $Qu = f$ with infinitely differentiable coefficients and its free term in $\mathscr{K}$.

From inequality (4.15) and condition (5.4) we get

$$K_+ = \int u_+^q \, dx \leqslant \mathfrak{A}^{2q} \varkappa^{-q} \int_0^{v(P_0)} [G(\mu)]^{q/p} \, d\mu,$$

where

$$G(\mu) = \Big(\int_\mu^{v(P_0)} t^{-2\alpha} \, dt \int_0^t f_+^*(\tau) \, d\tau \Big)^p.$$

Since $G(\mu)$ is a nonincreasing function,

$$\int_0^{v(P_0)} [G(\mu)]^{q/p} \, d\mu \leqslant \Big(\frac{p}{q}\Big)^{\frac{q}{p}-1} \int_0^{v(P_0)} \Big(\int_0^t G(\tau) \, \tau^{\frac{p-q}{q}} \, d\tau \Big)^{\frac{q-p}{p}} G(t) \, t^{\frac{p-q}{q}} \, dt$$

$$= \Big(\frac{p}{q}\Big)^{\frac{q}{p}} \Big[\int_0^{v(P_0)} G(\mu) \, \mu^{\frac{p}{q}-1} \, d\mu \Big]^{\frac{q}{p}}.$$

Consequently

$$K_+^{\frac{1}{q}} \leqslant p^{\frac{1}{p}} q^{\frac{1}{p'}} \mathfrak{A}^2 \varkappa^{-1} \Big\{ \int_0^{v(P_0)} \Big[\frac{1}{\mu} \int_0^\mu f_+^*(\tau) \, d\tau \Big]^p d\mu \Big\}^{\frac{1}{p}}.$$

Applying Hardy's inequality (1.5), we get

$$K_+^{\frac{1}{q}} \leqslant p^{\frac{1}{p}} q^{\frac{1}{p'}} \mathfrak{A}^2 \varkappa^{-1} \Big\{ \int_0^{v(P_0)} \Big[\frac{1}{\mu} \int_0^\mu f_+^*(\tau) \, d\tau \Big]^p d\mu \Big\}^{\frac{1}{p}},$$

which does not exceed

$$p' p^{\frac{1}{p}} q^{\frac{1}{p'}} \mathfrak{A}^2 \varkappa^{-1} \Big(\int_0^{v(P_0)} [f_+^*(\tau)]^p \, d\tau \Big)^{1/p}.$$

Thus

$$\Big(\int u_+^q \, dx \Big)^{\frac{1}{q}} \leqslant p' p^{\frac{1}{p}} q^{\frac{1}{p'}} \mathfrak{A}^2 \varkappa^{-1} \Big(\int_{P_0} |f|^p \, dx \Big)^{\frac{1}{p}}.$$

An analogous estimate is obtained for the integral

$$\int u_-^q \, dx.$$

The theorem is proved.

§6. Subareal mappings and a test of solvability conditions for the Neumann problem

We consider a one-to-one mapping $D \ni x \to \xi \in R_n$ that is locally Lipschitz in D. The image of an arbitrary set $A \subset D$ under the mapping ξ will be denoted by ξA. Let ξ_x' be the matrix $(\partial \xi_i / \partial x_k)$ and $\det \xi_x'$ be the Jacobian of the mapping ξ. The notations x_ξ' and $\det x_\xi'$ have analogous meanings.

DEFINITION 1. We call a mapping ξ *subareal* if there exists a constant K such that the following inequality holds for any locally Lipschitz $(n-1)$-dimensional manifold $V \subset D$:

$$s(\xi V) \leqslant K s(V). \tag{6.1}$$

In this section we study certain properties of subareal mappings and use them to estimate the function $\lambda(t)$.

LEMMA 14. *A mapping ξ is subareal if and only if for almost all $x \in D$*

$$|\det x'_\xi| \geqslant K \| x'_\xi \|, \tag{6.2}$$

where $\| \cdot \|$ is the norm of a matrix.

In proving this lemma we will make use of

LEMMA 15. *Let u be any function in $C^{(\infty)}(D)$ whose first derivatives are summable in D, and let E be any measurable subset of D. Then in order for the inequality*

$$\int_E |\nabla_x u| \, dx \geqslant K \int_{\xi E} |\nabla_\xi u| \, d\xi \tag{6.3}$$

with a constant K not depending on u and E to hold, it is necessary and sufficient that the mapping ξ satisfy condition (6.2). (*In addition, the validity of* (6.2) *follows if* (6.3) *holds for any ball in D.*)

SUFFICIENCY. Clearly,

$$\int_{\xi E} |\nabla_\xi u| \, d\xi = \int_E \left| \sum_{i=1}^n \frac{\partial u}{\partial x_i} \nabla_\xi x_i \right| |\det \xi'_x| \, dx = \int_E |\nabla_x u| \, |x'_\xi \boldsymbol{\alpha}| \, |\det x'_\xi|^{-1} \, dx,$$

where $\boldsymbol{\alpha} = |\nabla_x u|^{-1} \nabla_x u$. It but remains to take advantage of condition (6.2).

NECESSITY. We fix the vector $\boldsymbol{\alpha}$ of unit length and consider a ball $\sigma_\rho(x_0) \subset D$ of radius ρ with center at an arbitrary point $x_0 \in D$. We put $u(x) = \boldsymbol{\alpha}\mathbf{x}$. Then by virtue of (6.3)

$$\int_{\sigma_\rho(x_0)} dx \geqslant K \int_{\xi\sigma_\rho(x_0)} |\nabla_\xi u| \, d\xi = K \int_{\sigma_\rho(x_0)} |x'_\xi \boldsymbol{\alpha}| \, |\det x'_\xi|^{-1} \, dx.$$

Passing to the limit for $\rho \to 0$ and making use of the arbitrariness of $\boldsymbol{\alpha}$, we obtain inequality (6.3) for almost all $x_0 \in D$.

PROOF OF LEMMA 14. NECESSITY. Let ξ be a subareal mapping of D. We consider an arbitrary function $u \in L_2^{(1)} \cap C^{(\infty)}(D)$. As is known, the contour sets E_t for the function $|u|$ are locally infinitely differentiable $(n-1)$-dimensional manifolds for almost all $t \geqq 0$. Let σ be an arbitrary ball in D. By Lemma 1

$$\int_\sigma |\nabla_x u| \, dx = \int_0^\infty s(\sigma \cap E_t) \, dt,$$

$$\int_{\xi\sigma} |\nabla_\xi u| \, d\xi = \int_0^\infty s(\xi\sigma \cap \xi E_t) \, dt.$$

By the definition of a subareal mapping $s(\xi E_t) \leqq Ks(E_t)$. Consequently

$$\int_\sigma |\nabla_x u| \, dx \leqslant K \int_{\xi\sigma} |\nabla_\xi u| \, d\xi,$$

which together with Lemma 15 gives the required result.

SUFFICIENCY. Let V be an arbitrary $(n-1)$-dimensional locally Lipschitz manifold in D. We take any point $y \in V$ and a neighborhood U_y of it that is so small that the set $\overline{U_y \cap V}$ admits the explicit representation $x_n = f(x_1, \cdots, x_{n-1})$, where f is a function satisfying a Lipschitz condition. Let ϵ_0 be a fixed small positive number. The set of points $x \in U_y$ given by the equation $x_n = f(x_1, \cdots, x_{n-1}) + \epsilon$, where $\epsilon \in (0, \epsilon_0]$, is denoted by V_ϵ. We put

$$u_{\varepsilon_0}(x) = \begin{cases} 1 & \text{for } x_n \leqslant f(x_1, \ldots, x_{n-1}), \\ 1 - \varepsilon\varepsilon_0^{-1} & \text{for } x \in V_\varepsilon, \\ 0 & \text{for } x_n \geqslant f(x_1, \ldots, x_{n-1}) + \varepsilon_0. \end{cases}$$

Clearly, the function $u_{\epsilon_0}(x)$ satisfies a Lipschitz condition. By virtue of (6.2) and Lemma 15,

$$\int_{U_y} |\nabla_x u_{\varepsilon_0}| \, dx \geqslant K \int_{\xi U_y} |\nabla_\xi u_{\varepsilon_0}| \, d\xi.$$

By using Lemma 1, this inequality can be represented in the form

$$\varepsilon_0^{-1} \int_0^{\varepsilon_0} s(V_\varepsilon) \, d\varepsilon \geqslant K \varepsilon_0^{-1} \int_0^{\varepsilon_0} s(\xi V_\varepsilon) \, d\varepsilon.$$

Since the function $s(V_\epsilon)$ is continuous,

$$s(V \cap U_y) \geqslant K \varliminf_{\varepsilon_0 \to 0} \varepsilon_0^{-1} \int_0^{\varepsilon_0} s(\xi V_\varepsilon) \, d\varepsilon.$$

Therefore, taking the lower semicontinuity of area into consideration, we obtain

$$s(V \cap U_y) \geqslant Ks(\xi V \cap \xi U_y).$$

Since the point y is arbitrary, this result implies estimate (6.1).

The following theorem is useful for estimating the function $\lambda(t)$ introduced on page 152 in specific cases.

THEOREM 7. *Suppose there exists a subareal mapping ξ of the domain D onto a domain ξD which is a Lipschitz image of a ball. Let $\mu(t)$ denote*

$$\inf \int_E |\det \xi'_x| \, dx$$

on the set of all $E \subset D$ such that $D \cap \partial E$ is a locally Lipschitz manifold and

$$t \leqslant v_E \leqslant \frac{1}{2} v_D.$$

Then there exists a constant Q such that the function $\lambda(t)$ satisfies the inequality

$$Q\lambda(t) \geqslant [\mu(t)]^{\frac{n-1}{n}}. \tag{6.4}$$

PROOF. Let E be an arbitrary subset of D satisfying the condition $v_E \leqq v_D/2$ and such that $D \cap \partial E$ is a locally Lipschitz manifold. We put $M = \sup v_{\xi E}$ on the set of such E. Clearly, $M < v_{\xi D}$. Since ξD is the image of a ball under a bi-Lipschitz mapping, a constant C depending only on M can be found such that

$$v_{\xi E}^{\frac{n-1}{n}} \leqslant C s_{\xi(D \cap \partial E)}. \tag{6.5}$$

(This can be proved by taking advantage of the analogous inequality for a ball (see, for example, [**30**], p. 119).) Since the mapping ξ is subareal, we get from (6.5) that

$$v_{\xi E}^{\frac{n-1}{n}} \leqslant CK s_{D \cap \partial E}.$$

Therefore, if $v_E \geqq t$, then

$$[\mu(t)]^{\frac{n-1}{n}} \leqslant CK\lambda(t).$$

The theorem is proved.

EXAMPLE 1. We consider the n-dimensional "cone"

$$D_1 : \left\{x : \left(\sum_{i=1}^{n-1} x_i^2\right)^{\frac{1}{2}} < f(x_n),\ 0 < x_n < 1\right\},$$

where $f(t)$ is a nonnegative convex downwards function such that $f'(1) < \infty$ and $f(0) = 0$ (Figure 1). We will show that $\lambda(t)$ satisfies the following inequality for the domain D:

$$cf^{n-1}(\xi(t)) \leqslant \lambda(t) \leqslant f^{n-1}(\xi(t)), \tag{6.6}$$

where $c \in (0, 1)$ and $\xi(t)$ is the function defined by the equation

$$t = v_{n-1} \int_0^{\xi(t)} f^{n-1}(\tau)\, d\tau.$$

We consider the "cone" $E_\xi = D \cap (0, \xi)$, where ξ is subject to the condition

$$v_{E_\xi} = v_{n-1} \int_0^{\xi} f^{n-1}(\tau)\, d\tau \leqslant \frac{1}{2} v_D.$$

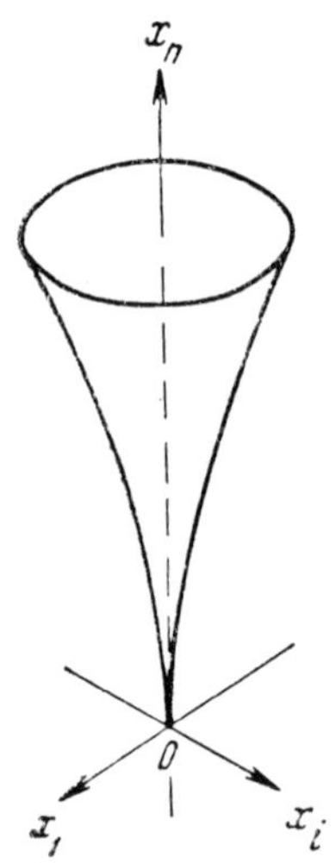

FIGURE 1

Clearly,

$$s_{D \cap \partial E_\xi} = v_{n-1} f^{n-1}(\xi).$$

Since by the definition of $\lambda(t)$

$$\lambda[v_{E_\xi}] \leqslant s_{D \cap \partial E_\xi},$$

the right inequality of (6.6) is proved.

The mapping

$$\xi : \{\xi_1 = x_1, \ldots, \xi_{n-1} = x_{n-1}, \ \xi_n = f(x_n)\}$$

takes D_1 into the cone

$$\xi D_1 = \left\{\xi : \sum_{i=1}^{n-1} \xi_i^2 < \xi_n^2, \ 0 < \xi_n < f(1)\right\}.$$

In the present case condition (6.2) of Lemma 14 is equivalent to the requirement that $f'(x_n)$ be bounded. Consequently the mapping ξ is subareal. It remains to estimate the function $\mu(t)$ defined in Theorem 7 from below. Since the modulus of the Jacobian of ξ is equal to $f'(x_n)$, it is necessary to estimate the integral

$$\int_E f'(x_n)\, dx \tag{6.7}$$

from below under the condition $v_D/2 \geqq v_E \geqq t$. Since $f'(x_n)$ is a nondecreasing function, of all the sets E the "cone" $E_{\xi(t)}$ yields the least value of the integral (6.7). Consequently

$$\int_E f'(x_n)\, dx \geqslant \int_0^{\xi(t)} f'(c)\, dc \int\limits_{\sum_{i=1}^{n-1} x_i^2 < f^2(x_n)} dx_1 \ldots dx_{n-1} = \frac{v_{n-1}}{n} f^n(\xi(t)).$$

Thus

$$\mu(t) \geqslant \frac{v_{n-1}}{n} f^n(\xi(t)).$$

From this result, applying Theorem 7, we deduce the left estimate of (6.6).

It directly follows from (6.6) that condition (5.4) is satisfied for the domain considered in this example if and only if

$$\varliminf_{\xi \to 0} g(\xi) \left(\int_0^{\xi} g(\tau)\, d\tau\right)^{-\alpha} > 0, \tag{6.8}$$

where $g(\xi) = f^{n-1}(\xi)$. In particular, if for some $\beta \in [1, \infty)$

$$\varliminf_{\xi \to 0} \xi^{-\beta} f(\xi) > 0, \tag{6.9}$$

then condition (5.4) is satisfied with exponent $\alpha = \beta(n-1)[\beta(n-1)+1]^{-1}$. Condition (5.8) for the boundedness of a solution is equivalent to the convergence of the integral

$$\int_0 \left(\int_0^x g(\tau)\, d\tau\right)^{p'} [g(x)]^{1-2p'} dx. \tag{6.10}$$

EXAMPLE 2. We consider the domain

$$D_2: \left\{0 < x_n < +\infty, \ \left(\sum_{i=1}^{n-1} x_i^2\right)^{\frac{1}{2}} < f(x_n)\right\},$$

where $f(t)$ is a nonnegative convex downwards function such that $f'(+0) > -\infty$ and $f(+\infty) = 0$ (Figure 2). We also assume that $v_D < \infty$, i.e. that

$$\int_0^\infty f^{n-1}(t)\, dt < \infty.$$

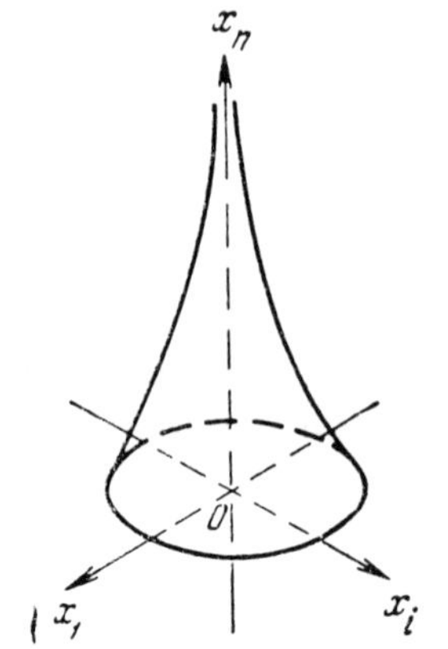

FIGURE 2

It can be shown that the following estimates are valid:

$$cf^{n-1}(\xi(t)) \leqslant \lambda(t) \leqslant f^{n-1}(\xi(t)), \tag{6.11}$$

where $c \in (0, 1)$ and $\xi(t)$ is the function defined by the equation

$$t = v_{n-1} \int_{\xi(t)}^\infty f^{n-1}(\tau)\, d\tau.$$

The proof is carried out analogously to Example 1. In this connection the role of the auxiliary subareal mapping is played by the mapping

$$\xi: \{\xi_1 = x_1, \ldots, \xi_{n-1} = x_{n-1}, \ \xi_n = f(x_n)\}$$

onto the cone

$$\left\{\xi: \sum_{i=1}^{n-1} \xi_i^2 < \xi_n^2, \ 0 < \xi_n < f(0)\right\}.$$

For the domain D_2 considered here conditions (5.5), (5.6) and (5.1) of Theorems 4 and 3 are equivalent to the corresponding conditions

$$\int^\infty \left(\int_x^\infty g(\tau)\, d\tau\right)^{\frac{1}{\alpha-1}} [g(x)]^{\frac{\alpha-2}{\alpha-1}}\, dx < \infty, \tag{6.12}$$

$$\int^\infty \left(\int_x^\infty g(\tau)\, d\tau\right)^{\frac{p-1}{1+p(\alpha-1)}} [g(x)]^{\frac{1+p(\alpha-2)}{1+p(\alpha-1)}}\, dx < \infty, \tag{6.13}$$

$$\int^\infty M\left[\frac{1}{g(x)}\right] g(x)\, dx < \infty, \tag{6.14}$$

where $g(x) = f^{n-1}(x)$.

EXAMPLE 3. We consider the plane helical domain D_3 defined in polar coordinates by the inequalities

$$1 - \varepsilon_1(\theta) > \rho > 1 - \varepsilon_2(\theta), \quad 0 < \theta < \infty,$$

where $0 < \varepsilon_1(\theta) < \varepsilon_2(\theta) < \varepsilon_1(\theta + 2\pi) < 1$

and ϵ_1 and ϵ_2 are functions satisfying a uniform Lipschitz condition on $[0, \infty)$ and such that the function $\epsilon_2 - \epsilon_1$ is convex downwards on $[0, \infty)$. We also assume that the area of D_3 is finite, i.e. that

$$\int^{\infty} (\varepsilon_2 - \varepsilon_1)\, d\theta < \infty.$$

Applying Theorem 7 to the subareal mapping ξ:

$$\xi_1 = 1 - \rho - \frac{1}{2}(\varepsilon_1(\theta) + \varepsilon_2(\theta)), \quad \xi_2 = \frac{1}{2}(\varepsilon_2(\theta) - \varepsilon_1(\theta))$$

of the domain D_3 onto the triangle

$$|\xi_1| < \xi_2, \ 0 < \xi_2 < \frac{1}{2}(\varepsilon_2(0) - \varepsilon_1(0))$$

and arguing in the same way as in Example 1, we obtain

$$c_1[\varepsilon_2(\theta(t)) - \varepsilon_1(\theta(t))] \leqslant \lambda(t) \leqslant c_2[\varepsilon_2(\theta(t)) - \varepsilon_1(\theta(t))], \tag{6.15}$$

where $\theta(t)$ is the function defined by the equality

$$t = \iint_{D_3 \cap \{(\rho,\theta):\theta(t)<\theta<\infty\}} \rho\, d\rho\, d\theta.$$

By using (6.15), one can easily rephrase conditions (5.1), (5.4)—(5.6) and (5.8) of Theorems 3—5 in terms of the function $\epsilon_2(\theta) - \epsilon_1(\theta)$.

§7. On the exactness of the solvability conditions for the Neumann problem

In conclusion we give some examples showing the sharpness of some of the sufficient conditions given above for the validity of the estimates for the solutions of the Neumann problem.

1°. Let D_4 be a domain of the type considered in Example 1. Suppose in addition that the function $f(t)$ satisfies the following condition: for $t \in [0,1]$

$$f(2t) \leqslant c f(t), \quad c = \text{const} > 1. \tag{7.1}$$

We also assume that $f(t)$ is linear on each interval $s_k = [2^{-k-1}, 2^{-k}]$, $k = 0, 1, \cdots$. We introduce the following notation: $g_k = f^{n-1}(2^{-k})$; $\rho = \rho(x)$ is the distance from a point $x = (x_1, x_2, \cdots, x_n)$, $x_n \in s_k$, to the point of intersection of the x_n axis with the straight line containing the interval $\{x_2 = 0, \cdots, x_{n-1} = 0, x_1 = f(x_n), x_n \in s_k\}$; $(0, \cdots, 0, \alpha_k)$ are the coordinates of the indicated point of intersection; $\rho_k = \rho(f(2^{-k}), 0, \cdots, 0, 2^{-k})$; and $C_k = \{x : x \in D_4, \rho(x) \in [\rho_{k+1}, 2^{-k} - \alpha_k]\}$.

We define a sequence of functions:

$$u_k = (2^{-k} - \alpha_k - \rho)^2 (\rho - \rho_{k+1})^2,$$

if $x \in C_k$, and $u_k = 0$ if $x \notin C_k$.

A simple calculation shows that

$$\left(\int_{D_4} |\Delta u|^p\, dx\right)^{\frac{1}{p}} \leqslant c 2^{-\left(2+\frac{1}{p}\right)k} g_k^{\frac{1}{p}},$$
$$\left(\int_{D_4} |\nabla u_k|^q\, dx\right)^{\frac{1}{q}} \geqslant c 2^{-\left(3+\frac{1}{q}\right)k} g_k^{\frac{1}{q}}, \tag{7.2}$$

where $p, q \in [0, \infty]$. Consequently

$$\| \nabla u_k \|_{L_q(D_4)} \| \Delta u_k \|^{-1}_{L_p(D_4)} \geqslant c 2^{-\left(\frac{1}{p'}+\frac{1}{q}\right)k} g_k^{\frac{1}{q}-\frac{1}{p}}. \tag{7.3}$$

We further note that by virtue of (7.1) condition (5.4) of Theorem 4, or what is the same thing, condition (6.8) for the domain D_4, is equivalent to inequality (6.9).

We assume that $f(\xi)$ satisfies the condition

$$\lim_{\xi\to 0} \xi^{-\beta} f(\xi) = 0, \tag{7.4}$$

and put in (7.3)

$$q = \frac{p}{1-p(1-\alpha)}, \quad \text{where } \alpha = \frac{\beta(n-1)}{1+\beta(n-1)}.$$

Then

$$2^{-\left(\frac{1}{p'}+\frac{1}{q}\right)k} g_k^{\frac{1}{q}-\frac{1}{p}} = [2^{k\beta} f(2^{-k})]^{\frac{1-n}{1+\beta(n-1)}} \xrightarrow[k\to\infty]{} \infty$$

and from (7.3) we get that estimate (5.7) is not correct in the case under consideration.

2°. The sharpness of Theorem 6 is shown by considering the domain $D_5 = D_4 \cup D_4^*$, where D_4^* is the specular reflection of D_4 with respect to the hyperplane $x_n = 1$. We extend the function u_k introduced in 1° to D_4^* as an odd function. Then, clearly,

$$\inf_{\gamma\in Z} \| u_k - \gamma \|_{L_q(D_5)} = \| u_k \|_{L_q(D_5)} \geqslant c 2^{-\left(4+\frac{1}{q}\right)k} g_k^{\frac{1}{q}}.$$

Taking (7.2) into consideration, we obtain

$$\inf_{\gamma\in Z} \| u_k - \gamma \|_{L_q(D_5)} \| \Delta u_k \|^{-1}_{L_p(D_5)} \geqslant c 2^{-\left(2-\frac{1}{p}+\frac{1}{q}\right)k} g_k^{\frac{1}{q}-\frac{1}{p}}. \tag{7.5}$$

Therefore, if the function $f(\xi)$ satisfies condition (7.4) and if

$$q = \frac{p}{1-2p(1-\alpha)}, \quad \text{where } \alpha = \frac{\beta(n-1)}{1+\beta(n-1)},$$

then

$$\lim_{k\to\infty} \inf_{\gamma\in Z} \| u_k - \gamma \|_{L_q(D_5)} \| \Delta u_k \|^{-1}_{L_p(D_5)} = c \lim_{k\to\infty} [2^{k\beta} f(2^{-k})]^{2\frac{1-n}{1+\beta(n-1)}} = \infty.$$

Thus estimate (5.12) is not correct in the case of the domain D_5.

3°. We consider the domain D_4 under the assumption that the integral

$$\int_0 \frac{x^{p'}}{[g(x)]^{p'-1}}\,dx \tag{7.6}$$

diverges. We first note that by virtue of (7.1) the convergence of this integral is equivalent to condition (6.10), which in turn is equivalent to condition (5.8) of Theorem 5.

We introduce the function

$$v_k(x) = (2^{-k} - \alpha_k - \rho)^2 (\rho - 2\rho_{k+1} + 2^{-k} - \alpha_k)^2 (2^{-k} - \alpha_k - \rho_k)^{-4}$$

which is defined for $x \in C_k$, and an arbitrary positive numerical sequence $h_0, h_1, \cdots$. We consider the sequence of continuous functions

$$u_N(x) = \begin{cases} h_0 v_0 & \text{for } x \in C_0, \\ h_k v_k + h_{k-1} + \ldots + h_0 & \text{for } x \in C_k,\ k = 1, \ldots, N \end{cases}$$

which are constant in the components $D_4 \setminus \bigcup_{j=1}^N C_j$ respectively. It is easily seen that

$$\int_{D_4} |\Delta u_N|^p\,dx \leqslant c \sum_{k=0}^N h_k^p 2^{(2p-1)k} g_k.$$

We now put

$$h_k = 2^{-(p'+1)k} g_k^{1-p'}.$$

Then

$$\int_{D_4} |\Delta u_N|^p\,dx \leqslant c \sum_{k=0}^N 2^{-(p'+1)k} g_k^{1-p'},$$

$$\operatorname*{osc}_{D_4} u_N = \sum_{k=0}^N h_k = \sum_{k=0}^N 2^{-(p'+1)k} g_k^{1-p'}.$$

By virtue of the divergence of the integral (7.6) we get

$$\lim_{N\to\infty} \operatorname*{osc}_{D_4} u_N \|\Delta u_N\|^{-1}_{L_p(D_4)} = \lim_{N\to\infty} \left(\sum_{k=0}^N 2^{-(p'+1)k} g_k^{1-p'}\right)^{\frac{1}{p'}} = \infty.$$

Consequently estimate (5.9) is not correct in the case under consideration.

4°. We also give an example showing the sharpness of condition (5.5) of Theorem 4. Let D_6 be a domain of the type considered in Example 2. Suppose in addition that the function $f(t)$ satisfies the following condition: for $t \in [0, \infty)$

$$f(t) \leqslant Kf(2t), \quad K = \text{const} > 1. \tag{7.7}$$

We also assume that $f(t)$ is linear on each interval $s_k = [2^k, 2^{k+1}]$, $k = 0, 1, \cdots$. We introduce the following notation: $g_k = f^{n-1}(2^k)$; $\rho = \rho(x)$ is the distance from a point $x = (x_1, \cdots, x_n)$, $x_n \in s_k$, to the point of intersection of the x_n axis with the straight line containing the interval $\{x_2 = 0, \cdots, x_{n-1} = 0, x_1 = f(x_n),\ x_n \in s_k\}$; $(0, \cdots, 0, \alpha_k)$ are the coordinates of the indicated point of intersection; $\rho_k = \rho(f(2^k), 0, \cdots, 0, 2^k)$; and $C_k = \{x : x \in D_6, \rho(x) \in$

$[\rho_{k+1}, \alpha_k - 2^k]\}$. We define a sequence of functions:

$$v_k(x) = (\alpha_k - 2^k - \rho)^2 (\rho - \rho_{k+1})^2,$$

if $x \in C_k$, and $v_k = 0$ if $x \notin C_k$.

A simple calculation shows that

$$\int_{D_6} |\Delta v_k|^p \, dx \leqslant c 2^{(2p+1)k} g_k,$$

$$\int_{D_6} |\nabla v_k|^q \, dx \geqslant c 2^{(3q+1)} g_k.$$

Let α_k be an arbitrary nonnegative numerical sequence. We define a sequence of functions: $w_N = \alpha_k v_k$ if $x \in C_k$ $(k = 1, 2, \cdots, N)$ and $w_N = 0$ if $x \in D_6 \backslash \bigcup_{k=1}^N C_k$. Clearly

$$\int_{D_6} |\Delta w_N|^p \, dx \leqslant c \sum_{k=1}^N \alpha_k^p 2^{(2p+1)k} g_k,$$

$$\int_{D_6} |\nabla w_N|^q \, dx \geqslant c \sum_{k=1}^N \alpha_k^q 2^{(3q+1)k} g_k,$$

where $q = p[1 + p(\alpha - 1)]^{-1}$, $1 + 1/p' \geqq \alpha > 1$, $\infty \geqq p > 1$. Suppose

$$\alpha_k = 2^{\frac{2p-3q}{q-p} k}.$$

Then

$$\| \nabla w_N \|_{L_q(D_6)} \| \Delta w_N \|_{L_p(D_6)}^{-1} \geqslant c \Big(\sum_{k=1}^N 2^{\frac{\alpha}{\alpha-1}} g_k \Big)^{\alpha - 1}.$$

The right side of this inequality is bounded uniformly in N if and only if

$$J = \int_0^\infty g(t) \, t^{\frac{1}{\alpha-1}} \, dt < \infty. \tag{7.8}$$

Using Hardy's inequality (1.5), the fact that $g(t)$ is decreasing and condition (7.7), we get

$$c_1 J \geqslant \int_0^\infty \Big(\int_x^\infty g^{\alpha-1}(t) \, dt \Big)^{\frac{1}{\alpha-1}} dx \geqslant \int_0^\infty [g(x)]^{\frac{\alpha-2}{\alpha-1}} \Big(\int_x^\infty g(t) \, dt \Big)^{\frac{1}{\alpha-1}} dx \geqslant c_2 J.$$

Hence (7.8) is equivalent to condition (6.12) or, what is the same thing, to condition (5.5) of Theorem 4. Thus, if (7.8) is not satisfied for the domain D_6, estimate (5.7) for the solutions of the Neumann problem is not generally correct.

Received 10 NOV 1967

BIBLIOGRAPHY

[1] E. Giorgi, *Sulla differenziabilità e l'analiticità delle estremali degli integrali multipli regolari,* Mem. Accad. Sci. Torino. Cl. Sci. Fis. Mat. Nat. (3) **3** (1957), 25-43. MR **20** # 172.

[2] J. F. Nash, Jr., *Continuity of solutions of parabolic and elliptic equations,* Amer. J. Math. **80** (1958), 931-954. MR **20** # 6592.

[3] C. B. Morrey, Jr., *Second order elliptic equations in several variables and Hölder continuity,* Math. Z. **72** (1959/60), 146-164. MR **22** # 11200.

[4] N. N. Ural'ceva, *Regularity of solutions of multidimensional elliptic equations and variational problems,* Dokl. Akad. Nauk SSSR **130** (1960), 1206-1209 = Soviet Math. Dokl. **1** (1960), 161-164. MR **23** # A4036.

[5] G. Stampacchia, *Problemi al contorno e ellittici, con dati discontinui, dotati di soluzioni hölderiane,* Ann. Mat. Pura Appl. (4) **51** (1960), 1-37. MR **23** # A3897.

[6] J. Moser, *A new proof of De Giorgi's theorem concerning the regularity problem for elliptic differential equations,* Comm. Pure Appl. Math. **13** (1960), 457-468. MR **30** # 332.

[7] A. G. Sigalov, *Two-dimensional problems of the calculus of variations in nonparametric form,* Trudy Moskov. Mat. Obšč. **2** (1953), 201-233. (Russian) MR **15**, 442.

[8] G. Stampacchia, *Contributi alla regolarizzazione delle soluzioni dei problemi al contorno per equazioni del secondo ordine ellittiche,* Ann. Scuola Norm. Sup. Pisa (3) **12** (1958), 223-245. MR **23** # A2616.

[9] ———, *Régularisation des solutions de problèmes aux limites elliptiques à données discontinues,* Proc. Internat. Sympos. Linear Spaces (Jerusalem, 1960), Jerusalem Academic Press, Jerusalem and Pergamon Press, Oxford, 1961, pp. 399-408. MR **26** # 4034.

[10] V. G. Maz'ja, *Some estimates of solutions of second-order elliptic equations,* Dokl. Akad. Nauk SSSR **137** (1961), 1057-1059 = Soviet Math. Dokl. **2** (1961), 413-415. MR **24** # A908.

[11] C. Miranda, *Alcune osservazioni sulla maggiorazione in L_p^ν delle soluzioni deboli delle equazioni ellittiche del secondo ordine,* Ann. Mat. Pura Appl. (4) **61** (1963), 151-169. MR **31** # 1451.

[12] H. F. Weinberger, *Symmetrization in uniformly elliptic problems,* Studies in Math. Anal. and Related Topics, Stanford Univ. Press, Stanford, Calif., 1962, pp. 424-428. MR **26** # 2726.

[13] N. G. Meyers, *An L^p-estimate for the gradient of solutions of second order elliptic divergence equations,* Ann. Scuola Norm. Sup. Pisa (3) **17** (1963), 189-206. MR **28** # 2328.

[14] J. Serrin, *Pathological solutions of elliptic differential equations,* Ann. Scuola Norm. Sup. Pisa (3) **18** (1964), 385-387. MR **30** # 335.

[15] W. Littman, G. Stampacchia and H. F. Weinberger, *Regular points for elliptic equations with discontinuous coefficients,* Ann. Scuola Norm. Sup. Pisa (3) **17** (1963), 43-77. MR **28** # 4228.

[16] H. L. Royden, *The growth of a fundamental solution of an elliptic divergence structure equation,* Studies in Math. Anal. and Related Topics, Stanford Univ. Press, Stanford, Calif., 1962, pp. 333-340. MR **26** # 2725.

[17] J. Serrin, *Local behavior of solutions of quasi-linear equations,* Acta Math. **111** (1964), 247-302. MR **30** # 337.

[18] E. M. Landis, *Some questions in the qualitative theory of second-order elliptic equations (case of several independent variables),* Uspehi Mat. Nauk **18** (1963), no. 1 (109), 3-62 = Russian Math. Surveys **18** (1963), no. 1, 1-62. MR **27** # 435.

[19] J. Serrin and H. F. Weinberger, *Isolated singularities of solutions of linear elliptic equations,* Amer. J. Math. **88** (1966), 258-272. MR **34** # 1697.

[20] V. G. Maz'ja, *The behavior near the boundary of the solution of the Dirichlet problem for an elliptic equation of the second order in divergence form,* Mat. Zametki **2** (1967), 209-220. (Russian) MR **36** # 2947.

[21] J. Moser, *On Harnack's theorem for elliptic differential equations,* Comm. Pure Appl. Math. **14** (1961), 577-591. MR **28** # 2356.

[22] A. D. Aleksandrov, *Majorants of solutions and uniqueness conditions for elliptic equations,* Vestnik Leningrad. Univ. **21** (1966), no. 7, 5-20; English transl., Amer. Math. Soc. Transl. (2) **68** (1968), 144-161. MR **34** # 1688.

[**23**] V. G. Maz′ja, *On the solvability of the Neumann problem,* Dokl. Akad. Nauk SSSR **147** (1962), 294-296 = Soviet Math. Dokl. **3** (1962), 1595-1597. MR **26** # 1606.

[**24**] ———, *On the theory of the higher-dimensional Schrödinger operator,* Izv. Akad. Nauk SSSR Ser. Mat. **28** (1964), 1145-1172. (Russian) MR **30** # 5070.

[**25**] G. H. Hardy, J. E. Littlewood and G. Pólya, *Inequalities,* Cambridge Univ. Press, New York, 1934; 2nd ed., 1952; Russian transl., IL, Moscow, 1948. MR **13**, 727; **18**, 722.

[**26**] A. Kronrod, *On functions of two variables,* Uspehi Mat. Nauk **5** (1950), no. 1 (35), 24-134. (Russian) MR **11**, 648.

[**27**] A. G. Vituškin, *On multidimensional variations,* GITTL, Moscow, 1955. (Russian) MR **17**, 718.

[**28**] H. Federer, *Curvature measures,* Trans. Amer. Math. Soc. **93** (1959), 418-491. MR **22** # 961.

[**29**] S. L. Sobolev, *On a theorem of functional analysis,* Mat. Sb. **4** (**46**) (1938), 471-497; English transl., Amer. Math. Soc. Transl. (2) **34** (1963), 39-68.

[**30**] Ju. D. Burago and V. G. Maz′ja, *Certain questions of potential theory and function theory for regions with irregular boundaries,* Zap. Naučn. Sem. Leningrad. Otdel. Mat. Inst. Steklov. (LOMI) **3** (1967). (Russian) MR **37** # 3031.

Translated by:
S. Smith

Trudy Moskov. Mat. Obšč.
Tom 20 (1969)

Trans. Moscow Math. Soc.
Vol. 20 (1969)

ON THE POINCARÉ BOUNDARY VALUE PROBLEM

M. B. MALJUTOV

Contents

INTRODUCTION 173

CHAPTER 1. Basic results on problem F 177
§1.1. Basic definitions 177
§1.2. Auxiliary estimates 178
§1.3. Construction of a solution of F. Estimate of its gradient 181
§1.4. Uniqueness of the solution 183
§1.5. Continuity of the solution of F in $\overline{D}$ 184

CHAPTER 2. Behavior of a solution of problem F near M 187
§2.1. Basic definitions 187
§2.2. Investigation of positive sides 188
§2.3. Behavior of a solution of F near M^- 190

CHAPTER 3. Investigation of problem P in domain D 191
§3.1. Passage to the limit alongside M^- 191
§3.2. Basic theorem 192

CHAPTER 4. Probabilistic analysis of problem P 194
§4.1. Probability interpretation of the preceding results 194
§4.2. Proof of Lemma 4.1 195
§4.3. Stochastic equations for trajectories of the process X 197
§4.4. Behavior of trajectories of X near M^+ 199

CHAPTER 5. Separation of classical solutions of problem P_0 in the 2-dimensional case 200
§5.1. Solutions continuous at null points 200
§5.2. Separation of classical solutions 203

BIBLIOGRAPHY 203

Introduction

While working on the theory of the tides, H. Poincaré [1] came to the following problem: find a solution $u(x)$ of the equation

$$Lu = f \tag{0.1}$$

in a domain D of Euclidean space E^n which satisfies the condition

$$\partial u / \partial l = \phi \tag{0.2}$$

on the boundary. Here $L = \sum a_{ij}(x)\,\partial^2/\partial x_i \partial x_j + \sum b_i(x)\,\partial/\partial x_i + C(x)$ is an elliptic operator, and $\partial u/\partial l$ is the derivative of the function $u(x)$ along the smooth vector field $l(x)$, $x \in \partial D$ on the boundary, and f and ϕ are given functions. Poincaré was not able to solve this problem. In 1936 it was shown by Giraud [**2**] that if the field $l(x)$ is not tangent to ∂D, we can get theorems analogous to the theorems on the classical Neumann problem. Later, for the case when L is a Laplace operator in a two-dimensional domain, A. Liénard [**3**] solved the problem with the field $l(x)$ being of general form by reducing it to the well-known problem of Hilbert in the theory of functions of a complex variable. His idea was that if the function $u(x_1, x_2)$ is harmonic, then the function $F_u(z) = \partial u/\partial x_1 - i\partial u/\partial x_2$ is an analytic function of the complex variable $z = x_1 + ix_2$, and moreover that between the real and imaginary parts of the function F_u there is, by (0.2), a linear relation on the boundary; this is indeed the problem of Hilbert. Then with the help of the theory of generalized analytic functions I. N. Vekua generalized the result of Liénard to the case of a general elliptic equation of second order. However it remained an essential requirement that the problem be two-dimensional.

In the following we will always limit ourselves to the case that $\phi = 0$ in condition (0.2). The general case can be reduced to this case. Denote the corresponding new condition by (0.2.0).

Another approach to Poincaré's problem in the two-dimensional case was proposed in [**5**]. The problem is solved in two stages. If we broaden the class of solutions of Poincaré's problem by admitting functions which can be discontinuous at certain exceptional points of the boundary, then it is possible to get an existence and uniqueness theorem which completely describes the class of such solutions. Then with the help of topological mappings it is possible to separate out of this broader class the class of solutions which are regular in the closed region.

The first stage, i.e. the construction of solutions with singularities on the exceptional set, was carried through without any special difficulty in [**6**] for the n-dimensional case of Poincaré's problem.[1)] Independently Ju. V. Egorov and V. A. Kondrat′ev [**7**], while developing some ideas of L. Hörmander, got some results on Poincaré's problem in n-dimensional space, but they could not get an existence theorem for the general case for the class of functions they were investigating. However, after becoming familiar with my results, they were able to improve their results, getting an existence theorem for the class of functions having singularities on some set of the boundary [**8**]. Nevertheless the classes of functions for which they and I got existence theorems are different, and neither result can be derived from the other. For some special cases the problem of Poincaré was investigated earlier by A. V. Bicadze [**9**], [**10**], [**11**] and by A. I. Janušauskas [**12**].

[1)] Although [**6**] deals with the 3-dimensional case, the methods carry over to the n-dimensional case.

Finding solutions to Poincaré's problem is equivalent to investigating the behavior of the trajectories of a certain diffusion process in the domain D. A whole series of results in this article can be formulated both in the language of differential equations and in the language of probability theory. Our presentation is such that the basic part of the article is understandable to specialists in differential equations who are not familiar with the special methods of the theory of stochastic processes. Only one auxiliary proposition is proved by probability considerations; and the purpose of this is merely to illustrate the use of probability methods. But at the same time it is necessary to point out that probability considerations were decisive in the first stage of the work on this problem. They led to an intuitive understanding of results which were later proved. And only later did we devise proofs which do not employ probability considerations.

The method of this article is based on results about the following mixed boundary value problem F for equation (0.1).

Suppose that condition (0.2.0) is satisfied on the smooth subregion Γ_1 of the boundary ∂D, and suppose that the Dirichlet condition $u|_{\Gamma_2} = g$ is satisfied on the remaining part Γ_2. In addition we impose the essential requirement that the vector field $l(x)$ be nowhere tangent to Γ_1. We prove the basic results about problem F in Chapter 1. An existence and uniqueness theorem for F was announced earlier in [**13**].

We consider the δ-neighborhood M_δ of the set M, on which the field $l(x)$ is tangent to ∂D, and we let $S_\delta = M_\delta \cap \partial D$. The idea is to pass to the limit as $\delta \to 0$ in the mixed boundary value problem F with Dirichlet conditions given on S_δ.

In probability language this is equivalent to investigating a diffusion process in D which "is reflected off ∂D along $l(x)$" and which breaks off at the moment of reaching S_δ. The concept of reflection off a boundary will be made more precise later; here we merely observe that the direction of reflection off the boundary is best described by the vector $l_1(x)$ which coincides with $l(x)$ if $l(x)$ is directed inside D and which equals $-l(x)$ in the opposite case. The vector $l_1(x)$, which is always directed inside the domain, shows the "primary direction of the reflection" of the diffusion process.

Let M consist of a finite number of smooth closed manifolds of dimension $n-2$, and suppose $l(x)$ is not tangent to these special manifolds. Then from the probability point of view there arises a natural classification of the behavior of the vector field near a manifold from M.

As a preliminary it is convenient to introduce in the tangent space T_x at the point $x \in \overline{D}$[2)] the new scalar product $(\xi, \eta)_L = \sum a_{ij}\xi_i\eta_j$, where the ξ_i (η_j) are the components of the vectors ξ (η), and the a_{ij} are coefficients of the operator L at x for second derivatives. We will refer to this metric as the L-metric. Sometimes we will investigate the $L(x)$-metric on some set B,

[2)] We denote the closure of B by $\overline{B}$.

i.e. for $y \in B$ we introduce a metric in T_y by the formula $(\xi, \eta)_{L(x)} = \sum a_{ij}(x)\xi_i\eta_j$. We will refer to the distance in the L-metric as the L-distance, and we will designate the scalar product by $(\cdot, \cdot)_L$.

The matrix $A(x)$ of coefficients of the principal part of L can be written in the form $A(x) = \sigma^2(x)$, because A is positive definite (ellipticity). We effect the linear mapping $\xi = \sigma(\alpha)x$. Then in the new coordinates A is a Laplace operator at α, and the metric $(\xi, \eta)_L$ in the new coordinates will be given by the formula $(\xi, \eta)_L = \sum \xi_i\eta_j$, where $\xi_i(\eta_j)$ are the coordinates of $\xi(\eta)$.

With the help of an additional orthogonal mapping we can preserve the above properties of L and $(\cdot, \cdot)_L$ and also arrange for the coordinate axes to be oriented more conveniently for purposes of computation.

If the limit of the vector $l_1(x)$, as x tends to $\alpha \in \gamma \subset M$ from one of the sides on which the manifold γ locally divides ∂D, is directed to the other side, i.e. if it forms an acute angle in the L-metric with the vector $\nu \in \hat{T}_\alpha$[3] which is L-normal to γ and directed to the other side of γ, then the given side of γ is said to be positive. The probability model[4] says that the diffusion process is attracted to the manifold from the positive side and is repelled from the negative side.

EXAMPLE. For Laplace's equation in the ball $x_1^2 + x_2^2 + x_3^2 \leqq 1$ and for the field $l(x)$ of vectors parallel to the x_3-axis, both sides of the circle $x_3 = 0$, $x_1^2 + x_2^2 + x_3^2 = 1$ are positive. For the same operator and the same vector field in the region $1 \leqq x_1^2 + x_2^2 + x_3^2 \leqq 2$, both sides of the above circle are negative.

For smoothness and simplicity of presentation we will henceforth deal with the 3-dimensional case of the problem. In this case M must consist of a finite number of closed curves. If we carefully look over the proofs of the theorems, we see that with obvious modifications they carry over to the n-dimensional case of the problem. A crucial factor in the proofs is the construction of a function in some neighborhood V of a singular manifold on the boundary; it plays a role analogous to the role of "barriers" in the investigation of the Dirichlet problem. These functions will always be chosen in the following manner: V is stratified into 2-dimensional planes, L-normal to the singular manifold, and on each of these planes we investigate the same function of two variables. In essence, then, the problem reduces to the 2-dimensional case. All difficulties concerning the fusing of these functions on different planes and the character of the singularity of the resulting function of n variables are the same in the n-dimensional case as in the 3-dimensional case.

Theorems for the general n-dimensional case can be obtained from the corresponding theorems for the 3-dimensional case by changing the word "curve" to "manifold of dimension $n-2$".

---

[3] T_α is a hyperplane tangent to ∂D at the point α.

[4] See Chapter 4.

The basic results of this article (Theorems 4.1 and 4.3) consist in making more precise the intuitive picture described above. It is proved that there exists a positive probability that the diffusion process described above reaches from the positive side an arbitrarily small interval γ_0 of the singular curve γ, and that this probability tends to 1 as the initial point approaches an interior point of this interval γ_0 from the positive side. It is also proved that certain functions describing this process form a complete class of solutions of Poincaré's problem with singularities on the negative sides of curves from M.

A certain property of repulsion from the negative side is contained in Theorem 2.2.

An equivalent formulation (cf. Theorem 3.1) is as follows: A bounded function $u(x)$ satisfying equation (0.1) in D, continuously differentiable everywhere inside the set M, and satisfying condition (0.2.0) inside M will be called a solution of problem $P = P_f$. Now enumerate the curves in M. To each positive side γ_k^i $(i = 1, 2)$ of the curves γ_k in M we assign continuous functions $g_k^i(x)$. Then there exists a unique solution $u(x)$ of P_f such that the limiting value of $u(x)$, as x approaches γ_k along the boundary from the positive side γ_k^i, is $g_k^i(x)$.

It is easy to prove that if both sides of the curve γ_k are positive and $g_k^1 = g_k^2$, then the solution to problem P is continuous in a 3-dimensional neighborhood of γ_k. If the curve γ has a negative side, then the solution, generally speaking, is discontinuous on γ_k even if all functions $g_k(x)$ are as smooth as we wish.

The description of solutions of P which are continuous in $\overline{D}$ is still an open problem.

Chapter 1 contains proofs of some necessary results on the mixed problem F. In Chapter 2 we prove some theorems on the behavior of solutions of the problem near singular curves; these theorems are used in Chapter 3 to prove the existence of a solution of P in the whole domain D. Chapter 4 is devoted to a probability analysis of P. Finally in Chapter 5 we describe the separation of continuous solutions in $\overline{D}$, and consequently of smooth ones, for the case that D is a simply connected 2-dimensional domain.

The author takes this opportunity to express his sincere thanks to E. B. Dynkin for his attention to this work and to M. I. Freĭdlin for his valuable advice.

CHAPTER 1

BASIC RESULTS ON PROBLEM F

§1.1. Basic definitions

Let D be a bounded domain of class C^3, and suppose that on the closure of a smooth subregion Γ_1 of the boundary ∂D there is a vector field $l(x)$ of class $C^3(\overline{\Gamma}_1)$ not tangent to ∂D. Let $f(x) \in C^3(\overline{D})$ and $g(x) \in C(\Gamma_2)$, where $\Gamma_2 = \partial D \setminus \Gamma_1$ is assumed to be nonempty. The bounded function $u(x)$ is called a solution of the problem $F[D, \Gamma_1, f, g]$[5] (or simply problem F) if 1) $u \in C^2(D)$ and $Lu = f$[6] in D); 2) $u \in C^1(\Gamma_1 \cap D)$ and $\partial u/\partial l = 0$ on Γ_1; 3) $u \in C(\Gamma_2 \backslash \partial \Gamma_2)$ and $u(x) = g(x)$ on $\Gamma_2 \backslash \partial\Gamma_2$.

The goal of this chapter is to prove an existence and uniqueness theorem for F. We also prove that the solution of F is continuous in $\overline{D}$, and we get an estimate of the gradient of the solution in any closed subregion Γ_1. We will make extensive use of the following proposition.

LEMMA 1.1 (MAXIMUM PRINCIPLE). *If $Lu \leqq 0$, and the function $u(x)$ $(x \in D)$ is not constant and has a minimum at $\xi \in \partial D$ and is differentiable in some neighborhood of the point ξ, then for any vector l forming an acute angle with the interior normal at ξ, $\partial u/\partial l > 0$ at this point.*

The proof of Lemma 1.1 may be found in [**14**], Chapter 14. In §1.2 we will prove a series of auxiliary propositions. In §1.3 we will construct a solution, prove that it is bounded, and get an estimate of the gradient of this solution; §1.4 is devoted to the uniqueness theorem. That the solution we construct is continuous will be proved in §1.5.

§1.2. Auxiliary estimates

Let $\gamma = \Gamma_1 \cap \Gamma_2$. Let N_β $(\beta \in \gamma)$ be a plane which is L-normal to γ at β. Fix the point $\alpha \in \gamma$, and let α_i, $i = 1, 2$, be points on the curve γ at a distance δ_2 from α. Let V_{δ_1, δ_2} be the part of the δ_1-neighborhood of γ bounded by the planes N_{α_i}, $i = 1, 2$. In those cases where there is no ambiguity we will use just V.

In the following discussion we will frequently make use of constants and auxiliary functions. We will designate these constants by Latin letters, sometimes with subscripts, and the functions by the letters u, v, w. The constants (and the functions designated by u, v, and w) in the following proof have no relation to the similarly designated constants (and functions) in the preceding proof. It is necessary to bear in mind that the constants do not depend on the dimensions δ_1 and δ_2 of domains to be investigated later. We associate with each point $\beta \in \gamma$ a natural system of coordinates $x_i^{(\beta)}$, $i = 1, 2, 3$, with origin at β. We will investigate the plane $\hat{T}_\beta$ tangent to ∂D at β. We will direct the axis $x_1^{(\beta)}$ along the tangent to γ at β; we will choose the axis $x_2^{(\beta)}$ in the plane $\hat{T}_\beta$, L-orthogonal to axis $x_1^{(\beta)}$, and we will direct it toward Γ_1. We direct $x_3^{(\beta)}$ along the interior L-normal to ∂D.

We will assign the triple of values (β, ξ_0, ξ_1) to the point $x \in N_\beta$ for which $x_2^{(\beta)} = \xi_0$, $x_3^{(\beta)} = \xi_1$. If δ_1 is sufficiently small, then in the neighborhood V_{δ_1, δ_2}

---

[5] The dependence of F on the operator L and the field l is understood.

[6] $L = \sum a_{ij}(x)\partial^2/\partial x_i \partial x_j + \sum b_i(x)\partial/\partial x_i + c(x)$, the coefficients belong to class $C^3(\overline{D})$, $\sum a_{ij}\lambda_i\lambda_j \geqq \theta\Sigma\lambda_i^2$, $\theta > 0$, $c(x) \leqq 0$.

the relation between the points x and the triples (β, ξ_0, ξ_1) is one-to-one. To each function $u(\xi_0, \xi_1)$ there corresponds a function $\hat{u}(x) = u(\xi_0(x), \xi_1(x))$ in the domain V_{δ_1,δ_2}, which we will refer to as the function generated by $u(\xi_0, \xi_1)$.

Because the coordinates $x_i^{(\beta)}$, $i = 1, 2, 3$, of the point x depend smoothly on β, the function $\hat{u}(x)$ will be smooth in V if $u(\xi)$ is smooth. Our most immediate goal is to evaluate the singularities of the derivatives of $\hat{u}(x)$ if the derivatives of $u(\xi)$ have singularities. For simplicity let L be constant on γ.

The estimate of the derivatives of $\hat{u}(x)$ is based on the geometrically obvious observation that the level surfaces of $\hat{u}(x)$ intersect N_β $L(\beta)$-orthogonally.[7] Hence $\partial \hat{u}/\partial x_1^{(\beta)} = 0$ for $x_1^{(\beta)} = 0$. Thus $\| \nabla \hat{u}(x) \| = \| \nabla u(\xi) \|$ for $x \in N_\beta$. Furthermore,

$$\partial \hat{u}/\partial x_1^{(\beta)}(x) \ll c_1 x_1^{(\beta)} \| \nabla u(\xi) \|, \tag{1.2.1}$$

where the constant c_1 depends on the curvature of the curve γ. In fact $\partial \hat{u}/\partial x_1^{(\beta)}$ is the projection of $\nabla \hat{u}$ on the axis $x_1^{(\beta)}$, and the cosine of the angle between the direction of $\nabla \hat{u}$ and the axis $x_1^{(\beta)}$ is less than $c_1 x_1^{(\beta)}$. Finally, differentiating with respect to x_1, we get

$$\partial^2 u/\partial x_i^{(\beta)} \partial x_1^{(\beta)} = 0 \quad \text{for} \quad x_1^{(\beta)} = 0, \quad i = 2, 3, \tag{1.2.2}$$

$$| \partial^2 \hat{u}/\partial x_1^{(\beta)} \partial x_1^{(\beta)} | \ll c_1 \| \nabla u \|. \tag{1.2.3}$$

Further, in the coordinates $x_i^{(\beta)}$ $(i = 1, 2, 3)$ we have $L = \Delta + \widetilde{L}$, where $\widetilde{L}$ contains first derivatives and derivatives of second order multiplied by some function of x which tends to 0 as $x \to \beta$ with at least a velocity which is linear in $x_2^{(\beta)}$ and $x_3^{(\beta)}$.

Let us assume that we have the estimates $\| \nabla u \| \leqq c_2 r^\theta$ and $| \partial^2 u/\partial \xi_i \partial \xi_j | < c_2 r^{\theta - 1}$ $(r = (\xi_0^2 + \xi_1^2)^{1/2})$. Then for $x \in N_\beta$ we get from (1.2.2) and (1.2.3) that

$$| L\hat{u}(x) - \Delta \hat{u}(x) | = | \widetilde{L} \hat{u} | \ll c_3 \hat{r}^\theta(x), \tag{1.2.4}$$

where $\hat{r}(x)$ is the function in V generated by $r(\xi)$.

Lemma 1.2. *Let* 1) $|f(x)| < c_1 \hat{r}^{\theta - 1}$ *for* $x \in \mathring{V}_{\delta_1,\delta_2}$, $0 \leqq \theta \leqq 1$, *and* $Lv = f$ *in* $\mathring{V}_{\delta_1,\delta_2}$; 2) $v \in C^1(\Gamma_1 \cap \partial V_{\delta_1,\delta_2})$ *and* $|\partial v/\partial l_1| < c_2 \delta_1^\theta$ *on* $\Gamma_1 \cap \partial V_{\delta_1,\delta_2}$; 3) $\lim_{x \to (\partial V_{\delta_1,\delta_2} \setminus \Gamma_1)} |v(x)| < c_3 \delta_1^{\theta+1}$.

Then, if δ_1 *is sufficiently small,* $\sup_{x \in \bar{V}_{\delta_1,\delta_2}} |v(x)| < c \delta_1^{\theta+1}$.

Proof. It is easy to verify that all second derivatives of the function $r^{\theta+1}(\xi) = (\xi_0^2 + \xi_1^2)^{(\theta+1)/2}$ satisfy the inequality

$$\left| \frac{\partial^2 r^{\theta+1}}{\partial \xi_i \partial \xi_j} \right| < (1 + \theta)^2 r^{\theta - 1}, \quad \Delta r^{\theta+1} = (1 + \theta)^2 r^{\theta - 1},$$

and the gradient $\nabla r^{\theta+1}$ satisfies the inequality

[7] In the cases we consider level surfaces make sense. In the general case the angle of intersection tends to a right angle as $x \to \beta$ linearly in $x_2^{(\beta)}$ and $x_3^{(\beta)}$. This is sufficient for (1.2.4).

$$\|\nabla r^{\theta+1}\| \leqslant c_4 r^{\theta}. \tag{1.2.5}$$

By (1.2.4) $L\hat{r}^{\theta+1}(x) > \frac{1}{2}(1+\theta)^2\hat{r}^{\theta-1}$ for sufficiently small δ_1.

We now utilize the fact that the vector field $l(x)$ is not tangent to ∂D. More precisely, it is necessary that $\partial/\partial l = \sum_{i=1}^{3} a_i(x)\,\partial/\partial x_i^{(\beta)}$ and also

$$|a_3|\,(a_1^2 + a_2^2 + a_3^2)^{-1/2} \geqslant \varkappa > 0 \text{ on } \Gamma_1 \cap \partial V. \tag{1.2.6}$$

From this and (1.2.5) it follows that there are constants c_5 and c_6 such that $\rho(x) = -\hat{r}^{\theta+1} + c_5\delta_1^{\theta+1} + c_6\delta_1^{\theta}\hat{x}_3$[8] will satisfy the conditions $\partial\rho/\partial l_1 \leqq 0$ on $\Gamma_1 \cap \partial V$, $L\rho < -\frac{1}{2}(1+\theta)^2\hat{r}^{\theta-1}$ and $\rho(x)$ will be nonnegative in the domain V.

From condition 2) of the lemma and (1.2.6) it follows that there is a constant c_7 such that $\partial/\partial l_1(v(x) + c_7\delta_1^{\theta}x_3) \geqq 0$. From this and conditions 1) and 3) of the lemma and the properties of the function $\rho(x)$ it follows that there are constants c_8 and c_9 such that the function

$$\tilde{\rho}(x) = \rho(x) - (v(x) + c_7\delta_1^{\theta}\hat{x}_3)\,c_8 + c_9\,\delta_1^{\theta+1}$$

will satisfy the following conditions: $L\tilde{\rho} \leqq 0$ in $\mathring{V}$, $\partial\tilde{\rho}/\partial l_1 \leqq 0$ on $\Gamma_1 \cap \partial V$ and $\lim_{x\to\partial V\setminus\Gamma_1}\tilde{\rho}(x) \geqq 0$. By Lemma 1.1 $\tilde{\rho} \geqq 0$, from which the estimate $v < c\delta_1^{\theta+1}$ immediately follows.

Analogously we get the estimate $v \geqq -c\delta_1^{\theta+1}$.

REMARK 1.1. From the minimum principle it easily follows that the estimates obtained in Lemma 1.2 for the domain V_{δ_1,δ_2} are also valid for the domain $\tilde{V} \subset V_{\delta_1,\delta_2}$.

LEMMA 1.3. *Let* 1) $Lw = 0$ *in* $\mathring{V}_{\delta_1,\delta_2}$; 2) $w \in C^1(\Gamma_1 \cap \partial V)$ *and* $\partial w/\partial l_1 = 0$ *on* $\Gamma_1 \cap \partial V$; 3) $\overline{\lim}_{x\to N_{\alpha_i}} w(x) \leqq 1$, $i = 1, 2$, *and* $\lim_{x\to G} w(x) \leqq 0$, *where* $G = \partial V \setminus (\Gamma_1 \cup N_{\alpha_1} \cup N_{\alpha_2})$. *Then on the plane* N_α *we have* $w(x) < K\delta_1\delta_2^{-1}$.

PROOF. We will show that the function w on N_α is evaluated almost exactly like the solution to Dirichlet's problem for a Laplace operator in the rectangle $|x_1| < \delta_2$, $|x_3| < \delta_1$.

We consider the function $w_0(x) = (x_1^{(\alpha)})^2 - (x_3^{(\alpha)})^2$ in the domain V_{δ_1,δ_2}. We have $Lw_0(\alpha) = 0$ because $Lw_0(\alpha) = \Delta w_0(\alpha) + \tilde{L}w_0(\alpha) = \tilde{L}w_0(\alpha) = 0$, since the first derivatives are zero at α.

Henceforth we will always assume that $\delta_2 > \delta_1$. Since the coefficients are smooth,

$$\sup_{x\in V} Lw_0(x) < c_1\delta_2 \tag{1.2.7}$$

and

$$\partial w_0/\partial l \leqslant c_2\delta_2 \quad \text{on} \quad \partial V \cap \Gamma_1. \tag{1.2.8}$$

By (1.2.6) and (1.2.8), if δ_2 is sufficiently small, then it is possible to select

---

[8] $\hat{x}_3$ denotes the function in V generated by the coordinate ξ_1. Analogously to the preceding it is proved that $\hat{x}_3$ and its derivatives up to the second order do not have singularities in V.

a constant c_3 such that $\partial w_1/\partial l_1 \leqq 0$, where $w_1(x) = w_0(x) + c_3\delta_2\hat{x}_3$. Moreover it is clear that $|Lw_1| < c_4\delta_2$; on the planes N_{α_i}, $i = 1, 2$,

$$w_1 \geqslant \delta_2^2 - c_5\delta_2\delta_1, \quad \min_{x\in V} w_1(x) \geqslant -c_5\delta_2\delta_1.$$

The function $w_2 = (w_1 + c_5\delta_1\delta_2)(\delta_2^2 - c_5\delta_1\delta_2)^{-1} \geqq 1$ on the planes N_{α_i}; $\inf_{x\in V} w_2(x) \geqq 0$, and on the plane N_α

$$w_2 < c_3\delta_1\delta_2(\delta_2^2 - c_5\delta_1\delta_2)^{-1} < c_6\delta_1\delta_2^{-1}, \tag{1.2.9}$$

if $\delta_1\delta_2^{-1}$ is sufficiently small. If $w(x)$ satisfies the conditions of Lemma 1.3, then $\delta_2(w(x) - w_2(x))$ satisfies condition 1) of Lemma 1.2 for $\theta = 1$. Thus $\delta_2(w(x) - w_2(x)) \leqq c\delta_1^2$ or $w(x) \leqq w_2(x) + c\delta_1\delta_2^{-1}$.

Recalling estimate (1.2.9), we have $w(x) < K\delta_1\delta_2^{-1}$ on N_α, where

$$K \leqslant c + c_6. \tag{1.2.10}$$

We proved (1.2.10) only under the assumption that δ_2 is small. But from the minimum principle it immediately follows that it is valid for any δ_2.

REMARK 1.2. In the 3-dimensional case it is possible to prove a substantially stronger result: $w(x) < \lambda^{\delta_2\delta_1^{-1}}$ for $x \in N_\alpha$ and some $\lambda < 1$. However the corresponding discussion makes use of the fact that γ is 1-dimensional and cannot be extended immediately to the n-dimensional case.

§1.3. Construction of a solution of F. Estimate of its gradient

The basic idea leading to the existence theorem for F in the work of M. I. Freĭdlin [**13**] consists of the following construction. Consider a second copy D' of D. All objects related to D' will be designated by a prime. Thus the point corresponding to x will be written as x'. We connect D and D' along Γ_1 and on the resulting manifold T we introduce a smooth structure such that the vector fields $l(x)$ and $-l'(x')$ coincide and form a vector field of class C^3. On the smooth manifold T there is defined an elliptic operator $\widetilde{L}$ which is equal to L on D and L' on D'. Finally the coefficients of the operator $\widetilde{L}$ have, generally speaking, discontinuities of the first kind on Γ_1.

On the manifold T with boundary $\partial T = \Gamma_2 \cup \Gamma_2'$ we investigate Dirichlet's problem

$$\widetilde{L}u = f, \quad u(x)|_{\Gamma_2} = g(x), \quad u(x')|_{\Gamma_2'} = g(x'). \tag{1.3.1}$$

In the work of I. V. Girsanov [**15**] there is an existence and uniqueness theorem for a solution of Dirichlet's problem in a smooth region $\hat{T}$ for an operator $\hat{L}$ the coefficients of which have discontinuities of the first kind on some smooth surface Γ, going out, perhaps, onto the boundary $d\hat{T}$ of $\hat{T}$. A small modification of problem (1.3.1) from the case analyzed by Girsanov consists in the fact that the boundary of the manifold T has a break on $\overline{\Gamma}_1 \cap \Gamma_2$.

Girsanov constructs a solution as the limit of solutions u_h for operators L_h with smoothed coefficients converging to $\hat{L}$ as $h \to 0$. In view of the extreme brevity of [**15**] it is perhaps worthwhile to describe this limiting process in detail. We will now do this for the case when $\hat{T}$ has a break on $\Gamma \cap d\hat{T}$.

Since $\hat{L}$ is nonsingular, u_h is bounded; it is also uniform in h (see for example [**13**], where probability methods are used for the proof).

From the uniform boundedness of u_h and Schauder estimates within the domain it follows that the derivatives of u_h up to the third order are uniformly bounded in $\hat{T} \backslash \Gamma$ with respect to h. Without loss of generality we can assume that in some neighborhood U of a point $\alpha \in \Gamma$ the surface Γ is given by the equation $x_1 = 0$.

Bernšteĭn's method of auxiliary functions [**16**] allows us to prove that the derivatives

$$\frac{\partial u_h}{\partial x_i}, \quad \frac{\partial^2 u_h}{\partial x_i \partial x_k}, \quad \frac{\partial^3 u_h}{\partial x_i \partial x_j \partial x_k}, \quad i, j, k \neq 1$$

are uniformly bounded in $\hat{T}$ with respect to h. We outline the proof that $\partial u_h / \partial x_i$, $i \neq 1$, is bounded. We consider the function $\zeta(x) \sum (\partial u_h / \partial x_i)^2 + cu = w_h$, where $\zeta(x)$ is an infinitely differentiable truncation function, equal to 1 on U and 0 within some neighborhood W of the boundary $\partial \hat{T}$. It can be verified that for sufficiently large c (depending on $\max u_h$, $\zeta(x)$, the coefficients of $\hat{L}_h$ and their derivatives in x_i, $i \neq 1$) $\hat{L}_h w_h > 0$ on $\hat{T} \backslash W$. Thus the maximum of w_h is reached on ∂W_h, where $w_h = cu$. Hence we get an estimate for $\sum_{i>1} (\partial u_h / \partial x_i)^2$. Now, using an analogous auxiliary function in which the squares of the second derivatives of u_h appear, it is possible to estimate the second derivatives and in an analogous manner estimate the third derivatives.

We now write the equation $\hat{L}_h u_h = f$ in the following form:

$$\sum a_{ij} \frac{\partial}{\partial x_i} \left(\frac{\partial u_h}{\partial x_1} \right) + b_1 \frac{\partial u_h}{\partial x_1} = \varphi, \tag{1.3.2}$$

carrying over into the right side all previously estimated expressions; (1.3.2) is a linear equation of the first order for $\partial u_h / \partial x_1$ the characteristics of which go out onto Γ because $\hat{L}$ is elliptic. Hence it follows easily that the function ∇u_h is uniformly bounded and uniformly continuous.

By Arzelà's theorem it is possible to distinguish a subsequence of the solutions u_h converging to the solution $u(x)$ of equation $\hat{L}u = f$. From the preceding estimates it follows that the convergence is uniform everywhere except $\Gamma \cap \partial \hat{T}$. With the help of barriers it is proved that on $\Gamma_2 \backslash \bar{\Gamma}_1$ Dirichlet's condition is satisfied. Hence $u(x)$ is the desired solution.

If $\tilde{\Gamma}$ is a closed subset of Γ lying inside of $\hat{T}$, then from the uniform boundedness of ∇u_h on $\tilde{\Gamma}$ it follows that the function ∇u is bounded on $\tilde{\Gamma}$.

§1.4. Uniqueness of the solution

If $z = \xi_0 + i\xi_1$, $r^2 = \xi_0^2 + \xi_1^2$ and $\phi(z)$ is the argument of z, then $\ln z = \ln r + i\phi(z)$. Hence the functions $\ln r$ and $\phi(z)$ are harmonic. From the formulas

$$\frac{\partial\varphi}{\partial\xi_k} = \frac{(-1)^{1-k}\xi_{1-k}}{r^2}, \quad \frac{\partial^2\varphi}{\partial\xi_k^2} = (-1)^k\frac{\xi_0\xi_1}{r^4}, \quad \frac{\partial^2\varphi}{\partial\xi_0\partial\xi_1} = \frac{\xi_1^2-\xi_0^2}{r^4}.$$

and analogous formulas for r we have:

1) $\|\nabla \ln r\| < c_1 r^{-1}$, $\|\nabla\varphi(z)\| < c_1 r^{-1}$;

2) $\left|\frac{\partial^2\varphi}{\partial\xi_k\partial\xi_l}\right| < c_2 r^{-2}$, $\left|\frac{\partial^2 \ln r}{\partial\xi_k\partial\xi_l}\right| < c_2 r^{-2}$, $k = 0, 1$;

3) if U is a smooth domain and ∂U is tangent to the real axis at 0, then on ∂U

$$\left|\frac{\partial\varphi}{\partial t}\right| < c_4, \quad \frac{\partial\varphi}{\partial n}(z) \sim c_3 r^{-1}(z),$$

$$\frac{\partial \ln r}{\partial t} \sim c_3 r^{-1}(z), \quad \left|\frac{\partial \ln r}{\partial n}\right| < c_4, .$$

where $\partial/\partial t$ ($\partial/\partial n$) is the derivative with respect to the unit tangent vector (with respect to the normal).

We consider the functions $\hat r(x)$ and $\hat\phi(x)$ in V_{δ_1,δ_2} generated by the functions $r(z)$ and $\phi(z)$. By (1.2.4) and subsections 1 and 2 we have

$$|L \ln \hat r(x)| < c\hat r^{-1}, \quad |L\hat\varphi(x)| < c\hat r^{-1}. \tag{1.4.1}$$

We consider on the plane the annulus $W: \delta^2 < \xi_0^2 + \xi_1^2 < \delta_1^2$, the boundaries of which we designate by S_δ and S_{δ_1}.

The solution u_δ of Dirichlet's problem for a Laplace operator with the conditions $u_\delta|_{S_\delta} = 1$, $u_\delta|_{S_{\delta_1}} = 0$ is, as is well known,

$$u_\delta(z) = [\ln(\delta_1/\delta)]^{-1}\ln(\delta_1/r(z)).$$

When $\delta_1 \to 0$

$$u_\delta \to 0 \tag{1.4.2}$$

at each point of the circle $\xi_0^2 + \xi_1^2 < \delta_1^2$ different from the origin of coordinates.

We investigate the connected component $\widetilde V_{\delta_1}$ of the δ_1-neighborhood γ; let $\overset{\circ}{V}_{\delta_1} = D \cap \widetilde V_{\delta_1}$ (this is a special case of $\overset{\circ}{V}_{\delta_1,\delta_2}$) and $\Gamma_\delta = V_\delta \cap \Gamma_1$. Let $w_\delta(x)$ be a solution of the problem $F[V_{\delta_1}, (\Gamma_1 \cap \partial V_\delta)\setminus\overline{\Gamma_\delta}, 0, g(x)]$, where $g(x) = 1$ on Γ_δ, $g(x)$ is continuous, $g(x) \leq 1$, and $g(x) = 0$ outside $\partial V_\delta \cap \Gamma_1$. In addition we assume that $w_\delta(x)$ is continuous on $\Gamma_\delta\setminus\gamma$, which by the minimum principle implies the uniqueness of a function satisfying these conditions. Obviously[9] $w_\delta \downarrow w$ when $\delta \downarrow 0$.

[9] The notation $x_\delta \downarrow x$ ($x_\delta \uparrow x$) means that x_δ converges to x from above (from below).

LEMMA 1.4. $w = 0$.

PROOF. Since the field $l(x)$ is not tangent to ∂D, and by property 3) of the derivatives of the functions $\phi(z)$ and $\ln r(z)$, there is a constant c_1 such that the function $\overline{w}_\delta = [\ln \hat{r}(x) + c_1\hat{\phi}(x)]$ will satisfy the condition $\partial\overline{w}/\partial l_1 \leqq 0$ on $(\partial V_\delta \cap \Gamma_1)\setminus\overline{\Gamma}_\delta$.

It is clear that $[\ln(\delta_1\delta^{-1})]^{-1}\overline{w}_\delta - w_\delta = \rho_\delta \geqq 0$ on ∂V_{δ_1}. By (1.4.1) and Lemma 1.2

$$\rho_\delta \geqslant -c_2^2\delta_1 (\ln(\delta_1\delta^{-1}))^{-1}. \tag{1.4.3}$$

From (1.4.2) and (1.4.3) we get the result of Lemma 1.4.

Let the function $u(x)$ be a solution of $F[D,\Gamma_1,0,0]$. By the minimum principle $\sup_{x\in D} u(x) = \overline{\lim}_{x\to\gamma} u(x) = A$. Obviously $u(x) \leqq A - \epsilon < A$ on $\partial V_{\delta_1} \cap D$. We consider the function $\tilde{u} = \epsilon^{-1}(u(x) - A + \epsilon)$ in V_{δ_1}. By the continuity of $\tilde{u}$ on $\Gamma_\delta\setminus\gamma$ and the minimum principle, for any δ we have $\tilde{u} < w_\delta$, where w_δ is the function investigated in Lemma 1.4. Hence $\tilde{u} \leqq 0$, i.e. $A \leqq 0$.

It is proved analogously that $\underline{\lim}_{x\to\gamma} u(x) \geqq 0$. The uniqueness of a solution of F follows immediately.

REMARK 1.3. The requirement that the domain D be smooth is clearly not essential. Everything in §§1.3 and 1.4 goes through for the case of a domain D with a finite number of breaks on Γ_2.

§1.5. Continuity of the solution of F in $\overline{D}$

In the general case of a domain $\hat{T}$ with a break on $\Gamma \cap \partial\hat{T}$ nothing can be said about the behavior of a solution of Dirichlet's problem (1.3.1) on $\Gamma \cap \partial\hat{T}$. If, however, we are dealing with problem F, investigation shows that a solution satisfies a Hölder condition on $\overline{\Gamma}_1 \cap \Gamma_2$ if $g(x)$ also has this property.

The value of the solution of F at $x \in D \cup \Gamma_1$ for a continuous boundary function g given on Γ_2 and null right side is the continuous linear functional of g which is given by Riesz's measure theorem:

$$u(x) = \int_{\Gamma_2} g(y)\,\mu(x, dy).$$

From the minimum principle it follows that the measure $\mu(x, B)$ is nonnegative. Obviously it is normalized:

$$\int_{\Gamma_2} \mu(x, dy) = 1.$$

We will call it the harmonic measure corresponding to the given problem.

Let $v(x)$ be a solution of $F[D,\Gamma,f,g]$.[10] We consider first the case when $g(x) \in H_\sigma$, i.e. satisfies a Hölder condition with index σ. We know that $v(x)$ is bounded. For concreteness let $0 \leqq v \leqq 1$. Let the L-distance from x_0 to Γ_2 be equal to h_1, and let $\hat{r}(x_0) = \delta_1$.

---

[10] In fact our discussion is fully applicable if $v(x)$ is a solution of the more general problem differing from F in that $\partial v/\partial l = \phi$, where ϕ is bounded.

LEMMA 1.5. *If* $\delta_1 \to 0$ *and* $h_1\delta_1^{-1} \to 0$, *then* $|v(x_0) - \phi(\alpha)| \leqq c_1 h_1 \delta_1^{-1} + c_2 \delta_1^{\sigma}$. *Here* α *is the point on the curve* γ *closest to* x_0.

PROOF. As is known, the solution $u(x)$ of Dirichlet's problem for the Laplace operator in the half-plane $x_2 > 0$ with the conditions $u(x_1, 0) = 0$ for $x_1 > 0$ and $u(x_1, 0) = \pi$ for $x_1 < 0$ is the angle between the vector x and the positive direction of the real axis. We now introduce the functions $\underline{v}(x)$ and $\bar{v}(x)$ as a solution of Dirichlet's problem for L with the Dirichlet condition $\bar{v}|_{\Gamma_2} = \underline{v}|_{\Gamma_2} = g$ and $\underline{v}|_{\Gamma_1} = 0$, $\bar{v}|_{\Gamma_1} = 1$. Obviously $\underline{v} \leqq v \leqq \bar{v}$.

Let $\tilde{g}(x)$ be a continuation of $g(x)$ on ∂D such that $\tilde{g}(x) \in H_\sigma(\partial D)$ and $0 \leqq \tilde{g} \leqq 1$. If $\tilde{v}(x)$ is a solution of Dirichlet's problem with boundary values $\tilde{g}(x)$, then, as is known, $\tilde{v}(x) \in H_\sigma(\bar{D})$; in particular,

$$|\tilde{v}(x_0) - g(\alpha)| < c_3 \delta_1^{\sigma}. \tag{1.5.1}$$

Obviously $\underline{v} \leqq \tilde{v} \leqq \bar{v}$. From the minimum principle it follows that

$$|\tilde{v} - v| \leqslant \bar{v} - \underline{v} = w. \tag{1.5.2}$$

On the other hand, w is a solution of Dirichlet's problem with boundary conditions $w|_{\Gamma_1} = 1$ and $w|_{\Gamma_2} = 0$. It is evaluated by an angle, as in the above example. We sketch a proof of this fact. The difference $\tilde{w} = w - \pi^{-1}\hat{\phi}(x)$ tends to zero as $x \to \gamma$, staying on ∂D, and is bounded on $\partial V_{\delta_1, \delta_2}$. Moreover, by (1.4.1) $|L\tilde{w}| < c_1 \hat{r}(x)^{-1} \leqq c_1 \tilde{r}(x)^{-1}$, where $\tilde{r}(x)$ is the distance from x to ∂D. Using this inequality and estimates of Green's function [**14**], it is not difficult to prove that $|\tilde{w}(x)| < c_2 \hat{r}(x) \ln[\hat{r}^{-1}(x)]$ in a neighborhood of γ. Hence $|w - \pi^{-1}\hat{\phi}(x)| < c_2 \hat{r}(x) \ln[\hat{r}^{-1}(x)]$.

Replacing the angle $\hat{\phi}$ by its tangent, which is possible because the tangent tends to 0 by hypothesis, we get the necessary estimate if $g(x)$ is constant. Combining the estimates (1.5.1) and (1.5.2), we get the result of the lemma.

We introduce the notation

$$B_\varepsilon = A_\varepsilon \cup (\Gamma_2 \cap \partial V_{\delta_1, \delta_2}), \tag{1.5.3}$$

where

$$A_\varepsilon = \{x : \hat{r}(x) = \delta_1,\ h_1(x) \hat{r}^{-1}(x) < \varepsilon\}.$$

Our immediate goal is the estimation of $\mu(x, B_\epsilon)$, after which it will be easy to establish a Hölder condition for the solution $v(x)$.

We pass a plane R_β through $l(\beta)$ and the tangent to γ at point β. Designate by $\pi\theta_\beta$ the angle between R_β and the real axis $x_2^{(\beta)}$ (cf. §1.2). Let $\theta < \min_{\beta \in \gamma} \theta_\beta$. We introduce the function $b_\theta(z) = r^\theta \sin \theta\phi$, where ϕ is the argument of z, $z = \xi_0 + i\xi_1$. We observe that $b_\theta(z) = \operatorname{Im} z^\theta$ is a harmonic function.

Consider the domain U_β on N_β bounded in the real coordinates by the level curve $b_\theta(z) = \delta_1^\theta \zeta$, the intersection $N_\beta \cap \partial D$, and the circle $|z| = \delta_1$. Observe that $\zeta = \sin\psi$, where ψ is the angle between the axis ξ_0 and the direction from the origin of coordinates to the point of intersection of the circle $|z| = \delta_1$ and the line $b_\theta(z) = \delta_1^\theta \zeta$.

Consider the union U of the domains U_β for all points $\beta \in \gamma$ which are a distance not greater than δ_2 from α in the L-metric.

We introduce the function $\hat{b}(x)$ in U generated by $b_\theta(z)$. The function $b_\theta(z)$ satisfies a Hölder condition with index θ. It can be verified that

$$\| \nabla b_\theta(z) \| < c_1 r^{\theta-1}, \quad \left| \frac{\partial^2 b_\theta(z)}{\partial \xi_k \, \partial \xi_l} \right| < c_2 r^{\theta-2}, \quad k = 0, 1.$$

Consequently $\hat{b}(x)$ satisfies a Hölder condition and by (1.2.4)

$$| L\hat{b}(x) | < c_3 \hat{r}^{\theta-1}. \tag{1.5.4}$$

Furthermore, for $z < 0$ we have $\cos \theta\pi \, \partial b_\theta / \partial \xi_\theta - \sin \theta\pi \, \partial b_\theta / \partial \xi_1 = 0$, whence $\partial \hat{b}(x) / \partial l_1 < c_4^2(\theta - \min_{\beta \in \gamma} \theta_\beta) \hat{r}^{\theta-1}(x)$. Finally, the function $b_\theta(z)$ is nonnegative in the upper half-plane and equal to 0 only on the real axis, whence it follows that

$$\min_{x \in \bar{U}} \hat{b}(x) \geqslant -c_5^2 \delta_1^{\theta+1}. \tag{1.5.5}$$

As in the previous lemmas the idea of the proof is that the function $v(x)$ cannot behave with respect to γ worse than $\hat{b}(x)$.

Note that the domain U is contained in $V_{\tilde{\delta}_1, \delta_2}$, where $\tilde{\delta}_1 = c(\theta)\delta_1$. Thus if we let $u(x)$ be a solution of $F[U, \Gamma_1 \cap \partial U, f, b(x)]$, then by Lemma 1.2

$$\hat{b}(x) - u(x) \geqslant -c_6^2 \delta_1^{\theta+1} \tag{1.5.6}$$

for sufficiently small δ_1. In fact the function $\hat{b}(x) - u(x)$ satisfies all conditions of the lemma except the condition that its derivative along the field l_1 be negative. But by the minimum principle such a function can be only still bigger. By Lemmas 1.2 and 1.3, estimates (1.5.5) and (1.5.6), and the definition of harmonic measure for problem $F[U, \Gamma_1 \cap \partial U, 0, g(x)]$, for $x \in N_\alpha$, we have, setting $\epsilon = \operatorname{tg} \arcsin \zeta$ in (1.5.3):

$$\begin{gathered} -c_5^2 \delta_1^{\theta+1} [\mu(x, B_\varepsilon) + c_7^2 \delta_1 \delta_2^{-1}] + \delta_1^\theta \zeta [1 - \mu(x, B_\varepsilon) - c_7^2 \delta_1 \delta_2^{-1}] \\ \leqslant u(x) \leqslant \hat{r}^\theta(x) \sin \theta \hat{\varphi} + c_8 \delta_1^{\theta+1}. \end{gathered} \tag{1.5.7}$$

If we take into consideration that $\delta_1 \delta_2^{-1} \to 0$ and $\zeta \to 0$ for $\delta_1 \to 0$ but $\zeta \delta_1^{-1} \to \infty$, then from (1.5.7) it follows that

$$\mu(x, B_\varepsilon) \geqslant 1 - c_9 \hat{r}^\theta(x) \delta_1^{-\theta} \zeta^{-1} - \delta_1 \zeta^{-1}. \tag{1.5.8}$$

We investigate the solution $w_k(x)$ of $F[u, \Gamma_1 \cap \partial U, f, g_k(x)]$, where $g_k(x) = v(x)$ on B_ϵ and $g_k(x) = k$ on $\partial U \setminus (B_\epsilon \cup \Gamma_1)$, $k = 0, 1$. We have $w_0(x) \leqq v(x) \leqq w_1(x)$. At the same time, taking Lemma 1.5 and (1.5.8) into account, we have

$$\begin{aligned} w_0(x) &\geqslant [\varphi(\alpha) - c_{10}\zeta - c_{11}\delta_1^\sigma] \, \mu(x, B_\varepsilon) \\ &\geqslant [\varphi(\alpha) - c_{10}\zeta - c_{11}\delta_1^\sigma] \, [1 - c_9 \hat{r}^\theta(x) \delta_1^{-\theta} \zeta^{-1} - \delta_1 \zeta^{-1}]. \end{aligned} \tag{1.5.9}$$

Let $\hat{r}(x) = \delta_1^{(2\sigma+\theta)/\theta}$ and $\zeta = \delta_1^{\sigma}$; then from (1.5.9) we get

$$|w_0(x) - \phi(\alpha)| < c_{12}\hat{r}^{\lambda}(x), \quad \text{where} \quad \lambda = \min\left[\frac{\sigma\theta}{2\sigma+\theta}, \frac{(1-\sigma)\theta}{2\sigma+\theta}\right].$$

Analogously we get the same sort of estimate for w_1. Thus we have proved the Hölder condition.

Because we can uniformly approximate a continuous function on Γ_2 by smooth ones, by the minimum principle the convergence of solutions will also be uniform, and thus we have continuity on $\overline{\Gamma}_1 \cap \Gamma_2$ also in the case of continuous Dirichlet conditions on Γ_2.

REMARK 1.4. If ∂D has a break relative to $\overline{\Gamma}_1 \cap \Gamma_2$ and the angle between the tangent planes $\hat{T}_i$ to Γ_i, $i = 1, 2$, at the points of $\overline{\Gamma}_1 \cap \Gamma_2$ is no greater than π, then all of the preceding discussion is valid, which shows the continuity of a solution on $\overline{\Gamma}_1 \cap \Gamma_2$.

REMARK 1.5. It is easy to get a more precise estimate of the index in the Hölder condition. For us this is not essential, and therefore we only outline a procedure. It is clear that the estimate of Lemma 1.5 is very crude, because in the proof of the lemma we used an estimate of a solution of Dirichlet's problem with discontinuous boundary values. Now that we know that $v(x) \in H_\lambda$ it is possible to strengthen the result of Lemma 1.5 in the following manner. If $h_1\delta_1^{-1} < \delta_1^{1-\lambda-\nu}$, $0 < \nu < 1 - \lambda$, then

$$|v(x) - \varphi(\alpha)| < c\delta_1^{\sigma - \frac{(\sigma-\lambda)\nu}{1-\lambda}}.$$

Moreover, again using the fact that $v(x) \in H_\lambda$, it is possible to estimate it a priori on the level line $b_\theta(x) = \delta_1^{\theta}\zeta$. Repeating the discussion of the lemma with the indicated improvements in precision, it is possible to prove that $v(x) \in H_{\lambda'}$, $\lambda' > \lambda$, and it is possible to continue this chain of estimates and thereby to improve substantially the result on the smoothness properties of $v(x)$.

CHAPTER 2

BEHAVIOR OF A SOLUTION OF PROBLEM *F* NEAR *M*

§2.1. Basic definitions

Let D be a bounded domain of class C^3 in the Euclidean space E^3. On its boundary ∂D let there be given a vector field $l(x)$ of class C^3 which is not mapped into 0. Designate by M the set of points at which $l(x)$ lies in the tangent plane $\hat{T}_x$ to ∂D.

Let

$$L = \sum a_{ij}(x)\frac{\partial^2}{\partial x_i \partial x_j} + \sum b_i(x)\frac{\partial}{\partial x_i} + c(x), \ x \in D,$$

be an elliptic operator, and let $\sum a_{ij}\lambda_i\lambda_j \geqq \theta \sum \lambda_i^2$, where $\theta > 0$, and the coefficients belong to class $C^3(\overline{D})$, where $\overline{D} = D \cup \partial D$ and $c(x) \in C^3(\overline{D})$, $c(x) \leqq 0$.

We assume that

1) M consists of a finite number of curves of class C^2;

2) if the curve $\gamma \subset M$, then $l(x)$ is not tangent to γ at any of the points of γ.

Let $n(x)$ be the vector of the interior normal to ∂D at the point x. We introduce the vector field

$$l_1(x) = \begin{cases} l(x), & \text{if } (l(x), n(x)) \geqslant 0, \\ -l(x) & \text{in the opposite case.} \end{cases}$$

Let M_δ be a δ-neighborhood of M and $S_\delta = M_\delta \cap \partial D$. If δ is sufficiently small, then S_δ consists of nonintersecting neighborhoods γ_δ of the curves γ (relative to ∂D) and the domain γ_δ is split by the curve γ into the parts γ_δ^1 and γ_δ^2. We will decompose each point α of the curve γ into points α^1 and α^2, and the term ϵ-neighborhood ($\epsilon > \delta$) of the point α^i will be used to designate the set of points of γ_δ^i which are at a distance less than ϵ from α. We will use the notation $\lim_{x \to \alpha^i} f(x)$ or simply $f(\alpha^i)$ to designate the limiting value of the function at the point α^i. The set of points α^i ($\alpha \in \gamma$) will be called the side γ^i of the curve γ. Let $n^i(\alpha)$ be the vector of the normal to γ at α in the sense of the metric $(\cdot,\cdot)_L$ (cf. the Introduction) which lies in the tangent plane $\hat{T}_\alpha$ to ∂D at α and which is directed to the side γ_δ^i. From (2.2) it follows that the scalar product $(l_1(\alpha^i), n^i(\alpha))_L \neq 0$ and that its sign is constant for all points of the side γ^i, $i = 1, 2$. We will say that the side γ^i of the curve γ is positive (respectively negative) if the scalar product is negative (positive). If both sides of γ are positive (negative), then we call γ positive (negative); if the sides of γ are of different signs, then γ is said to be null. The union of the positive (negative) sides will be designated by M^+ (M^-).

§2.2. Investigation of positive sides

Let U be a 3-dimensional neighborhood of point α on the singular curve γ which has positive side γ^1. Let $\partial U \cap \partial D$ split γ into parts γ_U^i ($i = 1, 2$) which adjoin the sides γ^i. We assume that $\partial U \in C^3$ and that ∂U does not contain points of other singular curves of M. Let $\Gamma_\delta = \gamma_\delta^1 \cap \partial U$. We designate by u_δ a solution of problem $F[\overset{\circ}{U}, \gamma_U^1 \backslash \Gamma_\delta, 0, g(x)]$, where $g(x) = 1$ on Γ_δ and $g(x) = 0$ on $\partial U \backslash \gamma_U^1$. The function u_δ decreases as $\delta \to 0$. This follows immediately from Lemma 1.1. In fact from this lemma it follows that for $\delta_1 < \delta_2$ the function $u_{\delta_2} - u_{\delta_1}$ cannot attain a minimum on $\gamma_U^1 \backslash \Gamma_{\delta_2}$. But at the remaining points of the boundary $u_{\delta_2} \geqq u_{\delta_1}$; thus everywhere $u_{\delta_2} \geqq u_{\delta_1}$. Let $u_0 = \lim_{\delta \to 0} u_\delta$.

THEOREM 2.1. $u_0(\alpha^1) = 1$.

PROOF. With the help of the minimum principle it is easy to prove that if the domain $\widetilde{U}$ is contained in U, then the solution of F for $\widetilde{U}$ will be no larger than the solution for U. We will take as $\widetilde{U}$ the neighborhood V_{δ_1,δ_2} of γ which was described in §1.2, only now γ^1 is a positive side. Let $u_{\delta_2}^{(\delta_1)}$ represent the corresponding solution. We will prove that

$$u_0^{(\delta_1)}(\alpha^1) \geqslant 1 - c\delta_1^{\nu}, \ \nu > 0. \tag{2.2.1}$$

The result of Theorem 2.1 follows from this estimate and the preceding remarks. Note that Lemmas 1.2 and 1.3 are valid for V_{δ_1,δ_2}.[11]

From property 3 of the function $\phi(x)$ introduced in §1.4 it follows that[12]

$$\left|\frac{\partial \widehat{\varphi}(x)}{\partial l}\right| < c_1 \text{ on } \gamma_V^1. \tag{2.2.2}$$

Let ν_δ be a solution of $F[\overset{\circ}{V}_{\delta_1,\delta_2}, \gamma_V^1 \setminus \Gamma_\delta, 0, g(x)]$, where $g(x) = \pi$ on N_{α_i}, $i = 1, 2$, and $g(x) = \hat{\phi}(x)$ on the remaining part of ∂V. Then from (1.2.4), (2.2.2), Lemmas 1.1 and 1.3, and Remark 2.1 it follows that

$$|\widehat{\varphi}(x) - v_\delta(x)| < c_2\delta_1 + c_3\delta_1\delta_2^{-1} < c_4\delta_1\delta_2^{-1} \tag{2.2.3}$$

on the plane N_α.

On the other hand, by the definition of the harmonic measure $\mu(x, \Gamma)$ of the problem $F[\overset{\circ}{V}_{\delta_1,\delta_2}, \gamma_V^1 \setminus \Gamma_\delta, 0, g(x)]$

$$v_\delta(x) = \int_{\partial V \setminus [\gamma_V^1 - \Gamma_\delta]} g(x)\, \mu(x, dy).$$

Let $\Gamma_{\delta,h} = \Gamma_\delta \cup \Phi_h$, where $\Phi_h = [\partial V \cap D] \cap \{x : \hat{\phi}(x) \leqq h\}$. We have

$$v_\delta(x) = \int_{\Gamma_{\delta,h}} g(y)\, \mu(x, dy) + \int_{\partial V \setminus [\gamma_V^1 \cup \Phi_h]} g(y)\, \mu(x, dy)$$

$$\geqslant \min_{\partial V \setminus [\gamma_V^1 \cup \Phi_h]} g(x)\, [1 - \mu(x, \Gamma_{\delta,h})].$$

Hence

$$v_\delta(x) \geqslant h[1 - \mu(x, \Gamma_{\delta,h})]. \tag{2.2.4}$$

Because $\hat{\phi}(x) \to 0$ as $x \to \alpha^1$, and by virtue of (2.2.3), we have

$$\overline{\lim_{x \to \alpha^1}}\, v_\delta(x) \leqslant c_4\delta_1\delta_2^{-1}. \tag{2.2.5}$$

From (2.2.4) and (2.2.5) we get

---

[11] REMARK 2.1. In the proof of Lemma 1.2 we used the inequality (1.2.6), which we got because $l(x)$ is not tangent to ∂D. Here (1.2.6) is not satisfied; however from 2.2 in §2.1 it follows that $|a_2|\,(a_1^2 + a_2^2 + a_3^2)^{-1/2} \geqq \varkappa > 0$ on $\Gamma_1 \cap V$. Thus if we replace $\hat{x}_3$ by $\hat{x}_2$ in the proof of Lemma 1.2, it is easy to see that Lemma 1.2 is still valid if γ is a singular curve.

[12] If $x \in N_\beta$, then $\hat{\phi}(x)$ is the angle in the $L(\beta)$-metric between the vector $x - \beta$ and the axis $x_2^{(\beta)}$ directed to γ_V^1.

$$\varliminf_{x\to\alpha^1}\lim_{\delta\to 0}\mu(x,\Gamma_{\delta,h}) \geqslant 1 - c_4\delta_1\delta_2^{-1}h^{-1}. \tag{2.2.6}$$

On the other hand, from the results of §1.5 it is easy to show that

$$\varliminf_{x\to\alpha^1}\lim_{\delta\to 0}\mu(x,\Gamma_\delta \cup \Psi_\varepsilon) \geqslant 1 - c_5\delta_1^\lambda, \tag{2.2.7}$$

$$\lambda > 0,\quad \Psi_\varepsilon = \partial V \cap \{x : \hat\varphi(x) > \pi - \varepsilon\}.$$

If we take into account that $h = \delta_1^{1/2}$ and $\delta_2 = \text{const}$, then it is clear that the estimates (2.2.6) and (2.2.7) can both be satisfied at the same time only in the case that for $\nu = \min(\lambda, 1/2)$ we have

$$\lim_{x\to\alpha^1}\lim_{\delta\to 0}\mu(x,\Gamma_\delta) \geqslant 1 - c_6\delta_1^\nu.$$

It remains only to note that $\mu(x,\Gamma_\delta) = u_\delta^{(\delta_1)}(x)$. Thus (2.2.1) is proved and the proof of Theorem 2.1 is complete.

§2.3. Behavior of a solution of F near M^-

We now consider the same problem as in §2.2 except that we now let γ^1 be the negative side. The behavior of the solution u_δ is quite different from its behavior in a neighborhood of the positive side.

THEOREM 2.2. 1) $\lim_{\delta\to 0} u_\delta \equiv 0$. 2) *Let $\tilde u_\delta$ be a solution of the same problem as for u_δ, but let $\tilde u_\delta|_{\gamma_V^2} = 1$. Then* $\varlimsup_{x\to\alpha^1}\tilde u_\delta(x) < 1$.

Assertion 1) of this theorem testifies to the stability of a solution of F in the presence of a disturbance of the solution on a small region near γ^1. Specifically, if we set the boundary value of the function equal to 1 on a small region and then let this region be retracted to γ^1, the limit of the solution will be the same as if we had set the boundary values equal to zero on this region and then retracted it to γ^1. The situation which we encountered in §1.4 obtains at those points of the boundary where the vector field is not tangent to the boundary. As we saw (Theorem 2.1) the situation changes on the positive sides of curves from M.

Assertion 2) shows the difference between negative sides and ordinary points at which the vector field is not tangent. In the last case the limiting function $\lim_{\delta\to 0}\tilde u_\delta$ would be continuous on γ in any case. Here this is not so. Dirichlet conditions imposed from the opposite side of γ exert only a weak influence on the value of the function from the negative side of the singular curve.

Assertion 1) has already been proved in §1.4 if in the discussion of §1.4 we let γ be a curve from M having negative side γ^1, and in this case the proof is simpler because it is obvious that $\partial \ln \hat r/\partial l_1 \leq 0$ and it is not necessary to introduce the function $\hat\phi$.

REMARK 2.2. By altering that same discussion only slightly, it is possible to prove the following fact. If both sides are negative and u'_δ is a solution of the problem $F[\overset{\circ}{V}, ({}_2V \cap \partial D) \backslash S_\delta, 0, g(x)]$, $g(x)|_{S_\delta} = 1$, and $g(x)|_{\partial V \cap D} = 0$, then $\lim_{\delta \to 0} u'_\delta \equiv 0$.

PROOF OF ASSERTION 2). Consider the function $\hat{\phi}(x)$ introduced in §1.4, except that now V_{δ_1,δ_2} is a neighborhood of the point α on the curve γ having negative side γ^1.

The function $\hat{\phi}(x)$ is discontinuous along γ; we will use this fact for the proof of 2). Consider the function v_δ, introduced in §2.2, in the domain $\overset{\circ}{V}_{\delta_1,\delta_2}$. If $\mu(x, \Gamma)$ is the harmonic measure of $F[\overset{\circ}{V}, \gamma_V^1 \backslash \Gamma_\delta, 0, g(x)]$, then

$$v_\delta(x) = \int_{\gamma_V^2} g(y)\, \mu(x, dy) + \int_{\Gamma_\delta \cup (\dot{D} \cup \partial V)} g(y)\, \mu(x, dy)$$

$$\geqslant A\mu(x, \gamma_V^2) + B(1 - \mu(x, \gamma_V^2)),$$

where

$$A = \inf_{y \in \gamma_V^2} g(y) > \frac{\pi}{2}, \qquad B = \inf_{(D \cap \partial V) \cup \Gamma_\delta} g(y) > -\frac{\pi}{2}.$$

By (2.2.3) and the fact that $\hat{\phi}(x) \to 0$ as $x \to \gamma^1$ there exists for any ϵ a neighborhood G_ϵ of α^1 having the following property: for $x \in G_\epsilon$ and any δ,

$$\varepsilon + c\delta_1 + K\delta_1\delta_2^{-1} \geqslant \mu(x, \gamma_V^2)(\pi - A) + (1 - \mu(x, \gamma_V^2))B,$$

whence

$$\mu(x, \gamma_V^2) \leqslant (A - B)^{-1}[\varepsilon + c\delta_1 + K\delta_1\delta_2^{-1} - B].$$

When $\delta_1 \to 0$ and $\delta_1\delta_2^{-1} \to 0$, this expression remains strictly less than 1, which proves 2) since $\mu(x, \gamma_V^2) = \tilde{u}_\delta(x)$.

CHAPTER 3

INVESTIGATION OF PROBLEM P IN DOMAIN D

§3.1. Passage to the limit alongside M^-

We consider the boundary value problem $P = P_f$:

$$Lu = f \text{ in } D \qquad (f \in C^3(\overline{D})) \tag{3.1.1}$$

$$\partial u / \partial l = 0 \text{ on } \partial D \backslash M. \tag{3.1.2}$$

A solution is a bounded function $u(x) \in C^2(D)$ such that $u \in C^1(\overline{D} \backslash M)$ and conditions (3.1.1) and (3.1.2) are satisfied.

In the preceding sections we investigated the behavior of solutions of problem F in small neighborhoods of singular curves. In this chapter, using the preceding results, we concern ourselves with solutions of problem P in the whole domain D.

First we prove an auxiliary proposition, from which the uniqueness theorem will follow.

Let T_δ be the union of the δ-neighborhood S_δ^- of the set M^- and a fixed neighborhood U of the set M^+. Consider the solution u_δ of $F[D, \partial D \setminus T_\delta, 0, g(x)]$, where $g(x) = A$ on S_δ^- and $g(x) = B$ on U.[13] (It is assumed that M^+ is not empty.) As $\delta \to 0$ the function u_δ decreases and the limit $\lim_{\delta\to 0} u_\delta = u$ exists. We will prove that $u = B$.

Now fix a neighborhood U^- of singular curves having a negative side such that it is possible to apply Theorem 2.2 to the neighborhood. Let $\Gamma = \overline{D \cap \partial U^-}$.

By Lemma 1.1, $\max_{x\in \overline{D}\setminus U^-} u_\delta(x) = \max_{x\in\Gamma} u_\delta(x)$ for all δ. By the estimate of the gradient of the solution of F, the sequence $u_\delta(x)$ converges uniformly on Γ to the function $u(x)$, which is thus continuous. Moreover on $\Gamma \cap \partial D$ the condition $\partial u/\partial l = 0$ is satisfied. In fact the sequence u_δ converges uniformly outside some neighborhood $\widetilde{U}^- \subset U^-$, and thus by Lemma 1.1 $u(x)$ coincides with the solution $u^*(x)$ of $F[D\setminus \widetilde{U}^-, \partial D \setminus \widetilde{U}^-, 0, u(x)]$.[14]

On the other hand, by Theorem 2.2 and the Remark to it

$$\sup_{x\in U^-} u(x) = \max_{x\in\Gamma} u(x).$$

So

$$\sup_{x\in D} u(x) = \max_{x\in\Gamma} u(x),$$

and this is possible, by Lemma 1.1, only when $u(x)$ is constant; hence $u = B$.

As an obvious consequence we have

LEMMA 3.1. *Suppose that we have a continuous function $\phi(x)$ on some neighborhood U^+ of the set M^+ (relative to ∂D). Consider the solution v_δ of $F[D, \partial D \setminus [U^+ \cup \overline{S}_\delta], 0, g_\delta(x)]$, where $g_\delta = \phi$ on M^+, $g_\delta = \phi_\delta(x)$ on S_δ^-, and $\phi_\sigma(x)$ is bounded by a constant not depending on δ. Then the limit $\lim_{\delta\to 0} v_\delta(x) = v(x)$ exists and does not depend on the choice of ϕ_δ; moreover*

$$\min_{x\in U^+} \varphi(x) < v < \max_{x\in U^+} \varphi(x).$$

§3.2. Basic theorem

In this section we concern ourselves with the construction of solutions of problem P in the whole domain D.

We enumerate the positive sides and let γ_k^i $(i = 1, 2)$ be the ith side of the kth positive singular curve.

THEOREM 3.3. *Let $f \in C^3(\overline{D})$. Let a continuous function $g_k^i(\alpha)$ correspond to each positive side γ_k^i. There exists a unique solution $u(x)$ of P_f such that*

---

[13] A and B are constants; for concreteness suppose $A > B$.

[14] $u^*(x) - u_\delta(x)$ has an extremum on $\overline{D \cap \widetilde{U}}$ and therefore converges uniformly to 0 as $\delta \to 0$.

$\lim_{x\to\alpha^i} u(x) = g_k^i(\alpha)$ *for any point* α^i *of the positive side* γ_k^i. *If* M^+ *is empty, then every solution of* P_0 *is a constant.*

PROOF. Let $\tilde{g}_k^i(x)$ be a continuous extension of the function $g_k^i(x)$ to a δ-neighborhood of M^+. Let there be continuous functions uniformly bounded in δ on neighborhoods of negative sides; consider the corresponding solution $u_{\epsilon,\delta}$ of F and let ϵ go to 0. By the preceding lemma the limit of u_δ exists and does not depend on the values that were assigned from the negative sides.

Now let δ go to 0.

Let α^i be a point of the positive side γ_k^i. Let V be a neighborhood of α of the form investigated in Theorem 2.1. Let u_δ (respectively $\bar{u}_\delta$) be a solution of $F[\overset{\circ}{V}, \gamma_V^1\backslash\Gamma_\delta, 0, \underline{g}_k^i(x)]$ $(F[\overset{\circ}{V}, \gamma_V^1\backslash\Gamma_\delta, 0, \bar{g}_k^i(x)])$, where $\underline{g}_k^i$ $(\bar{g}_k^i)$ equals g_k^i on Γ_δ and $\underline{g}_k^i(x) = \min_{i,k,x}\tilde{g}_k^i(x)$ $(\bar{g}_k^i(x) = \max_{i,k,x}\tilde{g}_k^i(x))$ on the remaining part of ∂V. By Theorem 2.1

$$\lim_{x\to\alpha^i}\lim_{\delta\to 0} \underline{u}_\delta(x) = \lim_{x\to\alpha^i}\lim_{\delta\to 0} \bar{u}_\delta(x) = g_k^i(\alpha).$$

But

$$\underline{u}_\delta(x) \leqslant u_\delta(x) \leqslant \bar{u}_\delta(x.)$$

From this and the minimum principle it is clear that the sequence $u_\delta(x)$ converges uniformly as $\delta\to 0$ near M^+, and consequently in the whole region $\bar{D}$. The limiting function $u(x)$ is clearly continuous in M^+, $Lu = 0$, and the condition $\partial u/\partial l|_{\partial D\backslash M} = 0$ is satisfied, which is proved exactly as in Lemma 3.1.

Because the convergence of the functions $u_\delta(x)$ to $u(x)$ is uniform, we can get the following minimum principle for P from Lemma 3.1 by passing to the limit when the neighborhood of M_δ^+ is retracted to M^+.

If the solution of P_0 is continuous in M^+, then its minimum is attained on M^+.

The uniqueness of a solution of P_f follows immediately. If M^+ is empty and $u(x)$ is a solution of P_0, then from Theorem 2.2 it follows that $u(x)$ has an extremum on $\partial M_\delta \cap D$, and from the minimum principle it follows that $u(x) = \text{const}$.

In order to get the existence and uniqueness theorem for P_f, it remains to construct a particular nonhomogeneous solution of P_f. It is sufficient to consider the case when $f \leqq 0$ in (3.1.1). Let $f \in C^3(\bar{D})$. Consider a solution F_δ of $F[D, \partial D\backslash S_\delta, f, 0]$.

In a corollary to Lemma 4.2 it will be proved that the family F_δ is uniformly bounded. From the minimum principle it is clear that $F_\delta(x) \uparrow F(x)$. We will prove that $F(x)$ is the desired solution of P. Let $\Gamma_\epsilon = D \cap \partial M_\epsilon$ and let $G_\delta(x)$ be a solution of $F[D\backslash\bar{M}_\epsilon, \partial D\backslash S_\epsilon, 0, F_\delta(x)]$ (ϵ fixed). Clearly, within $D\backslash M_\epsilon$ we have $F_\delta = G_\delta + R$, where R is a solution of $F[D\backslash M_\epsilon, \partial D\backslash S_\epsilon, f, 0]$, and thus R does not depend on δ.

On the other hand the functions $G_\delta(x)$ converge to a solution of $F[D\backslash\bar{M}_\delta, \partial D\backslash S_\delta, 0, F(x)]$. (This is proved exactly as the corresponding result in Lemma 3.1.) Because ϵ is arbitrary, $F(x)$ is a solution of P.

The assertion $F(x) \to 0$ as $x \to M^+$ is a consequence of Lemma 4.1. In fact $\min_{x\in\bar{D}} fw(x) \leqq F(x) \leqq \max_{x\in\bar{D}} fw(x)$ (the function $w(x)$ is introduced in the formulation of the theorem), and $w(x) \to 0$ as $x \to M^+$.

Note that we go through the proof of Lemma 4.1 in the next chapter in order to illustrate the use of probability methods. Everything proved there can be done without the use of probability techniques, but then the presentation would be less intuitive.

CHAPTER 4

PROBABILISTIC ANALYSIS OF PROBLEM P

§4.1. Probability interpretation of the preceding results

A homogeneous Markov process in a domain D defines a semigroup of operators T_t $(0 \leqq t \leqq \infty)$ acting on the space $C(D)$ of continuous functions on D by the formula

$$T_t f(x) = \int_D \mathrm{P}(t, x, dy) f(y).$$

Here $\mathrm{P}(t, x, \Gamma)$ is the hitting probability of the trajectory of the process from the point x into the set Γ in time t.

The semigroup T_t is completely characterized by its infinitesimal operator $A: Af(x) = \lim_{t\to 0} t^{-1}(T_t f(x) - f(x))$ defined on the domain of definition D_A, which is dense in $C(D)$ and which consists of functions for which this limit exists in the sense of uniform convergence.

In [13] M. I. Freĭdlin constructed a diffusion process X^δ with continuous trajectories in D the infinitesimal operator of which coincides with the elliptic differential operator L on smooth functions in D_A; and in the domain of definition there are only those smooth functions for which the derivative along the field l on $\partial D \backslash S_\delta$ equals 0 and the limiting value on S_δ equals 0.

The process X^δ is called a reflection process along the field l on $\partial D\backslash S_\delta$, breaking off at the moment of hitting S_δ. In [13] it is proved that X^δ is strictly Markovian.

In [17] there is a construction of a Markov process in D such that the processes X^δ are its parts, i.e., until the moment ζ^δ when X^δ breaks off, all its characteristics coincide with the characteristics of X. Moreover $\zeta^\delta \uparrow \zeta$, where ζ is the moment when X breaks off. The process X breaks off at the moment of hitting the set M.

The connection between X and problem F is based on the following theorem, which comes from [13].

THEOREM 4.1. *If problem $F[D, \partial D\backslash S_\delta, f, g]$, $f \in C(\bar{D})$, has a solution $u(x)$, then*

$$u(x) = \int_0^{\zeta^\delta} f(x_t)\,dt + \mathrm{M}_x g(x_{\zeta^\delta}). \tag{4.1.1}$$

Here x_t is the state of the trajectory of X at time t, and $\mathrm{M}_x\xi$ denotes the mathematical expectation of the quantity ξ given that the process X starts from the point x.

As we know, if $f \in C^3(\overline{D})$, then problem F has a solution which is given by formula (4.1.1).

THEOREM 4.2. *Let V be the domain described in Theorem 2.1, and let γ^1 be the positive side of the curve γ. Then the probability that the trajectory of X breaks off before hitting $\partial V \backslash \gamma_V^1$ tends to 1 when $x \to \alpha^1$.*

PROOF. Let $\tau_\delta = \inf(t: x_t \in \Lambda_\delta)$, where $\Lambda_\delta = \Gamma_\delta \cup (\partial V \backslash \Gamma_V^1)$ (Γ_δ is introduced in Theorem 2.1).

From Lemma 1.2 (for $\theta = 1$) and Theorem 4.1 it follows that $\mathrm{M}_x\tau_\delta$ is uniformly bounded in δ. Hence $\tau = \lim_{\delta\to 0}\tau_\delta < \infty$ with probability 1. Thus the section of the trajectory of X until time τ is a compact set.

By Theorem 4.1 the assertion of Theorem 2.1 is equivalent to the assertion that the probability $u(x)$ of hitting the trajectory of X on all sets Γ_δ, $\delta > 0$, before hitting $\partial V \backslash \gamma_V^1$ tends to 1 as the initial point x of the trajectory of X tends to α^1. Hence by the compactness of the trajectory the distance up to γ^1 is equal to zero up to the moment of hitting $\partial V \backslash \gamma_V^1$ with probability $u(x)$ arbitrarily close to zero, which was to be proved.

Theorem 2.2 means in probability language that the probability of reaching the negative side of the curve equals 0; moreover, the probability that, starting arbitrarily close to γ from the negative side of γ, the trajectory will hit the neighborhood of the opposite side of γ is less than a constant which in turn is less than 1.

Finally, by Theorem 4.1, if $f \in C^3(\overline{D})$, then $\mathrm{M}_x \int_0^{\zeta^\delta} f(x_t)\,dt + \mathrm{M}_x g(x_{\zeta^\delta}) = F_\delta(x)$ $(f \geqq 0)$ is a solution of $F[D, \partial D \backslash S_\delta, f, g]$. If $\delta \to 0$, then $\zeta^\delta \to \zeta$; if $g(x)$ is continuous on M^+, then $\mathrm{M}_x g(x_{\zeta^\delta}) \to \mathrm{M}_x g(x_\zeta)$ by Theorem 4.2; and by Lebesgue's theorem on passing to the limit under the integral sign we get $\mathrm{M}_x \int_0^{\zeta^\delta} f(x_t)\,dt \to \mathrm{M}_x \int_0^{\zeta} f(x_t)\,dt$. On the other hand, as we proved in Chapter 3, F_δ tends to the solution $F(x)$ of P_f.

We have proved

THEOREM 4.3. *The function $F(x) = \mathrm{M}_x[\int_0^\zeta f(x_t)\,dt + g(x_\zeta)]$ is the unique solution of the problem P_f: $F|_{M^+} = g$.*

§4.2. Proof of Lemma 4.1

Here we will prove the assertion used in §3.2.

LEMMA 4.1. $\sup_x \mathrm{M}_x\zeta < K < \infty$; $\lim_{x\to M^+} \mathrm{M}_x\zeta = 0$.

PROOF. As we know, the moment ζ^δ of reaching the set S_δ tends to ζ. It is also known [13] that $\mathrm{M}_x\zeta^\delta$ are bounded functions of x. We will prove that $\mathrm{M}_x\zeta^\delta$ is less than a constant not dependent on δ, and we will show that $\mathrm{M}_x\zeta$ is bounded.

Now fix two systems U^i, $i = 1, 2$, of neighborhoods of the curves from M which have positive sides, and let $U^1 \subset U^2$. We enumerate the curves from

$\overline{M}$ and let U_k^i be the connected component of U^i containing γ_k. Let $\Gamma_k^i = \overline{D \cap U_k^i}$, $\Gamma^i = \bigcup_k \Gamma_k^i$, $\Gamma_k = \Gamma_k^1 \cup \Gamma_k^2$, and let them have the following properties:

a) There exists a set $V_k^1 \subset \Gamma_k^1$ such that the probability $\pi_k(x)$ of hitting M^+ before exiting on U_k^2, starting from $x \in V_k^1$, satisfies the following inequality:

$$\pi_k(x) > 1/2.$$

b) Let $\tau_1 = \min(t : x_t \in U^1)$. Then[15] $P_x\{x_{\tau_1} \in V_k^1\} > \sigma > 0$ if $x \in \Gamma_k^2$.

The existence of such neighborhoods follows from Theorem 2.1. Let $A_\delta = \max_{x \in \Gamma_k^2} M_x \zeta^\delta$, let $B = \max_{x \in \Gamma^2} M_x \tau_1$, and let τ_2 be the moment at which the trajectory hits $\Gamma^2 \cup S_\delta$; moreover $M_x \tau_2 < c$ for $x \in \Gamma^1$. Let $x \in \Gamma_k^2$ and $M_x \zeta^\delta = A_\delta$. Then by the strict Markov property of X^δ

$$A_\delta = M_x \tau_1 + M_x M_{x_{\tau_1}} \zeta^\delta \leqslant B + M_x [\chi_{E_1} M_{x_{\tau_1}} (\zeta^\delta \chi_{E_2})]$$

$$+ M_x [\chi_{\overline{E}} M_{x_{\tau_1}} \zeta^\delta] = B + \mathrm{I} + \mathrm{II},$$

where χ_A is the indicator function[16] of the event A; $\overline{E}$ is the negation of E, $E = E_1 E_2$, $E_1 = \{x_{\tau_1} \in V_k^1\}$, $E_2 = \{\zeta^\delta < \theta_{\tau_1} \tau_2\}$, and $\theta_{\tau_1}\tau_2$ is the value of τ_2 for a trajectory for which the section from 0 to τ_1 is thrown out; so here $\mathrm{I} \leqq M_x \chi_E \cdot c \leqq c$; $\mathrm{II} \leqq (A_\delta + c)(1 - \sigma/2)$, because $M_{x_{\tau_1}} \zeta^\delta \leqq M_{x_{\tau_1}} \tau_2 + M_{x_{\tau_1}} M_{x_{\tau_2}} \zeta^\delta \leqq c + A_\delta$, and $M_x \chi_{\overline{E}} = P_x(\overline{E}) = 1 - P_x(E) = 1 - M_x \chi_{E_1} P_{x_{\tau_1}}(E_2) \leqq 1 - \frac{1}{2} P_x(E_1) \leqq 1 - \sigma/2$. So

$$A_\delta \leqslant B + c + \left(1 - \frac{\sigma}{2}\right)(A_\delta + c),$$

whence

$$A_\delta \leqslant \frac{2B + 4c}{\sigma}.$$

We see that $M_x \zeta^\delta$ is bounded by a constant not dependent on δ on Γ^2, but by the strict Markov property

$$M_x \zeta^\delta = M_x \tau_1 + M_x M_{x_{\tau_1}} \zeta^\delta \leqslant M_x \tau_1 + \frac{2B + 4c}{\sigma}.$$

As we know, $M_x \tau_1$ is bounded; thus the assertion is proved for all $x \in D \setminus \overline{M}$.

Let $T_\epsilon = \overline{M_\epsilon \cap D}$ and let γ^1 be the positive side. Let τ be the moment that the trajectory hits the set $\gamma_\epsilon^2 \cup T_\epsilon$. Then from Lemma 1.3, Theorem 2.1, and their probabilistic interpretations in §4.1 we have for $x \in M_\epsilon$:

$$M_x \tau \leqslant c\varepsilon^2; \qquad \lim_{x \to \gamma^1} P_x \{\zeta = \tau\} = 1.$$

Again using the strict Markov property of X^δ, we get

[15] $P_x(A)$ designates the probability of event A if the trajectory of X starts from the point x.

[16] The indicator function of A is a function equal to 1 on A and equal to 0 on the complement of A.

$$M_x\zeta \leqslant M_x\tau + M_x[\chi_{(\tau<\zeta)}\theta_\tau\zeta] \leqslant c\varepsilon^2 + M_x\chi_{(\tau<\zeta)}M_{x_\tau}\zeta$$
$$\leqslant c\varepsilon^2 + \max_{x\in\overline{D}\setminus M} M_x\zeta\cdot P_x\{\tau<\zeta\}.$$

When $x\to\gamma^1$, the second member of the right side tends to 0, and because in the left side there is a quantity not dependent on ϵ, $\lim_{x\to\gamma^1}M_x\zeta = 0$, which was to be proved.

By the results of §4.1 we now have

COROLLARY. *Let $w_\delta(x)$ be a solution of $F[D,\partial D\setminus S_\delta,-1,0]$. Then the family $w_\delta(x)$, $\delta>0$, is uniformly bounded by a constant not dependent on x. The limit $\lim_{\delta\to0}w_\delta(x)=w(x)$ exists, and $w(x)\to0$ as $x\to M^+$.*

§4.3. Stochastic equations for trajectories of the process X

In this section I describe briefly some additional results on the behavior of trajectories of the process X. In order that the presentation be intuitive and also brief, I limit myself to the 2-dimensional case and let L be the Laplace operator. The following results extend easily to the general case.

First of all I justify the assertion that the field $l(x)$ is the "primary direction of reflection" of the process X off the boundary.

Let D be the upper half-plane, and let the field be constant, i.e. $\partial/\partial l = a\,\partial/\partial x + \partial/\partial y$. We introduce the functional $\Delta_t^\eta = \int_0^t\delta(\eta_s)\,ds$ of the Wiener process η_t by the formula $\Delta_t^\eta = \text{l.i.m.}(2\epsilon)^{-1}\int_0^t\chi_\epsilon(\eta_s)\,ds$, where χ_ϵ is the indicator function of the set $\{x:|x|<\epsilon\}$. Our definition coincides with one of the definitions of this functional in [**18**] (there it is denoted by $t^+(t)$), where it is proved that the limit exists not only in the mean but with probability 1, and that Δ_t^η is almost certainly continuous, increases only at times when η_t stays at the point 0, and has the same distribution as $|\eta_t|$.

Consider the process X, the components x_t, y_t of which are given by the equations

$$x_t = \xi_t + a\Delta_t^\eta,$$
$$y_t = |\eta_t|,$$

where ξ_t and η_t are one-dimensional Wiener processes and Δ_t^η is a functional of η_t. The measure of P_x is defined by the process ξ_t and η_t.

It is not difficult to prove that X is a Markov process with continuous trajectories. In the proof we can utilize the joint distribution of Δ_t^η and η_t described in [**18**].

Since Δ_t^η increases only at times when y_t stays at 0 and because X in the upper half-plane before going out onto the boundary has the same characteristics as a Wiener process, it follows easily that the infinitesimal operator A of the process X is the Laplace operator on smooth functions from D_A.

The standard proof of this fact is based on a theorem of E. B. Dynkin ([**19**], Russian p. 204) on the relationship between an infinitesimal and a local characteristic operator.

Finally we establish a boundary condition for smooth functions from D_A. We have

$$M_z f(x_t, y_t) - f(z) = \frac{\partial f}{\partial x}(z) M_z [\xi_t - x + a\Delta_t^\eta]$$

$$+ \frac{\partial f}{\partial y}(z) M_z |\eta_t| + O(t), \quad (z = (x, 0)).$$

Hence it is clear that the limit of $t^{-1}(T_t f - f)$ can exist only when $a\, \partial f/\partial x(z) M_x \Delta_t^\eta + \partial f/\partial y(z) M_x |\eta_t| = 0$; but, as we know, Δ_t^η has the same distribution as $|\eta_t|$; therefore necessarily $a\, \partial f/\partial x(z) + \partial f/\partial y(z) = 0$. So we have described a process X with reflection along a constant vector field.

Now let Ξ be a process with reflection relative to the normal in D the infinitesimal operator of which is the Laplace operator (for its construction see, for example, [**20**]). For it the functional $\Delta_t = \text{l.i.m.}(2\epsilon)^{-1} \int_0^t \chi_\epsilon(z_u)\, du$ is defined, and $z_u = (\xi_u^1, \xi_u^2)$ is the state of the trajectory Ξ at the moment u; χ_ϵ is the indicator function of an ϵ-neighborhood of ∂D. In a neighborhood of $x \in \partial D$ we introduce rectangular coordinates x^1 and x^2, letting x^2 be directed according to the interior normal and letting

$$dx_t^1 = d\xi_t^1 + d\Delta_t \cdot a(x),$$

$$dx_t^2 = d\xi_t^2,$$

analogously to the preceding case in these coordinates. In these equations it is necessary, as usual, to take the integral equation for the corresponding differential equation. Also $a(x) = \operatorname{tg} \phi(x)$, and $\phi(x)$ is the angle between the vector $l(x)$ and the normal to ∂D at x. This system of equations can be solved by the method of successive approximations, and the resulting process will be a Markov diffusion process in D.

The infinitesimal operator of the process X on smooth functions coincides with the Laplace operator, and the boundary condition for a smooth function $f \in D_A$ has the form

$$\partial f / \partial l = 0.$$

The proof of these assertions, in a form somewhat different from the description given here, is contained in [**21**]. This construction is suitable only if $a(x) < \infty$ or, equivalently, the field $l(x)$ is not tangent. If the field $l(x)$ is tangent to the boundary, then the corresponding process is obtained by passing to the limit as described in §4.1.

We now deal with an equation connected with the process X. For simplicity let D be the upper half-plane.

Following [22], we introduce a new scale of time $\tilde{t}_t = \Delta_t^\eta$, and in this new time $\tilde{t}$ we investigate the process X only at random moments when the trajectory of X hits the boundary. It is possible to show that we get a Markov process $\widetilde{X}$ which is a solution of the stochastic equation

$$d\tilde{x}_{\tilde{t}} = a(\tilde{x}_{\tilde{t}})\, d\tilde{t} + dk_{\tilde{t}}.$$

Here $dk_{\tilde{t}}$ is the increment of Cauchy's process, i.e. a process with independent increments such that the distribution of the increments has the form

$$\mathrm{P}_x\{k_{\tilde{t}} - x < y\} = \frac{1}{\pi}\int_{-\infty}^{y} \frac{\tilde{t}\, dy}{\tilde{t}^2 + y^2}.$$

In this way we get a Markov process on the boundary constructed relative to Cauchy's process just as a diffusion process with a unitary coefficient of diffusion is constructed relative to Wiener's process.

The equations described above for trajectories of the process are very useful for an intuitive understanding of the behavior of X, but when I used them to prove theorems on the behavior of X, I was not able to obtain proofs as simple as those obtained by the use of partial differential equations.

§4.4. Behavior of trajectories of X near M^+

After we proved that the trajectories of X are in M^+, the question of their behavior near M^+ arises naturally.

THEOREM 4.4. *Trajectories of X are almost certainly tangent to ∂D at the time ζ.*

PROOF. Let γ be a curve on ∂D having positive side γ^1; let V_{δ_1,δ_2} be a neighborhood of $\alpha \in \gamma$ and $\hat{\phi}(x)$ the function considered in §2.2. Let $W^h_{\delta_1,\delta_2}$ be that part of V_{δ_1,δ_2} consisting of points x for which $\tilde{\phi}(x) < h$. Let $\tilde{\phi}(x) = h^{-1}\hat{\phi}(x)$ if $x \in W^h_{\delta_1,\delta_2}$ and $\tilde{\phi}(x) = 1$ on $\overline{D}\setminus W^h_{\delta_1,\delta_2}$. Let A_ϵ be the event that the trajectory of X before the time ζ hits γ^1_ϵ, and let $\zeta_0^\epsilon = \inf(t: x_t \in \gamma^1_\epsilon)$ if A_ϵ occurs and $\zeta_0^\epsilon = \infty$ in the opposite case. Let $\tau_\epsilon = \inf(t: t > \zeta_0^\epsilon, x_t \notin W^h_{\delta_1,\delta_2})$. By the strict Markov property of the process X^ϵ we have[17] (a.s.) $\mathrm{M}_x[\tilde{\phi}(x_{\tau_\epsilon}) \mid A_\epsilon] = \mathrm{M}_{x_{\zeta_0}}\tilde{\phi}(x_{\tau_\epsilon})$. Consequently (a.s.) $\mathrm{M}_x[\tilde{\phi}(x_{\tau_\epsilon}) \mid A_\epsilon] \leqq \max_{x \in \gamma^1_\epsilon} \mathrm{M}_x\tilde{\phi}(x_{\tau_\epsilon})$.

It is well known that the probability that the trajectory of X will hit a fixed smooth curve at the moment of first reaching ∂D is equal to 0. Hence it follows easily that with probability 1 the trajectories of X which reach γ^1 should before this be tangent to any neighborhood γ^1_ϵ of side γ^1. In other words, for $\epsilon \downarrow 0$ we have $A_\epsilon \downarrow A$, where the event A designates the reaching of γ^1. From (2.2.3) and (4.1.1) it follows that (a.s.)

$$\overline{\lim_{\varepsilon \to 0}}\, \mathrm{M}_x[\tilde{\varphi}(x_{\tau_\varepsilon}) \mid A_\varepsilon] \leqslant \overline{\lim_{\varepsilon \to 0}} \max_{x \in \gamma^1_\varepsilon} \tilde{\varphi}(x) + ch^{-1}\delta_1\delta_2^{-1} = ch^{-1}\delta_1\delta_2^{-1} \quad (x \in \overline{D}\setminus M).$$

---

[17] $\mathrm{M}_x[\xi \mid A]$ designates the conditional expectation of the quantity ξ given the event A.

From this inequality it is clear that for any fixed h (a.s.)

$$\mathrm{P}_x\{\overline{\lim_{t\to\zeta}}\,\hat{\varphi}(x_t)<h\mid A\}=\lim_{\varepsilon\to 0}\mathrm{P}_x\{\overline{\lim_{t\to\zeta}}\,\hat{\varphi}(x_t)<h\mid A_\varepsilon\}>1-ch^{-1}\delta_1\delta_2^{-1}. \qquad (4.4.1)$$

Because of the decrease of $\delta_1\delta_2^{-1}$ the right side of (4.4.1) can be made as close to 1 as desired, and the probability under consideration does not depend on δ_1 or δ_2. Hence $\mathrm{P}_x\{\overline{\lim}_{t\to\zeta}\hat{\phi}(x_t)<h|A\}=1$ (a.s.). Because this equation is satisfied for any h, the theorem is proved.

Chapter 5

SEPARATION OF CLASSICAL SOLUTIONS OF PROBLEM P_0 IN THE 2-DIMENSIONAL CASE

§5.1. Solutions continuous at null points

We saw that in order to get general theorems for problem P it is necessary to broaden the class of solutions, including in it functions discontinuous on some subset of the boundary. The problem of separating out the solutions which are smooth in D arises naturally.

In the n-dimensional case we encounter difficulties which have as yet not been overcome. Thus it is of interest to analyze this problem in the 2-dimensional case, where a solution is possible (cf. [**5**]). It is necessary, however, to point out that in the 2-dimensional case smooth solutions of P can be found without investigating beforehand a broader class of discontinuous functions (cf. [**23**]).

We assume that D is a domain in the plane with smooth boundary ∂D and that on ∂D we have a vector field $l(x)$ tangent to ∂D at a finite set of singular points of M. Analogously to Chapter 2 we introduce the concept of a negative and a positive side of a singular point α. For example, let $t_1(\alpha)$ be the tangent vector of ∂D at the singular point α, oriented in the positive orientation of ∂D. We will say that $x_n\to\alpha^1$ if the points $x_n\in\partial D$ and $x_n\to\alpha$ as $n\to\infty$ and remains in the positive orientation from α.

If $(l_1(\alpha^1),t_1(\alpha))>0$, then the side α^1 of the singular point α is said to be negative. In the opposite case the side is said to be positive.

A point both sides of which are positive (negative) is said to be positive (negative); if the sides are of both signs, then the point is said to be null.

In D we consider the problem P: $Lu=f$ in D, $\partial u/\partial l=0$ on $\partial D\backslash M$, where L is an elliptic operator with smooth coefficients. For this problem it is not difficult to prove theorems analogous to those proved earlier for the 3-dimensional case. We will refer to theorems proved earlier, assuming that their formulation has been altered in the obvious manner in order that they be applicable to the 2-dimensional case of P.

Let $u(x)$ be a solution of the problem $P_0(x=x_1,x_2)$. We introduce $K_u(z)=\partial u/\partial x_1-i\partial u/\partial x_2$, $z=x_1+ix_2$. From an estimate of derivatives of the kernel yielding a solution of Dirichlet's problem in D [**14**] and from the boundedness of $u(x)$, it follows that the function $K_u(z)$ can only have a power singularity: $|K_u(z)| < |z-\alpha|^{-2}$ at singular points. We will use the following results proved in [**23**] (Russian pp. 251-253 and 316).

LEMMA 5.1. *If $u(x)$ is a solution of problem P_0, then $K_u(z) = R(z)w(z)$, where $R(z)$ is a rational function and $w(z)$ is a continuous function not vanishing on $\overline{D}$.*

LEMMA 5.2. *If $u(x)$ is a solution of P_0 continuous in $\overline{D}$, then the number of zeros N_D of $K_u(z)$ within D and the number of zeros $N_{\partial D}$ on the boundary are related by the equation $2N_D + N_{\partial D} = 2n$, where the multiplicities have been taken into consideration. Here n is the number of full revolutions made by the field $l(x)$ as the point x moves along ∂D, assuming that D remains on the left. A revolution is considered to have a plus sign if it is counterclockwise and a minus sign in the opposite case. The number n is called the index of $l(x)$.*

It is obvious that the zeros of $K_u(z)$ are in one-to-one correspondence with the saddle points of $u(x)$ ($\nabla u = 0$), for which the equation

$$2N_D + N_{\partial D} = 2n \tag{5.1.1}$$

is also valid.

The above assertions are proved with the help of the theory of generalized analytic functions. But note that (5.1.1) for the number of saddle points of a solution can be gotten from simple topological considerations (from the equation of Morse) without using the theory of analytic functions of a complex variable. A discussion of the derivation is in [**24**], and condition (A) on which it is based can be easily derived by using the minimum principle.

As an easy corollary to Lemma 5.1 we have

THEOREM 5.1.[18] *Values of the solution $u(x)$ of P_0 on ∂D cannot have discontinuities of the second kind at singular points of M.*

PROOF. Assume the contrary. Then there is a sequence of pairs of points a_k, b_k, with $a_k \to \alpha$, $b_k \to \alpha$, where a_k is a local minimum and b_k is a local maximum of the function $u(x)$ on ∂D.

By the minimum principle (Lemma 1.1) a_k (b_k) cannot be a local minimum (maximum) in a 2-dimensional neighborhood of the point a_k (b_k). Thus a_k, $k = 1, 2, \cdots$, are saddle points of $u(x)$ at which $\partial u/\partial x_1 = \partial u/\partial x_2 = 0$. But we know that there are only a finite number of saddle points of u, so that we have arrived at a contradiction.

The positive (negative) side of a null point will be designated by α^+ (α^-).

LEMMA 5.3. *If a nontrivial solution of P_0 has an extremum from the positive side α^+ of a null point α, then $u(x)$ is discontinuous at α.*

[18] This theorem can be proved without using the results of [**23**].

PROOF. Let U be a neighborhood of α not containing other singular points, and let $\Gamma = \partial U \cap D$. From the conditions of the lemma it follows that for some $\epsilon > 0$ we have $\sup_{x \in \Gamma} u(x) + \epsilon \leqq \max_{\bar{D}} u(x) = u(\alpha^+)$.[19] On the other hand, by Theorem 2.2 $\underline{\lim}_{x \to \alpha^-} \mu(x, \Gamma) \geqq c > 0$, where $\mu(x, A)$ is the harmonic measure (cf. Chapter 1) for problem F in which Γ_2 is replaced by $\gamma_U^+ \cup \Gamma$.[20] We have

$$u(x) \leqslant \sup_{x \in \Gamma} u(x) \mu(x, \Gamma) + u(\alpha^+)(1 - \mu(x, \Gamma))$$

$$\leqslant (1 - c) u(\alpha^+) + c(u(\alpha^+) - \varepsilon) \leqslant u(\alpha^+) - c\varepsilon,$$

which was to be proved.

COROLLARY. *If $u(x)$ is a solution of problem P_0 which is continuous at null and positive points and equal to 0 at positive points, then $u(x) \equiv 0$.*

PROOF. As we know, an extremum of a solution of P_0 is attained at null or positive points. Because $u(x) = 0$ at positive points, it follows from Lemma 5.3 and the continuity of $u(x)$ that $u(x) \equiv 0$.

THEOREM 5.2. *For given values at positive points there exists a unique solution of P_0 which is continuous at all null points.*

PROOF. Enumerate the null points. Let $p_k(x)$, $k = 1, \cdots, n_0$, be a solution of P_0 such that $p_k(\alpha_k^+) = 1$ and $p_k(\alpha_j^+) = 0$, $j \neq k$, $1 \leqq j \leqq n_0$,[21] and $p_k = 0$ at positive points. Let

$$r_{jk} = p_k(\alpha_j^+) - p_k(\alpha_j^-), \quad j, k = 1, \ldots, n_0.$$

We enumerate the positive points and let $\tilde{p}_k(x) = 1$ at the kth positive point and 0 on the remaining part of M^+. Let

$$c_{jk} = \tilde{p}_k(\alpha_j^+) - \tilde{p}_k(\alpha_j^-), \ j = 1, \ldots, n_0, \ k = 1, \ldots, n_+.$$

Let $u(\alpha_k^+) = x_k$, $k = 1, \cdots, n_0$, and $u(\alpha_i) = a_k$, $k = 1, \cdots, n_+$, be the values of a solution $u(x)$ of P_0 from positive sides of null points and at positive points respectively.

In order that $u(x)$ be continuous at null points for given a_k, it is necessary that it satisfy the system of equations

$$u(\alpha_j^+) - u(\alpha_j^-) = c_j + \sum_{k=1}^{n_0} r_{jk} x_k = 0, \tag{5.1.2}$$

$$j = 1, \ldots, n_0 \ \left(\text{here } c_j = \sum_{k=1}^{n_0} a_k c_{jk}\right).$$

---

[19] For concreteness we consider the case when a maximum of $u(x)$ is attained at α^+.

[20] Analogously to §2.1 we let γ_U^+ represent the part of $\partial D \cap \partial U$ which adjoints α^+.

[21] We use n_0, n_+ and n_- to designate the number of null, positive and negative points respectively.

By the Corollary to Lemma 5.3 the corresponding homogeneous system has only a trivial solution. Thus the solution of (5.1.2) exists and is unique.

We observe that the analog of Theorem 5.2 is valid also in the n-dimensional case, although the following results do not lend themselves to generalization.

§5.2. Separation of classical solutions

Let $\pi_k(x)$ be a solution of P_0 which is continuous at positive and null points and which is equal to 1 at the kth positive point and 0 at the remaining points. They form a basis for the space B of solutions of P_0 which are continuous outside the set $\widetilde{M}^-$ of negative points.

We enumerate the negative points and let $R_{jk} = \pi_k(\alpha_j^1) - \pi_k(\alpha_j^2), j = 1, \cdots, n_-$, $k = 1, \cdots, n_+$, where α_k^i, $i = 1, 2$, are the sides of the kth negative point.

In order to find solutions of P_0 which are continuous in $\overline{D}$, it is necessary to solve the system of equations $\sum_{k=1}^{n_+} x_k R_{jk} = 0$, $j = 1, \cdots, n_-$, with respect to the x_k. It is clear that there exists at least an $(n_+ - n_-)$-dimensional space C of continuous solutions.

If D is not simply connected, then the dimension of C can be greater than $n_+ - n_-$. So if D is the annulus $1 < x_1^2 + x_2^2 + x_3^2 < 2$ and the field $l(x)$ is directed along the axis x_1, then $n_+ - n_- = 0$ when there exist harmonic functions not dependent on x_1: $ax_2 + b$ (a and b are arbitrary constants).

THEOREM 5.3. *If the domain D is simply connected, then for $n \geqq 0$ the dimension of C is equal to $n_+ - n_-$.*

PROOF. The formula $n_+ - n_- = 2n + 2$, valid in the simply connected case, follows easily from the definition of the index n.

By Lemma 5.2 the number of saddle points $2N_D + N_{\partial D} = 2n$.

We will say that two positive points are adjacent if no positive or negative points lie between them.

It is not difficult to show that there are $2n + 1$ pairs of adjacent points. The subspace S of functions from B the values of which coincide at the $2n + 1$ adjacent points is of dimension $n_+ - 2n - 1$. A nonconstant function from $S \cap C$ must have an extremum on ∂D between adjacent points, which by Lemma 1.2 will be a saddle point. But by (5.1.1) the number of saddle points on the boundary cannot be greater than $2n$.

Thus $S \cap C$ consists of constants, i.e. is 1-dimensional. We have

$$1 = \dim (S \cap C) \geqslant \dim C - (\dim B - \dim S) = \dim C - 2n - 1,$$

whence $\dim C \leqslant 2n + 2 = n_+ - n_-$.

Received 20 NOV 1967

BIBLIOGRAPHY

[1] H. Poincaré, *Leçons de mécanique céleste*, Tome III, *Théorie de marées*, Gauthier-Villars, Paris, 1910.

[2] G. Giraud, *Équations à intégrales principales d'ordre quelconque*, Ann. Sci. École Norm. Sup. (3) **53** (1936), 1-40.

[3] A. Liénard, *Problème plan de la dérivée oblique dans la théorie du potentiel*, J. École Polytech. **144** (1938), 35-96, 97-158, 177-226; *Nouveau mémoire sur le problème de la dérivée oblique dans la théorie du potential*, ibid. **145** (1939), 55-84, 85-137.

[4] I. N. Vekua, *New methods for solving elliptic equations*, OGIZ, Moscow, 1948; English transl., Series in Appl. Math., vol. 1, North-Holland, Amsterdam and Interscience, New York, 1967. MR **11**, 598; **35** # 3243.

[5] M. B. Maljutov, *Brownian motion with reflection and the problem of the oblique derivative*, Dokl. Akad. Nauk SSSR **156** (1964), 1285-1287 = Soviet Math. Dokl. **5** (1964), 822-825. MR **29** # 2867.

[6] ———, *Oblique derivative problem in three-dimensional space*, Dokl. Akad. Nauk SSSR **172** (1967), 283-286 = Soviet Math. Dokl. **8** (1967), 87-90. MR **35** # 6993.

[7] Ju. V. Egorov and V. A. Kondrat′ev, *On an oblique derivative problem*, Dokl. Akad. Nauk SSSR **170** (1966), 770-772 = Soviet Math. Dokl. **7** (1966), 1271-1273. MR **34** # 4668.

[8] ———, *Existence of a solution to a problem with oblique derivative*, Uspehi Mat. Nauk **22** (1967), no. 1 (133), 165-167. (Russian) MR **34** # 7950.

[9] A. V. Bicadze, *The problem of the oblique derivative with polynomial coefficients*, Dokl. Akad. Nauk SSSR **157** (1964), 1273-1275 = Soviet Math. Dokl. **5** (1964), 1102-1104. MR **29** # 3662.

[10] ———, *A special case of the problem of the oblique derivative for harmonic functions in three-dimensional domains*, Dokl. Akad. Nauk SSSR **155** (1964), 730-731 = Soviet Math. Dokl. **5** (1964), 477-478. MR **30** # 1257.

[11] ———, *The homogeneous problem for the oblique derivative for harmonic functions in three-dimensional regions*, Dokl. Akad. Nauk SSSR **148** (1963), 749-752 = Soviet Math. Dokl. **4** (1963), 156-159. MR **26** # 5173.

[12] A. Janušauskas, *On the problem of the oblique derivative*, Dokl. Akad. Nauk SSSR **164** (1965), 753-756 = Soviet Math. Dokl. **6** (1965), 1303-1306. MR **33** # 297.

[13] M. I. Freĭdlin, Dissertation, Steklov Institute of Mathematics, Moscow, 1962. (Russian)

[14] C. Miranda, *Equazioni alle derivate parziali di tipo ellittico*, Ergebnisse der Mathematik und ihrer Grenzgebiete, Heft 2, Springer-Verlag, Berlin, 1955; Russian transl., IL, Moscow, 1957. MR **19**, 421.

[15] I. V. Girsanov, *The solution of certain boundary problems for parabolic and elliptic equations with discontinuous coefficients*, Dokl. Akad. Nauk SSSR **135** (1960), 1311-1313 = Soviet Math. Dokl. **1** (1960), 1373-1375. MR **23** # A1938.

[16] S. N. Bernšteĭn, *Collected works*. Vol. 3: *Differential equations, calculus of variations and geometry* (1903-1947), Izdat. Akad. Nauk SSSR, Moscow, 1960. (Russian) MR **23** # A32.

[17] E. B. Dynkin, *Foundations of the theory of Markov processes*, Fizmatgiz, 1959; English transl., Prentice-Hall, Englewood Cliffs, N. J. and Pergamon Press, Oxford, 1961. MR **24** # A1745; **24** # A1747.

[18] K. Itô and H. P. McKean, *Diffusion processes and their sample paths*, Die Grundlehren der math. Wissenschaften, Band 125, Academic Press, New York and Springer-Verlag, Berlin, 1965; Russian transl., "Mir", Moscow, 1967. MR **33** # 8031.

[19] E. B. Dynkin, *Markov processes*, Fizmatgiz, Moscow, 1963; English transl., Vols. I, II, Academic Press, New York and Springer-Verlag, Berlin, 1965. MR **33** # 1886; 1887.

[20] M. I. Freĭdlin, *Diffusion processes with reflection and an oblique derivative problem on a manifold with boundary*, Teor. Verojatnost. i Primenen. **8** (1963), 80-88. (Russian) MR **27** # 810.

[21] N. Ikeda, *On the construction of two-dimensional diffusion processes satisfying Wentzell's boundary conditions and its application to boundary value problems*, Mem. Coll. Sci. Univ. Kyoto Ser. A **33** (1960-61), 367-427. MR **23** # A4177.

[22] K. Sato and T. Ueno, *Multi-dimensional diffusion and the Markov process on the boundary*, J. Math. Kyoto Univ. **4** (1964/65), 529-605. MR **33** # 6702.

[23] I. N. Vekua, *Generalized analytic functions*, Fizmatgiz, Moscow, 1959; English transl., Pergamon Press, London and Addision-Wesley, Reading, Mass., 1962. MR **21** # 7283; **27** # 321.

[24] M. Morse, *Topological methods in the theory of functions of a complex variable*, Ann. of Math. Studies, no. 15, Princeton Univ. Press, Princeton, N. J., 1947; Russian transl., IL, Moscow, 1951. MR **9**, 20.

Translated by H. N. Parker

Trudy Moskov. Mat. Obšč.
Tom 20 (1969)

Trans. Moscow Math. Soc.
Vol. 20 (1969)

MIXED PROBLEMS FOR A CERTAIN CLASS OF PARTIAL DIFFERENTIAL EQUATIONS

Ju. A. DUBINSKIĬ

Contents

INTRODUCTION 205

§1. Mixed problem for strictly elliptico-parabolic equations 209

§2. Generalized mixed problem for elliptico-parabolic equations. The regularizer 214

§3. Elliptico-parabolic equations. Normal solvability of the generalized mixed problem 223

§4. Mixed problem for elliptico-hyperbolic equations 224

§5. Generalized mixed problem for elliptico-hyperbolic equations. The regularizer. Normal solvability 228

§6. Structure of the subspace of right sides of the mixed problem 232

§7. Smoothness theorem 234

§8. Examples. Remarks 237

LITERATURE 241

Introduction

1. This paper is devoted to studying mixed problems for equations of the form

$$\mathfrak{A}(u) \equiv \mathscr{P}_s\left(\frac{\partial}{\partial t}\right) u + \mathscr{L}(x, D)\, u = h(x, t), \tag{0.1}$$

where $\mathscr{P}_s(\partial/\partial t)u \equiv \sum_{q=0}^{s} a_q u^{(q)}$; $\mathscr{L}(x,D)u$ is a $2m$th order elliptic differential operator.

Let $Q = G \times [0, \infty)$ be the cylinder with base $G \subset \mathbf{R}^n$ and lateral surface $S = \partial G \times [0, \infty)$. We first consider the homogeneous mixed problem for equation (0.1). This means that we look for a solution of equation (0.1) which satisfies the additional conditions

$$u(x, 0) = 0, \; \ldots, \; u^{(s-1)}(x, 0) = 0, \quad B_j(x, D)\, u\,|_S = 0, \quad j = 1, \ldots, m, \tag{0.2}$$

where the $B_j(x, D)u$ are the boundary differential operators. Obviously, for $s > 2$ the mixed problem is automatically correct in the sense of Hadamard and Petrovskiĭ. In particular, this means that there does not exist a solution $u(x,t) \in H_\gamma^{s,2m}(Q)$ of problem (0.1), (0.2) for every right side $h(x,t) \in \mathscr{L}_{2,\gamma}(Q)$ (see below for notation).

Moreover, even in the case when a solution exists for a function $h(x,t)$, it may be extremely unstable relative to arbitrarily small variations of $h(x,t)$.

In this connection we note that in recent years papers by Tihonov [**14**], [**15**], Lavrent′ev [**11**], Ivanov [**9**] and others have created a theory of problems which had been called incorrect problems. From this point of view, the mixed problem for equation (0.1), which is generally incorrect in the sense of Hadamard and Petrovskiĭ, is correct in the sense of Tihonov. It became convenient here to divide equations (0.1) into two classes, which are called elliptico-parabolic and elliptico-hyperbolic (the terminology is justified because the three classical equations—Laplace's equation, the heat conduction equation, and the inverse heat conduction equation—belong to the first class, while the wave equation belongs to the second).

At first glance these definitions depend on the parameter $\gamma \in \mathbf{R}^1$ which determines the weight function $e^{-\gamma t}$ in the spaces $H_\gamma^{s,2m}(Q)$. In fact, the real axis $-\infty < \gamma < +\infty$ is divided into a countable number of intervals, and among them there is one for which the mixed problem (0.1), (0.2) is correct in the sense of Tihonov. In this case the operator $\mathfrak{A}(u)$ is called strictly elliptico-parabolic or elliptico-hyperbolic.

For all the other intervals, the mixed problem is correct in the sense of Tihonov if the right side $h(x,t)$ is orthogonal to a finite-dimensional subspace $\mathscr{L}_{2,\gamma}(Q)$, where the dimension of this subspace increases with $|\gamma|$.

As we know, an important role in the theory of incorrect problems is played by the notion of the regularizer of an incorrect problem, i.e. by the correct problem with the property that if the data of this problem is arbitrarily close to that of the incorrect problem, then its solution will be arbitrarily close to that of the original incorrect problem.

For the mixed problem (0.1), (0.2) it turned out that the regularizer is the following generalized mixed problem: for a given function $h(x,t)$ find a system of functions $u(x,t), \rho_1(x), \cdots, \rho_k(x)$, such that

$$\mathscr{P}_s\left(\frac{\partial}{\partial t}\right)u + \mathscr{L}(x, D)\,u = h(x, t) + \mathscr{P}_{k-1}(t), \tag{0.3}$$

$$u(x, 0) = 0, \ldots, u^{(s-1)}(x, 0) = 0, \quad B_j(x, D)\,u\,|_S = 0, \tag{0.4}$$

where

$$\mathscr{P}_{k-1}(t) = \rho_1(x) + \rho_2(x)\,t + \ldots + \rho_k(x)\,t^{k-1},$$

and the number k is uniquely determined by the type of the equation and, to within ± 1, equals $[s/2]$.

In connection with the statement of the generalized mixed problem (0.3), (0.4), we call attention to the fact that the appearance of the new unknown functions $\rho_1(x), \cdots, \rho_k(x)$, which depend only on x, is to some extent analogous to the introduction of additional boundary potentials for elliptic problems in the work of Višik and Èskin (see, for example, [**8**]).

We then have equivalence of the generalized mixed problem and the correct problem, called the boundary value problem, which does not contain additional unknown functions. The boundary value problem consists in finding a solution of equation (0.1) under the additional conditions

$$u(x, 0) = 0, \ldots, u^{(s-1-k)}(x, 0) = 0, \quad B_j(x, D)|_S = 0. \tag{0.5}$$

We note that Lions [12] examined a special case of problem (0.5).

Lemma 2 of §2 on the solvability of ordinary differential equations plays a fundamental role in the proof of unique solvability of the boundary value problem. This lemma implies that the geometric meaning of the strict elliptico-parabolicity condition

$$\sum_{q=0}^{s} a_q \sum_{p=0}^{\left[\frac{q}{2}\right]} \beta_{2p,q} (-1)^p \tau^{2p} + \lambda_0 > 0,$$

$$\beta_{2p,q} = \frac{q!}{(q-2p)!\,(2p)!} \left(\frac{\gamma}{2}\right)^{q-2p}$$

is that for any $\lambda > \lambda_0$ (λ_0 is the first eigenvalue of the operator $\mathscr{L}(x, D)u$) the characteristic polynomial

$$\mathscr{P}_s(\mu) + \lambda \equiv \sum_{q=0}^{s} a_q \mu^q + \lambda$$

has precisely $s - 1 - k$ roots μ such that $\operatorname{Re}\mu < \gamma$.

As we noted before, the operator $\mathfrak{A}(u)$ is generally strictly elliptico-parabolic only on some set Γ_0 of the real axis. The desire to study equation (0.1) in the spaces $H_\gamma^{s,2m}(Q)$ for any γ naturally leads to the definition of an elliptico-parabolic operator (see Definition 1 of §3 and Remark 8.2 at the end of the paper). But then the previous problems for γ not in the correctness set Γ_0 will with certainty be only normally solvable, i.e. solvable if the right side $h(x,t)$ is subject to a finite number of conditions.[1] Here the kernel of the problem may be a finite-dimensional subspace.

All these problems satisfy smoothness theorems analogous to those for elliptic and parabolic equations.

Returning to the original mixed problem, we show, on the one hand, that the complete image of the space $H_\gamma^{s,2m}(Q)$ under the map $\mathfrak{A}(u)$ is a subspace $K \subset \mathscr{L}_{2,\gamma}(Q)$ defined by infinitely many conditions. This means that the mixed problem (0.1), (0.2) is solvable if and only if the right side $h(x,t)$ belongs to a subspace of infinite codimension in $\mathscr{L}_{2,\gamma}(Q)$.

On the other hand, some questions of mathematical physics lead to problems of this form (see, for example, [11]). This gives reason to hope that problems involving partial differential equations with infinite-dimensional defect subspaces can also be successfully studied.

---

[1] These conditions have the character of conditions of orthogonality to a finite-dimensional subspace.

It should be noted in conclusion that the fact that $\mathscr{L}(x,D)u$ is a differential operator does not play any role in the basic considerations of this paper, so that everything can be carried over to, say, the differential-operator equations

$$\mathscr{P}_s\left(\frac{d}{dt}\right)u+\mathscr{L}(u)=h(t),$$

where $\mathscr{L}(u)$ is an unbounded operator which acts in a Hilbert space. In particular, this solves a problem of Hille (see [**16**]) on finding the defect of an abstract Cauchy problem. The author hopes to discuss these questions in another paper.

In addition, in this paper the boundary and initial conditions are always assumed to be homogeneous. The case of nonhomogeneous conditions can be similarly examined.

In this paper we do not consider boundary value problems for equation (0.1) in the half-space $\mathbf{R}_{n+1}^+=\{t\geqq 0; x\in\mathbf{R}_n\}$. In the case of constant coefficients, this was done earlier in terms of the Fourier transform by Dikopolov, Palamodov and Šilov. A survey of the results of these authors is given in the book [**19**], where references can also be found.

2. Notation. Let the function $u(x)$ be defined in a region $G\subset\mathbf{R}^n$ with a smooth boundary Γ. We let $H^s(G)$ designate the Sobolev space $W_2^{(s)}(G)$ (see [**13**]). The norm in the space $H^s(G)$ equals

$$\|u\|_s^2=\sum_{|\alpha|\leqslant s}\int_G|D^\alpha u|^2\,dx,$$

where $D^\alpha u\equiv\partial^{|\alpha|}u/\partial x_1^{\alpha_1}\cdots\partial x_n^{\alpha_n}$, $\alpha=(\alpha_1,\cdots,\alpha_n)$ and $|\alpha|=\alpha_1+\cdots+\alpha_n$. Further, the scalar product (u,v) is the integral

$$(u,v)\equiv\int_G u(x)\,v(x)\,dx,$$

possibly understood in the sense of generalized functions.

For functions $u(x,t)$ defined in the cylinder $Q=G\times[0,\infty)$, we introduce the following Sobolev-Slobodeckiĭ spaces:

(1) $$\mathscr{L}_{2,\gamma}(Q)=\left\{u(x,t)\;|\;|||u|||_{0,\gamma}^2=\int_0^\infty\|u\|_0^2e^{-\gamma t}\,dt<\infty\right\},\quad\gamma\in\mathbf{R}^1;$$

(2) $$H_\gamma^{s,l}(Q)=\left\{u(x,t)\;|\;|||u|||_{s,l,\gamma}^2=\int_0^\infty(\|u^{(s)}\|_0^2+\|u\|_l^2)\,e^{-\gamma t}\,dt<\infty\right\};$$

(3) $\overset{\circ}{H}{}_\gamma^{s,l}(Q)$ is the subspace of $H_\gamma^{s,l}(Q)$ which consists of the functions $u(x,t)$ for which $u(x,0)=0,\cdots,u^{(s-1)}(x,0)=0$;

(4) if $u(x,t)\equiv u(t)$, then the spaces $\mathscr{L}_{2,\gamma}(Q)$, $H_\gamma^{s,0}(Q)$ are designated by $\mathscr{L}_{2,\gamma}(0,\infty)$ and $H_\gamma^s(0,\infty)$, respectively.

§1. Mixed problem for strictly elliptico-parabolic equations

1. Definition of a strictly elliptico-parabolic operator. Let $G \subset \mathbf{R}^n$ be a bounded region with a smooth boundary Γ. In the region G we consider the elliptic operator

$$\mathcal{L}(x, D)u \equiv \sum_{|\alpha| \leq 2m} a_\alpha(x) D^\alpha u. \tag{1.1}$$

We recall that the operator $\mathcal{L}(x, D)u$ is called elliptic if for any point $x \in G$ and any $\xi \in \mathbf{R}^n \setminus \{0\}$ we have

$$\mathcal{L}_{2m}(x, -i\xi) \neq 0,$$

where $\mathcal{L}_{2m}(x, D)u \equiv \sum_{|\alpha|=2m} a_\alpha(x) D^\alpha u$ is the principal part of the operator $\mathcal{L}(x, D)u$.

We shall assume the coefficients $a_\alpha(x)$ to be sufficiently smooth functions in the region G.

Further, let

$$B_j(x, D)u \equiv \sum_{|\beta| \leq m_j} b_{\beta j}(x) D^\beta u, \quad j = 1, \ldots, m, \tag{1.2}$$

be a system of differential boundary operators of orders $m_j \leqq 2m - 1$. We let $H^{2m}(\{B_j\})$ designate the closure in the $H^{2m}(G)$-norm of the smooth functions which satisfy the conditions

$$B_j u|_\Gamma = 0, \quad j = 1, \ldots, m.$$

We suppose that the operator $\mathcal{L}(x, D)u$, considered on $H^{2m}(\{B_j\})$ is selfadjoint.

REMARK. In connection with the last assumption we note that the question of existence and characterization of the selfadjoint boundary operators $B_j(x, D)u$ has been recently studied in detail by many authors (see papers by Schechter [**17**], [**18**], Aronszajn and Milgram [**5**], and others).

We know that a selfadjoint elliptic operator has a purely point spectrum $\lambda_0 \leqq \lambda_1 \leqq \cdots$ of finite multiplicity, where the corresponding eigenfunctions $\omega_0(x), \omega_1(x), \cdots$ form a complete system in $H^{2m}(\{B_j\})$ (see, for example, [**2**] or [**6**]). We shall assume that the eigenfunctions of the operator $\mathcal{L}(x, D)u$ are orthonormal.

Further, let

$$\mathcal{P}_s\left(\frac{\partial}{\partial t}\right)u \equiv \sum_{q=0}^{s} a_q \frac{\partial^q u}{\partial t^q} \tag{1.3}$$

be a differential operator with constant real coefficients.

DEFINITION 1.1. The operator

$$\mathfrak{A}(u) \equiv \mathcal{P}_s\left(\frac{\partial}{\partial t}\right)u + \mathcal{L}(x, D)u$$

is called *strictly elliptico-parabolic* on the set[2] Γ_0 of the real axis if for any $\gamma \in \Gamma_0$ the polynomial

$$A(\gamma;\tau)+\lambda_0 \equiv \sum_{q=0}^{s} a_q \sum_{p=0}^{\left[\frac{q}{2}\right]} (-1)^p \beta_{2p,q}\tau^{2p}+\lambda_0>0,$$

where

$$\beta_{2p,q}=\frac{q!}{(q-2p)!\,(2p)!}\left(\frac{\gamma}{2}\right)^{q-2p},$$

and λ_0 is the first eigenvalue of the operator $\mathscr{L}(x,D)u$.

2. Tihonov correctness of a mixed problem. In the cylinder $Q=G\times[0,\infty)$ with lateral surface $S=\partial G\times[0,\infty)$ we consider the mixed problem

$$\mathscr{P}_s\left(\frac{\partial}{\partial t}\right)u+\mathscr{L}(x,D)\,u=h(x,t), \tag{1.4}$$

$$u(x,0)=0,\ \ldots,\ u^{(s-1)}(x,0)=0, \tag{1.5}$$

$$B_j(x,D)\,u\,|_S=0,\quad j=1,\ldots,m. \tag{1.6}$$

We easily see that for $s>2$ problem (1.4)—(1.6) is incorrect in the sense of Hadamard and Petrovskiĭ. The purpose of this section is to prove that problem (1.4)—(1.6) is Tihonov correct.

DEFINITION 1.2 (SEE, FOR EXAMPLE, [**11**]). A problem is called *Tihonov correct* if the following conditions are fulfilled:

a) for some class K of data the problem has a solution which belongs to a given class M (we emphasize that the class K is a priori unknown, while, on the other hand, M is defined);

b) the solution is unique in the class M;

c) continuous perturbations of the data of the problem which do not take the solution out of M result in continuous perturbations of the solution of the problem.

It is known that several problems of mathematical physics occur in precisely this form [**11**].

THEOREM 1.1. *Let $K\subset\mathscr{L}_{2,\gamma}(Q)$ be the subspace of right sides $h(x,t)$ of problem* (1.4)—(1.6) *which have a solution in the space $H_\gamma^{s,2m}(Q)$. Then for $\gamma\in\Gamma_0$ (see Definition* 1.1) *the mixed problem is Tihonov correct, and the following estimate holds here*:

$$\int_0^\infty \|u^{(s)}\|_0^2 e^{-\gamma t}\,dt+\int_0^\infty \|u\|_{2m}^2 e^{-\gamma t}\,dt \leq C\int_0^\infty \|h\|_0^2 e^{-\gamma t}\,dt, \tag{1.7}$$

where $C>0$ is a constant which depends on γ.

[2] It will be natural to call Γ_0 the *correctness set* of the mixed problem for the operator $\mathfrak{A}(u)$. It is clear here that $A(\gamma;\tau)+\lambda_0\equiv \operatorname{Re}\mathscr{P}_s(\gamma/2+i\tau)+\lambda_0$.

PROOF. Since the system of eigenfunctions $\omega_0(x), \omega_1(x), \cdots$ forms a complete system, the space $H \equiv H^{2m}(\{B_j\})$ can be represented as the orthogonal sum of the two invariant subspaces

$$H = H^+ \oplus H^-,$$

spanned by the eigenfunctions of the operator $\mathscr{L}(x, D)u$ with eigenvalues $\lambda_i > 0$ and $\lambda_i \leqq 0$, respectively. The subspace H^- is obviously finite dimensional.

We let $\mathscr{L}^+$ designate the restriction of the operator $\mathscr{L}(x, D)u$ to the subspace H^+. The operator $\mathscr{L}^+$ is a positive selfadjoint operator, so that the operator $M^+u = \sqrt{\mathscr{L}^+u}$ exists.

To derive estimate (1.7) we shall need the following two well-known inequalities.

LEMMA 1.1. *Any function $u(x) \in H^{2m}(\{B_j\})$ satisfies the inequalities*

(a) $(\mathscr{L}u, u) \geqq \lambda_0 \| u \|_0^2$, *where λ_0 is the first eigenvalue of the operator $\mathscr{L}$;*

(b) $(\mathscr{L}^+u, \mathscr{L}^+u) \geqq \mu_1 \| M^+u \|_0^2$, *where μ_1 is the first positive eigenvalue of the operator $\mathscr{L}$.*

PROOF. We set $u(x) = \sum_{i=0}^{\infty} c_i \omega_i(x)$. Then

$$(\mathscr{L}u, u) = \sum_{i=0}^{\infty} \lambda_i c_i^2 \geqslant \lambda_0 \sum_{i=0}^{\infty} c_i^2 \equiv \lambda_0 \| u \|_0^2,$$

which constitutes inequality (a).

We similarly prove inequality (b).

We proceed to the derivation of estimate (1.7). We first note that the decomposition of the space $H^{2m}(\{B_j\})$ into the sum

$$H^{2m}(\{B_j\}) = H^+ \oplus H^-$$

corresponds to the orthogonal decomposition of the space $\overset{\circ}{H}{}_\gamma^{s,2m}(Q)$[3] into the sum

$$\overset{\circ}{H}{}_\gamma^{s,2m}(Q) = \overset{\circ}{H}{}_\gamma^+ \oplus \overset{\circ}{H}{}_\gamma^-,$$

where

$$\overset{\circ}{H}{}_\gamma^+ = \{u^+ \mid u^+ = \sum_{\lambda_i > 0} c_i(t)\, \omega_i(x),\ c_i(t) \in H_\gamma^s(0, \infty),\ c_i(0) = \ldots = c_i^{(s-1)}(0) = 0\},$$

$\overset{\circ}{H}{}_\gamma^-$ is the same for $\lambda_i \leqq 0$.

We now set $u(x, t) = u^+(x, t) + u^-(x, t)$, where $u^+ \in \overset{\circ}{H}{}_\gamma^+$, $u^- \in \overset{\circ}{H}{}_\gamma^-$, and we first obtain estimate (1.7) for u^+. To do this, we multiply equation (1.4) by $\mathscr{L}^+u^+e^{-\gamma t}$ and integrate over the cylinder Q. We obtain

---

[3] It will be assumed everywhere that the functions $u(x, t) \in H_\gamma^{s,2m}(Q)$ satisfy the conditions $B_j(x, D)u|_S = 0$, $j = 1, \cdots, m$.

$$\int_0^\infty \left(\mathscr{P}_s\left(\frac{\partial}{\partial t}\right)u^+,\ \mathscr{L}^+u^+\right)e^{-\nu t}\,dt + \int_0^\infty (\mathscr{L}^+u^+,\ \mathscr{L}^+u^+)\,e^{-\nu t}\,dt = \int_0^\infty (h,\ \mathscr{L}^+u^+)\,e^{-\nu t}\,dt \tag{1.8}$$

where $(u, v) = \int_G uv dx$.

By the definition of the operator M^+u we have

$$\left(\mathscr{P}_s\left(\frac{\partial}{\partial t}\right)u^+,\ \mathscr{L}^+u\right) = \left(\mathscr{P}_s\left(\frac{\partial}{\partial t}\right)v^+,\ v^+\right),$$

where $v^+ = M^+u^+$. It is not hard to see that

$$v^{+(q)}e^{-\frac{\gamma}{2}t} = \sum_{r=0}^{q} \beta_{rq}\,(v^+e^{-\frac{\gamma}{2}t})^{(r)},$$

where

$$\beta_{rq} = \frac{q!}{(q-r)!\,r!}\left(\frac{\gamma}{2}\right)^{q-r}.$$

Hence we obtain that

$$\begin{aligned}\int_0^\infty \left(\mathscr{P}_s\left(\frac{\partial}{\partial t}\right)u^+,\ \mathscr{L}^+u^+\right)e^{-\nu t}\,dt &= \int_0^\infty \left(\mathscr{P}_s\left(\frac{\partial}{\partial t}\right)v^+,\ v^+\right)e^{\nu - t}\,dt \\ &= \sum_{q=0}^{s} a_q \sum_{r=0}^{q} \beta_{rq} \int_0^\infty ((v^+e^{-\frac{\gamma}{2}t})^{(r)},\ v^+e^{-\frac{\gamma}{2}t})\,dt \\ &= \sum_{q=0}^{s} a_q \sum_{p=0}^{\left[\frac{q}{2}\right]} \beta_{2p,q} \int_0^\infty ((v^+e^{-\frac{\gamma}{2}t})^{(2p)},\ v^+e^{-\frac{\gamma}{2}t})\,dt,\end{aligned} \tag{1.9}$$

since $v^+(x, 0) = 0, \cdots, v^{+(s-1)}(x, 0) = 0$. Using Parseval's equality, we find from (1.9) that

$$\int_0^\infty \left(\mathscr{P}_s\left(\frac{\partial}{\partial t}\right)u^+,\ \mathscr{L}^+u^+\right)e^{-\nu t}\,dt = \int_0^\infty A(\gamma;\tau)\,\|\widetilde{v^+e^{-\frac{\gamma}{2}t}}\|_0^2\,dt,$$

where

$$A(\gamma;\tau) = \sum_{q=0}^{s} a_q \sum_{p=0}^{\left[\frac{q}{2}\right]} \beta_{2p,q}\,(-1)^p\,\tau^{2p}.$$

(Here and later $\tilde{z}(\tau)$ designates the Fourier transform of the function $z(t)$ extended by zero for $t \leq 0$.)

Further, by Lemma 1.1 we have

$$\int_0^\infty (\mathcal{L}^+u^+, \mathcal{L}^+u^+)\, e^{-\gamma t}\, dt \geqslant \mu_1 \int_0^\infty \| v^+ \|_0^2\, e^{-\gamma t}\, dt,$$

where $\mu_1 > 0$ is the first positive eigenvalue of the operator $\mathcal{L}$.

Taking this into account, from (1.8) we obtain the inequality

$$\varepsilon \int_0^\infty (\mathcal{L}^+u^+, \mathcal{L}^+u^+)\, e^{-\gamma t}\, dt + \int_0^\infty (A(\gamma; \tau) + (1-\varepsilon)\mu_1) \| \widetilde{v^+ e^{-\frac{\gamma}{2}t}} \|_0^2\, dt \leqslant \int_0^\infty (h, \mathcal{L}^+u^+)\, e^{-\gamma t}\, dt.$$

If $\epsilon > 0$ is sufficiently small, it follows from Definition 1.1 of an elliptico-parabolic problem that for any fixed $\gamma \in \Gamma_0$ we have

$$A(\gamma; \tau) + (1-\varepsilon)\mu_1 > 0,$$

so that

$$\int_0^\infty (\mathcal{L}^+u^+, \mathcal{L}^+u^+)\, e^{-\gamma t}\, dt \leqslant \frac{1}{\varepsilon} \int_0^\infty (h, \mathcal{L}^+u^+)\, e^{-\gamma t}\, dt.$$

This and well-known inequalities for the solutions of elliptic equations (see, for example, [**3**]) imply that

$$\int_0^\infty \| u^+ \|_{2m}^2\, e^{-\gamma t}\, dt \leqslant C \int_0^\infty \| h \|_0^2\, e^{-\gamma t}\, dt, \tag{1.10}$$

where $C = C(\gamma)$ is a constant which may depend on $\gamma \in \Gamma_0$.

Further, projecting equation (1.4) onto H_γ^+, we find that $u^+(x,t)$ satisfies the equation

$$\mathcal{P}_s\left(\frac{\partial}{\partial t}\right) u^+ + \mathcal{L}(x, D)\, u^+ = h^+(x, t),$$

where $h^+(x,t) + h^-(x,t) = h(x,t)$. By (1.10), we find from this equation that

$$\int_0^\infty \| u^{+(s)} \|_0^2\, e^{-\gamma t}\, dt \leqslant C \int_0^\infty \| h \|_0^2\, e^{-\gamma t}\, dt,$$

which, together with (1.10), gives estimate (1.7) for the function $u^+(x,t)$.

We now estimate $u^-(x,t)$. For it (1.4) implies

$$\mathcal{P}_s\left(\frac{\partial}{\partial t}\right) u^- + \mathcal{L}(x, D)\, u^- = h^-(x, t). \tag{1.11}$$

We multiply (1.11) by $u^- e^{-\gamma t}$ and integrate over Q:

$$\int_0^\infty \left(\mathcal{P}_s\left(\frac{\partial}{\partial t}\right) u^-, u^-\right) e^{-\gamma t}\, dt + \int_0^\infty (\mathcal{L}u^-, u^-)\, e^{-\gamma t}\, dt = \int_0^\infty (h^-, u^-)\, e^{-\gamma t}\, dt.$$

Repeating this reasoning and taking into account inequality (a) of Lemma 1.1, we arrive at the estimate

$$\int_0^\infty (A(\gamma;\tau)+\lambda_0)\|\widetilde{u^- e^{-\frac{\gamma}{2}t}}\|_0^2\,dt \leqslant \int_0^\infty (h^-, u^-)\,e^{-\gamma t}\,dt. \tag{1.12}$$

By Definition 1.1, for any $\gamma \in \Gamma_0$

$$A(\gamma;\tau)+\lambda_0 \geqslant \delta > 0,$$

which by (1.12) implies that

$$\int_0^\infty \|u^-\|_0^2\,e^{-\gamma t}dt \leqslant \frac{1}{\delta^2}\int_0^\infty \|h\|_0^2 e^{-\gamma t}\,dt. \tag{1.13}$$

We now note that for every $t \in [0,\infty)$ the function $u^-(x,t)$ is generated by a finite number of eigenfunctions $\omega_i(x)$ with eigenvalues $\lambda_i \leqq 0$. Since in a finite-dimensional space all norms are equivalent, there is a constant $C>0$ such that

$$\|u^-\|_{2m} \leqslant C\|u^-\|_0.$$

We then derive estimate (1.7) for the function $u^-(x,t)$ from (1.13) and equation (1.11):

$$\int_0^\infty \|u^{-(s)}\|_0^2\,e^{-\gamma t}\,dt + \int_0^\infty \|u^-\|_{2m}^2\,e^{-\gamma t}dt \leqslant C\int_0^\infty \|h\|_0^2\,e^{-\gamma t}dt.$$

These estimates for the functions $u^+(x,t)$ and $u^-(x,t)$ imply estimate (1.7) for the function $u(x,t)=u^+(x,t)+u^-(x,t)$. Theorem 1.1 is proved.

REMARK. It will later be important to note that to derive estimate (1.7) it is sufficient to require that

$$u(x,0)=0,\ \ldots,\ u^{(l)}(x,0)=0,$$

where $l=s-1-k$, with the number k defined as follows:

a) $k=s/2$ if s is even,

b) $k=[s/2]$ if s is odd and $(-1)^{[s/2]}a_s>0$,

c) $k=[s/2]+1$ if s is odd and $(-1)^{[s/2]}a_s<0$.

This is proved in exactly the same way, except that in (1.9) we must first integrate by parts over t and then use Parseval's equality. For details, see the proof of Lemma 2.1 at the end of §2.

§2. Generalized mixed problem for elliptico-parabolic equations. The regularizer

1. Generalized mixed problem. In this section we study the correct problem which is the regularizer of the mixed problem examined in §1.

DEFINITION 2.1. By the *generalized mixed problem* for the equation

$$\mathscr{P}_s\left(\frac{\partial}{\partial t}\right) u + \mathscr{L}(x, D)\, u = h(x, t) \tag{2.1}$$

we mean the problem of finding a function $u(x,t)$ which satisfies the conditions

$$u(x, 0) = 0, \ldots, u^{(s-1)}(x, 0) = 0, \quad B_j(x, D)\, u|_S = 0,$$

and a system of functions $\rho_1(x), \rho_2(x), \cdots, \rho_k(x)$, such that

$$\mathscr{P}_s\left(\frac{\partial}{\partial t}\right) u + \mathscr{L}(x, D)\, u = h(x, t) + \mathscr{P}_{k-1}(t), \tag{2.3}$$

where

$$\mathscr{P}_{k-1}(t) = \rho_1(x) + \rho_2(x)\, t + \ldots + \rho_k(x)\, t^{k-1}.$$

Here the number k of new unknown functions is defined as follows:
a) $k = s/2$ if s is even,
b) $k = [s/2]$ if s is odd and $(-1)^{[s/2]} a_s > 0$,
c) $k = [s/2] + 1$ if s is odd and $(-1)^{[s/2]} a_s < 0$.

The generalized mixed problem is closely connected with the following boundary value problem:

$$\mathscr{P}_s\left(\frac{\partial}{\partial t}\right) u + \mathscr{L}(x, D)\, u = h(x, t), \tag{2.4}$$

$$u(x, 0) = 0, \ldots, u^{(s-1-k)}(x, 0) = 0, \quad B_j(x, D)\, u|_S = 0, \tag{2.5}$$

$$j = 1, \ldots, m.$$

The following two theorems are equivalent.

THEOREM 2.1. *For any function $h(x,t) \in H_\gamma^{k,0}(Q)$, where γ belongs to the correctness set Γ_0 of equation (2.4), the generalized mixed problem has a unique solution*

$$u(x, t), \rho_1(x), \rho_2(x), \ldots, \rho_k(x),$$

where

$$\int_0^\infty (\| u^{(s+k)} \|_0^2 + \| u^{(k)} \|_{2m}^2)\, e^{-\gamma t} dt + \sum_{i=1}^k \| \rho_i \|_0^2 \leqslant C \int_0^\infty \| h^{(k)} \|_0^2 e^{-\gamma t} dt.$$

THEOREM 2.2. *For any function $h(x,t) \in H_\gamma^{k,0}(Q)$, where $\gamma \in \Gamma_0$, the boundary value problem*

$$\mathscr{P}_s\left(\frac{\partial}{\partial t}\right) u + \mathscr{L}(x, D)\, u = h^{(k)}(x, t), \tag{2.6}$$

$$u(x, 0) = 0, \ldots, u^{(s-1-k)}(x, 0) = 0, \quad B_j(x, D)\, u|_S = 0, \tag{2.7}$$

$$j = 1, \ldots, m,$$

has a unique solution $u(x,t) \in H_\gamma^{s,2m}(Q)$, where

$$\int_0^\infty \| u^{(s)} \|_0^2 e^{-\gamma t} dt + \int_0^\infty \| u \|_{2m}^2 e^{-\gamma t} dt \leqslant C \int_0^\infty \| h^{(k)} \|_0^2 e^{-\gamma t} dt.$$

In fact, if the system of functions $u(x,t)$, $\rho_1(x), \cdots, \rho_k(x)$ is a solution of the generalized mixed problem, then, differentiating equation (2.3) k times with respect to t, we find that the function $v(x,t) = u^{(k)}(x,t)$ is a solution of the boundary value problem (2.6), (2.7).

Conversely, if the function $u(x,t)$ is a solution of the boundary value problem, then, setting

$$v(x, t) = \frac{1}{(k-1)!} \int_0^t (t-\tau)^{k-1} u(x, \tau)\, d\tau,$$

we find from equation (2.6) that

$$\left(\mathcal{P}_s\left(\frac{\partial}{\partial t}\right) v + \mathcal{L}(x, D) v - h(x, t)\right)^{(k)} \equiv 0,$$

i.e.

$$\mathcal{P}_s\left(\frac{\partial}{\partial t}\right) v + \mathcal{L}(x, D) v = h(x, t) + \mathcal{P}_{k-1}(t),$$

where

$$(i-1)!\, \rho_i(x) = \mathcal{P}_s\left(\frac{\partial}{\partial t}\right) v^{(k-i)}(x, 0) - h^{(k-i)}(x, 0),$$

$$i = 1, \ldots, k.$$

The equivalence is proved.

We now prove Theorem 2.3, which obviously implies Theorem 2.2.

THEOREM 2.3. *For any function $h(x,t) \in \mathcal{L}_{2,\gamma}(Q)$ there is a unique solution $u(x,t) \in H_\gamma^{s,2m}(Q)$ of the problem*

$$\mathcal{P}_s\left(\frac{\partial}{\partial t}\right) u + \mathcal{L}(x, D) u = h(x, t), \tag{2.8}$$

$$u(x, 0) = 0, \ldots, u^{(s-1-k)}(x, 0) = 0, \quad B_j(x, D) u|_S = 0, \tag{2.9}$$

$$j = 1, \ldots, m,$$

where the number k is defined in Theorem 2.1. The following inequality holds here:

$$\int_0^\infty \| u^{(s)} \|_0^2 e^{-\gamma t} dt + \int_0^\infty \| u \|_{2m}^2 e^{-\gamma t} dt \leqslant C \int_0^\infty \| h \|_0^2 e^{-\gamma t} dt.$$

PROOF. Let $\omega_0(x), \omega_1(x), \cdots$ be an orthonormal system of eigenfunctions of the operator $\mathcal{L}(x,D)u$, considered in the space $H^{2m}(\{B_j\})$. We shall look for a solution in the form of a series

$$u(x, t) = \sum_{i=0}^\infty c_i(t)\, \omega_i(x). \tag{2.10}$$

Substituting this series in equation (2.8), we find that the functions $c_i(t)$ satisfy the ordinary differential equations

$$\mathscr{P}_s\left(\frac{d}{dt}\right)c_i(t) + \lambda_i c_i(t) = h_i(t), \tag{2.11}$$

$$c_i(0) = 0, \ldots, c_i^{(s-1-k)}(0) = 0, \tag{2.12}$$

where the λ_i are the eigenvalues of the operator $\mathscr{L}(x, D)u$, $h_i(t) = (h, \omega_i)$.

LEMMA 2.1. *Every equation* (2.11) *has a unique solution in the space* $H_\gamma^s(0, \infty)$, $\gamma \in \Gamma_0$, *which satisfies the initial conditions* (2.12).

This lemma is proved at the end of the section. We show that the series (2.10) defined in this way converges and gives a solution of problem (2.8), (2.9). By (2.11), for the partial sum $u_n = \sum_{i=0}^n c_i(t)\omega_i(x)$ we have the relation

$$\mathscr{P}_s\left(\frac{\partial}{\partial t}\right)u_n + \mathscr{L}(x, D)u_n = P_n h(x, t), \tag{2.13}$$

where

$$P_n h(x, t) = \sum_{i=0}^{n} h_i(t)\omega_i(x).$$

The functions $u_n(x,t)$ satisfy the following estimate:

$$\int_0^\infty \|u_n^{(s)}\|_0^2 e^{-\gamma t}dt + \int_0^\infty \|u_n\|_{2m}^2 e^{-\gamma t}dt \leqslant C\int_0^\infty \|P_n h\|_0^2 e^{-\gamma t}dt.$$

The proof repeats the derivation of estimate (1.7) of the previous section (see the Remark at the end of §1).

It immediately follows from this estimate that the sequence of partial sums of the series (2.10) is fundamental in $H_\gamma^{s,2m}(Q)$. Hence this series converges and gives a solution of problem (2.8), (2.9). Theorem 2.3 is proved.

Theorem 2.1 is thus also proved, i.e. we have proved the Hadamard-Petrovskiĭ correct solvability of the generalized mixed problem.

2. The regularizer. We show that the generalized mixed problem is the regularizer of an incorrect problem with Cauchy data at $t = 0$.

THEOREM 2.4. *Suppose that* $h(x,t) \in K \cap K_\gamma^{k,0}(Q)$ *and* $u \in H_\gamma^{s+k,2m}(Q)$ *is a solution of the mixed problem*

$$\mathscr{P}_s\left(\frac{\partial}{\partial t}\right)u + \mathscr{L}(x, D)u = h(x, t), \tag{2.14}$$

$$u(x, 0) = 0, \ldots, u^{(s-1)}(x, 0) = 0, \quad B_j(x, D)u|_S = 0,$$
$$j = 1, \ldots, k.$$

Further let $h_\epsilon(x,t) \in H_\gamma^{k,0}(Q)$, *and let the system of functions* $u_\epsilon(x,t), \rho_{1\epsilon}(x), \cdots, \rho_{k\epsilon}(x)$ *be a solution of the generalized mixed problem*

$$\mathscr{P}_s\left(\frac{\partial}{\partial t}\right)u_\varepsilon + \mathscr{L}(x, D)u_\varepsilon = h_\varepsilon(x, t) + \mathscr{P}_{k-1,\varepsilon}(t), \tag{2.15}$$

where

$$\mathscr{P}_{k-1,\varepsilon}(t) = \rho_{1\varepsilon}(x) + \rho_{2\varepsilon}(x)\,t + \ldots + \rho_{k\varepsilon}(x)\,t^{k-1},$$

with the number k defined on p. 215. Then

$$\int_0^\infty \| (u - u_\varepsilon)^{(k+s)} \|_0^2 e^{-\gamma t} dt + \int_0^\infty \| (u - u_\varepsilon)^{(k)} \|_{2m}^2 e^{-\gamma t} dt \leqslant C \int_0^\infty \| (h - h_\varepsilon)^{(k)} \|_0^2 e^{-\gamma t} dt.$$

PROOF. Differentiating equations (2.14) and (2.15) k times with respect to t, we find that the difference $v_\varepsilon(x,t) = u^{(k)}(x,t) - u_\varepsilon^{(k)}(x,t)$ is a solution of the boundary value problem

$$\mathscr{P}_s\left(\frac{\partial}{\partial t}\right) v_\varepsilon + \mathscr{L}(x, D)\, v_\varepsilon = h^{(k)} - h_\varepsilon^{(k)},$$

$$v_\varepsilon(x, 0) = 0, \ldots, v_\varepsilon^{(l)}(x, 0) = 0, \quad B_j(x, D)\, v_\varepsilon|_S = 0,$$

where $l = s - 1 - k$, $j = 1, \cdots, m$. By Theorem 2.3

$$\int_0^\infty \| v_\varepsilon^{(s)} \|_0^2 e^{-\gamma t} dt + \int_0^\infty \| v_\varepsilon \|_{2m}^2 e^{-\gamma t} dt \leqslant C \int_0^\infty \| h^{(k)} - h_\varepsilon^{(k)} \|_0^2 e^{-\gamma t} dt,$$

which constitutes the inequality of the theorem. Theorem 2.4 is proved.

REMARK. In Theorem 2.4 we assumed that for $h(x,t) \in H_\gamma^{k,0}(Q)$ the solution $u(x,t)$ of the mixed problem is such that $u(x,t) \in H_\gamma^{s+k,2m}(Q)$. It will be shown in §7 that this is actually so.

3. Proof of Lemma 2.1. We consider the ordinary differential equation

$$\mathscr{P}_s\left(\frac{d}{dt}\right) u + \lambda u = h(t), \tag{2.16}$$

where the polynomial $\mathscr{P}_s(d/dt)$ satisfies the strict elliptico-parabolicity condition. We recall that this means that there is a correctness set Γ_0 such that for $\gamma \in \Gamma_0$ and for all $\tau \in \mathbf{R}^1$ we have the inequality

$$A(\gamma; \tau) \equiv \sum_{q=0}^{s} a_q \sum_{p=0}^{\left[\frac{q}{2}\right]} \beta_{2p,q} (-1)^p \tau^{2p} + \lambda > 0,$$

where

$$\beta_{2p,q} = \frac{q!}{(q-2p)!\,(2p)!} \left(\frac{\gamma}{2}\right)^{q-2p}.$$

We formulate again the lemma which we want to prove.

LEMMA 2.1. *For any right side* $h(t) \in \mathscr{L}_{2,\gamma}(0, \infty)$ *there exists a unique solution* $u(t) \in H_\gamma^s(0, \infty)$ *of equation* (2.16) *satisfying the following conditions*:

$$u(0) = 0, \ldots, u^{(l)}(0) = 0,$$

where $l = s - 1 - k$, *with the number k defined in Theorem* 2.1.

PROOF. a) First let $s = 2k$, $k \geqq 1$. We are then required to find a solution of equation (2.16) satisfying the initial conditions $u(0) = 0, \cdots, u^{(k-1)}(0) = 0$.

On the semiaxis $(0, \infty)$ we choose a system $v_1(t), v_2(t), \cdots$ of finite smooth functions which is complete in the space

$$\overset{\circ}{H}{}^k(0, \infty) = \left\{u(t) \,\Big|\, \int_0^\infty |u^{(k)}|^2\, dt < \infty,\ u(0) = 0, \ldots, u^{(k-1)}(0) = 0\right\}.$$

We shall look for an approximate solution $u_n(x, t)$ in the form

$$e^{-\gamma t} u_n(x, t) = \sum_{i=1}^n c_{in} v_i(t),$$

where the constants c_{in} are determined from the system of algebraic equations

$$\int_0^\infty \left[\mathcal{P}_s\left(\frac{d}{dt}\right) u_n + \lambda u_n\right] v_i\, dt = \int_0^\infty h v_i\, dt, \quad i = 1, \ldots, n. \tag{2.17}$$

We show that the following a priori inequality holds:

$$\int_0^\infty \left[\mathcal{P}_s\left(\frac{d}{dt}\right) u_n + \lambda u_n\right] u_n e^{-\gamma t}\, dt \geqslant a_0 \int_0^\infty |u_n^{(k)}|^2 e^{-\gamma t}\, dt, \tag{2.18}$$

where $a_0 > 0$ is a constant.

In fact, the product $u_n^{(q)}(t) e^{-\gamma t/2}$ can be represented in the form

$$u_n^{(q)}(t)\, e^{-\frac{\gamma}{2} t} = \sum_{r=0}^q \beta_{rq} \left(u_n(t)\, e^{-\frac{\gamma}{2} t}\right)^{(r)},$$

where

$$\beta_{rq} = \frac{q!}{(q-r)!\, r!} \left(\frac{\gamma}{2}\right)^{q-r}.$$

In this case we obtain that

$$\int_0^\infty \mathcal{P}_s\left(\frac{d}{dt}\right) u_n u_n e^{-\gamma t}\, dt = \sum_{q=0}^s a_q \sum_{r=0}^q \beta_{rq} \int_0^\infty \left(u_n e^{-\frac{\gamma}{2} t}\right)^{(r)} \left(u_n e^{-\frac{\gamma}{2} t}\right) dt$$

$$= \sum_{q=0}^s a_q \sum_{p=0}^{\left[\frac{q}{2}\right]} (-1)^p \beta_{2p,q} \int_0^\infty \left| \left(u_n e^{-\frac{\gamma}{2} t}\right)^{(p)} \right|^2 dt,$$

since $u_n(t)$ is finite on the semiaxis $(0, \infty)$. Using Parseval's equality, we find that

$$\int_0^\infty \left[\mathcal{P}_s\left(\frac{d}{dt}\right) u_n + \lambda u_n\right] u_n e^{-\gamma t}\, dt \equiv \int_{-\infty}^{+\infty} A(\gamma; \tau)\, |v_n(\tau)|^2\, d\tau,$$

where $v_n(\tau)$ is the Fourier transform $u_n(t) e^{-\gamma t/2}$.

We now note that, since $A(\gamma;\tau) > 0$ for $\gamma \in \Gamma_0$ and for any $\tau \in \mathbf{R}^1$, it follows that the coefficient $b_{2k} = (-1)^k a_{2k}$ of the leading term τ^{2k} is positive, i.e. $b_{2k} > 0$.

We write $A(\gamma;\tau)$ in the form

$$A(\gamma;\tau) \equiv \varepsilon b_{2k}\tau^{2k} + A_1(\gamma;\tau),$$

where $A_1(\gamma;\tau) \equiv A(\gamma;\tau) - \epsilon b_{2k}\tau^{2k}$. Obviously, for sufficiently small $\epsilon > 0$ the polynomial

$$A_1(\gamma;\tau) > 0$$

for $\gamma \in \Gamma_0$ and for all τ. Consequently

$$\int_0^\infty \left[\mathscr{P}_s\left(\frac{d}{dt}\right)u_n + \lambda u_n\right] u_n e^{-\gamma t}dt \geqslant \varepsilon b_{2k}\int_0^\infty \tau^{2k}|v_n(\tau)|^2\,d\tau,$$

which implies the a priori inequality (2.18).

In the first place, this inequality implies uniqueness, and hence solvability, of system (2.17) in $c_{1n}, \cdots, c_{nn}$, and, in the second place, it implies the following estimate for approximate solutions:

$$\int_0^\infty |u_n^{(k)}|^2 e^{-\gamma t}dt \leqslant C\int_0^\infty |h|^2 e^{-\gamma t}dt, \qquad (2.18')$$

where C is a constant which may depend on γ.

By the weak compactness of a bounded set in a Hilbert space, there is a subsequence (which we designate again by $u_n(t)$) such that $u_n(t) \to u(t)$ in $\overset{\circ}{H}{}^k_\gamma(0,\infty)$. This function is a generalized solution of the original equation: to see this, it suffices to pass to the limit as $n \to \infty$ in the moment equations (2.17) for any fixed i, and then consider the closure with respect to $v(t)$.

This solution $u(t)$ satisfies the initial conditions $u(0) = 0, \cdots, u^{(k-1)}(0) = 0$ and, in addition, $u(t) \in H^s_\gamma(0,\infty)$, since $h(t) \in \mathscr{L}_{2,\gamma}(0,\infty)$.

Uniqueness follows from inequality (2.18′) for the function $u(t)$.

Thus the lemma is proved for $s = 2k$.

b) Now let $s = 2k+1$ and $(-1)^k a_{2k+1} < 0$.

We are then required to find a solution $u(t) \in H^s_\gamma(0,\infty)$ of equation (2.16) satisfying the initial conditions

$$u(0) = 0, \ldots, u^{(k-1)}(0) = 0. \qquad (2.19)$$

Existence is proved exactly as in case a). We prove uniqueness. Namely, we show that for any function $u(t)$ which is a solution of equation (2.16) with conditions (2.19) we have the inequality

$$|u^{(k)}(0)|^2 + \int_{-\infty}^\infty A(\gamma;\tau)\,|\widetilde{ue^{-\frac{\gamma}{2}t}}|^2\,d\tau \leqslant C\int_0^\infty h^2 e^{-\gamma t}dt. \qquad (2.20)$$

In fact, we have

$$\int_0^\infty \mathscr{P}_s\left(\frac{d}{dt}\right) u \cdot ue^{-\gamma t} dt = \sum_{q=0}^{s} a_q \int_0^\infty u^{(q)} ue^{-\gamma t} dt$$

$$= \sum_{q=0}^{s} a_q \sum_{r=0}^{q} \beta_{rq} \int_0^\infty (ue^{-\frac{\gamma}{2}t})^{(r)} (ue^{-\frac{\gamma}{2}t})\, dt,$$

where

$$\beta_{rq} = \frac{q!}{(q-r)!\, r!} \left(\frac{\gamma}{2}\right)^{q-r}.$$

Hence

$$\int_0^\infty \left[\mathscr{P}_s\left(\frac{d}{dt}\right) u + \lambda u\right] ue^{-\gamma t} dt = \frac{1}{2} (-1)^{k+1} a_{2k+1} \,|\, u^{(k)}(0)\,|^2$$

$$+ \int_{-\infty}^{\infty} A(\gamma; \tau) \,|\, \widetilde{ue^{-\frac{\gamma}{2}t}} \,|^2 d\tau.$$

Since $(-1)^{k+1} a_{2k+1} > 0$, it follows from this inequality that any solution $u(t)$ satisfies estimate (2.20). Uniqueness is proved.

c) We finally consider the case when $s = 2k+1$ and $(-1)^k a_{2k+1} > 0$. Here we must find a solution $u(t) \in H_\gamma^s(0, \infty)$ of equation (2.16) under the conditions

$$u(0) = 0, \ldots, u^{(k)}(0) = 0. \tag{2.21}$$

We choose a system of basis functions $v_1(t), v_2(t), \cdots$ which are finite as $t \to +\infty$ and satisfy the initial conditions

$$v_i(0) = 0, \ldots, v_i^{(k-1)}(0) = 0, \quad i = 1, 2, \ldots.$$

We look for an approximate solution in the form

$$e^{-\gamma t} u_n(t) = \sum_{i=1}^{n} c_{in} v_i(t),$$

where the constants $c_{1n}, \cdots, c_{nn}$ are determined from the following system of equations:

$$[\mathfrak{A}(u_n), v_i] \equiv (-1)^{k+1} a_{2k+1} \int_0^\infty u_n^{(k)} v_i^{(k+1)} dt + \int_0^\infty Q_{s-1}\left(\frac{d}{dt}\right) u_n \cdot v_i\, dt = \int_0^\infty h v_i\, dt,$$

$$i = 1, \ldots, n. \tag{2.22}$$

Here

$$Q_{s-1}\left(\frac{d}{dt}\right) u \equiv \mathscr{P}_s\left(\frac{d}{dt}\right) u - a_{2k+1} u^{(2k+1)} + \lambda u.$$

We show that the following a priori inequality holds:

$$[\mathfrak{A}(u_n), u_n e^{-\gamma t}] \geqslant \frac{1}{C}(|u_n^{(k)}(0)|^2 + \int_{-\infty}^{\infty} A(\gamma;\tau)|\widetilde{u_n e^{-\frac{\gamma}{2}t}}|^2 d\tau). \tag{2.23}$$

In fact,

$$(-1)^{k+1}a_{2k+1}\int_0^\infty u_n^{(k)}(u_n e^{-\gamma t})^{(k+1)}\,dt = (-1)^k a_{2k+1}|u_n^{(k)}(0)|^2$$

$$+ a_{2k+1}\int_0^\infty u_n^{(2k+1)} u_n e^{-\gamma t}\,dt.$$

Hence

$$[\mathfrak{A}(u_n), u_n e^{-\gamma t}] = (-1)^k a_{2k+1}|u_n^{(k)}(0)|^2 + \int_0^\infty \left[\mathscr{P}_s\left(\frac{d}{dt}\right)u_n + \lambda u_n\right] u_n e^{-\gamma t}\,dt$$

$$= (-1)^k a_{2k+1}|u_n^{(k)}(0)|^2 + \sum_{q=0}^{2k+1} a_q \sum_{r=0}^{q} \beta_{rq}\int_0^\infty (u_n e^{-\frac{\gamma}{2}t})^{(r)}(u_n e^{-\frac{\gamma}{2}t})\,dt$$

$$+ \lambda\int_0^\infty u_n^2 e^{-\gamma t}\,dt.$$

In particular, $\beta_{2k+1,2k+1} = 1$. Hence, integrating by parts and using Parseval's equality, we obtain that

$$[\mathfrak{A}(u_n), u_n e^{-\gamma t}] = \frac{(-1)^k a_{2k+1}}{2}|u_n^{(k)}(0)|^2 + \int_{-\infty}^{\infty} A(\gamma;\tau)|\widetilde{u_n e^{-\frac{\gamma}{2}t}}|^2 d\tau,$$

which implies inequality (2.23).

Using this inequality just as in case a), we find a function $u(t) \in H_\gamma^s(0, \infty)$, which satisfies the initial conditions

$$u(0) = 0, \ldots, u^{(k-1)}(0) = 0$$

and is a solution of equation (2.16) in the sense that for any smooth function $v(t), v(0) = 0, \cdots, v^{(k-1)}(0) = 0$ which is finite as $t \to +\infty$ we have the identity

$$[\mathfrak{A}(u), v] \equiv \int_0^\infty hv\,dt. \tag{2.24}$$

It remains to show that $u^{(k)}(0) = 0$. To do this, we choose functions which are also finite at $t = 0$ for $v(t)$ in (2.24) and find that

$$\left[\mathscr{P}_s\left(\frac{d}{dt}\right)u + \lambda u, v\right] \equiv [h, v]. \tag{2.25}$$

Since $u(t) \in H_\gamma^s(0, \infty)$, because $h(t) \in \mathscr{L}_{2,\gamma}(0, \infty)$, we can close this equation for any function $v(t)$ which is finite as $t \to \infty$ and is such that $v(0) = 0, \cdots, v^{(k-1)}(0) = 0$.

Recalling that such functions satisfy relation (2.24) and comparing this relation with (2.25), we find that

$$\int_0^\infty u^{(k)} v^{(k+1)}\, dt = (-1)^{k+1} \int_0^\infty u^{(2k+1)} v\, dt.$$

Since $v^{(k)}(0)$ is arbitrary, this means that $u^{(k)}(0) = 0$. Thus existence is proved. Uniqueness follows from a suitable a priori estimate. Lemma 2.1 is completely proved.

§3. Elliptico-parabolic equations. Normal solvability of the generalized mixed problem

In the previous section we considered the generalized mixed problem and the boundary value problem for the equation

$$\mathfrak{A}(u) \equiv \mathscr{P}_s\left(\frac{\partial}{\partial t}\right) u + \mathscr{L}(x, D)\, u = h(x, t)$$

in the spaces H_γ, where $\gamma \in \Gamma_0$ (Γ_0 is the correctness set). It is now natural to study these problems in spaces H_γ for arbitrary γ.

DEFINITION 3.1. We call an operator $\mathfrak{A}(u)$ on the set Γ *elliptico-parabolic* if for any $\gamma \in \Gamma$ there is a constant $\alpha(\gamma) \geqq 0$ such that the polynomial

$$A(\gamma; \tau) + \alpha(\gamma) > 0.$$

THEOREM 3.1. *If the operator $\mathfrak{A}(u)$ is elliptico-parabolic, then for any $\gamma \in \Gamma$ the generalized mixed problem* (2.2), (2.3), *and also the boundary value problem* (2.4), (2.5), *is normally solvable. This means that for any right side $h(x,t) \in \mathscr{L}_{2,\gamma}(Q)$ which satisfies a finite number of conditions, these problems are solvable in the space $H_\gamma^{s,2m}(Q)$. In addition, the solution space of the corresponding homogeneous problem is finite dimensional.*[4)]

PROOF. Because Theorems 1.1 and 2.2 are equivalent, it suffices to consider the boundary value problem (2.4), (2.5). Let λ_N be the first eigenvalue of the operator $\mathscr{L}(x, D)u$ which is greater than $\alpha(\gamma)$, i.e. $\lambda_N \geqq \alpha(\gamma)$. We consider the decomposition of the space $H_\gamma \equiv H_\gamma^{s,2m}(Q)$ into the orthogonal sum

$$H_\gamma = H_\gamma^+ \oplus H_\gamma^-,$$

where

$$H_\gamma^+ = \left\{u^+(x, t) \,|\, u^+(x, t) = \sum_{i=N}^{\infty} c_i(t)\, \omega_i(x),\ c_i(t) \in H_\gamma^s(0, \infty)\right\},$$

$$H_\gamma^- = \left\{u^-(x, t) \,|\, u^-(x, t) = \sum_{i=0}^{N-1} c_i(t)\, \omega_i(x),\ c_i(t) \in H_\gamma^s(0, \infty)\right\}.$$

---

4) Here we have the estimate $\| u \|_{s,2m,\gamma} \leq C(\| h \|_{0,0,\gamma} + \| u \|_{0,0,\gamma})$.

We accordingly write the space of right sides $\mathscr{L}_{2,\gamma}(Q)$ in the form

$$\mathscr{L}_{2,\gamma} = \mathscr{L}_{2,\gamma}^{+} \oplus \mathscr{L}_{2,\gamma}^{-}.$$

Since these decompositions are orthogonal, it follows that the equation

$$\mathfrak{A}(u) = h$$

decomposes into a system of two equations of the form

$$\mathscr{P}_s\left(\frac{\partial}{\partial t}\right) u^+ + \mathscr{L}(x, D)\, u^+ = h^+(x, t) \tag{3.1}$$

and

$$\mathscr{P}_s\left(\frac{\partial}{\partial t}\right) u^- + \mathscr{L}(x, D)\, u^- = h^-(x, t), \tag{3.2}$$

where $u = u^+ + u^-$ and $h = h^+ + h^-$. Thus, solving the boundary value problem is equivalent to finding the components $u^+(x,t)$ and $u^-(x,t)$.

Since $\lambda_N \geqq \alpha(\gamma)$, it follows that equation (3.1) is strictly ellipticoparabolic in the sense of Definition 1.1. Consequently for any $h^+(x,t) \in \mathscr{L}_{2,\gamma}^{+}(Q)$ there is a unique solution $u^+(x,t) \in H_\gamma^+(Q)$, where

$$\int_0^\infty \| u^{+(s)} \|_0^2 e^{-\nu t}\, dt + \int_0^\infty \| u^+ \|_{2m}^2 e^{-\nu t}\, dt \leqslant C \int_0^\infty \| h^+ \|_0^2 e^{-\nu t}\, dt.$$

Further, equation (3.2) is obviously equivalent to the system of ordinary differential equations

$$\mathscr{P}_s\left(\frac{d}{dt}\right) c_i(t) + \lambda_i c_i(t) = h_i(t), \quad i = 0, 1, \ldots, N-1,$$
$$c_i(0) = 0, \ldots, c_i^{(l)}(0) = 0,$$

where $u^-(x,t) = \sum_{i=1}^{N-1} c_i(t)\,\omega_i(x)$. Since both the kernel and the cokernel of a system of ordinary differential equations are always finite dimensional, the theorem is proved.

§4. Mixed problem for elliptico-hyperbolic equations

1. Definition of a strictly elliptico-hyperbolic operator. In the cylinder $Q = G \times [0, \infty)$ with lateral surface S we consider the mixed problem:

$$\mathscr{P}_s\left(\frac{\partial}{\partial t}\right) u + \mathscr{L}(x, D)\, u = h(x, t), \tag{4.1}$$

$$u(x, 0) = 0, \ldots, u^{(s-1)}(x, 0) = 0, \quad B_j(x, D)\, u|_S = 0, \quad j = 1, \ldots, m, \tag{4.2}$$

where

$$\mathscr{P}_s\left(\frac{\partial}{\partial t}\right) u \equiv \sum_{q=0}^{s} a_q \frac{\partial^q}{\partial t^q} u$$

(the a_q are constants), and

$$\mathscr{L}(x, D)u \equiv \sum_{|\alpha|\leqslant 2m} a_\alpha(x) D^\alpha u$$

is an elliptic operator of order $2m$.

As in §1, we require that the boundary operators $B_j(x, D)u$ determine a selfadjoint boundary value problem in the region G for the operator $\mathscr{L}(x, D)u$. We let $\omega_0(x), \omega_1(x), \cdots$ designate a system of eigenfunctions of the operator $\mathscr{L}(x, D)u$ with eigenvalues $\lambda_0, \lambda_1, \cdots$.

DEFINITION 4.1. The operator

$$\mathfrak{A}(u) \equiv \mathscr{P}_s\left(\frac{\partial}{\partial t}\right)u + \mathscr{L}(x, D)u$$

is called *strictly elliptico-hyperbolic* on the set Γ_0 of the real axis if for any $\gamma \in \Gamma_0$ the polynomial

$$\bar{A}(\gamma; \tau) + \frac{\gamma}{2}\lambda_0 \equiv \left(\tau^2 + \frac{\gamma^2}{4}\right)\sum_{q=0}^{s-1} a_{q+1} \sum_{p=0}^{\left[\frac{q}{2}\right]} (-1)^p \beta_{2p,q}\tau^{2p} + \frac{a_0\gamma}{2} + \frac{\gamma}{2}\lambda_0 > 0,^{5)}$$

where λ_0 is the first eigenvalue of the operator $\mathscr{L}(x, D)u$ and

$$\beta_{2p,q} = \frac{q!}{(q-2p)!\,(2p)!}\left(\frac{\gamma}{2}\right)^{q-2p}.$$

2. Tihonov correctness of a mixed problem.

THEOREM 4.1. *Let $K \subset H_\gamma^{1,0}(Q)$ be the subspace of right sides $h(x,t)$ of problem* (4.1), (4.2) *for which there is a solution $u(x,t) \in H_\gamma^{s,2m}(Q)$. Further, let the operator $\mathfrak{A}(u)$ be strictly elliptico-hyperbolic in the sense of Definition* 4.1. *Then for any $\gamma \in \Gamma_0$ the mixed problem is Tihonov correct, and in addition the following inequality holds*:

$$\int_0^\infty \|u^{(s)}\|_0^2 e^{-\gamma t}dt + \int_0^\infty \|u\|_{2m}^2 e^{-\gamma t}dt \leqslant C\int_0^\infty \|(he^{-\gamma t})'\|_0^2 e^{\gamma t}dt. \qquad (4.3)$$

PROOF. It obviously suffices to prove estimate (4.3). The derivation of this estimate is to a large extent analogous to that of estimate (1.7) in the proof of Theorem 1.1 in §1. Let

$$H_\gamma^{s,2m}(Q) = H_\gamma^+ \oplus H_\gamma^-$$

be the orthogonal decomposition of the space $H_\gamma^{s,2m}(Q)$ into the subspaces H_γ^+ and H_γ^- generated by the eigenfunctions of the operator $\mathscr{L}(x, D)u$ with positive and negative eigenvalues, respectively (see §3).

Then for $u(x,t) \in H_\gamma^{s,2m}(Q)$ we have the decomposition

---

5) It is easily seen that

$$\bar{A}(\gamma; \tau) + \frac{\gamma}{2}\lambda_0 \equiv \mathrm{Re}\left(\frac{\gamma}{2} - i\tau\right)\left[\mathscr{P}_s\left(\frac{\gamma}{2} + i\tau\right) + \lambda_0\right]$$

$$u(x,t) = u^+(x,t) + u^-(x,t),$$

where $u^+ \in H_\gamma^+$, $u^- \in H_\gamma^-$.

LEMMA. *Any function $u(x,t) \in H_\gamma^{s,2m}(Q)$ satisfies the following inequalities:*

a) $$\int_0^\infty (\mathscr{L}(x,D)u, u')e^{-\gamma t}\,dt \geqslant \frac{\gamma}{2}\lambda_0 \int_0^\infty \|u\|_0^2 e^{-\gamma t}dt,$$

where λ_0 is the first eigenvalue of the operator $\mathscr{L}(x,D)u$;

b) $$\int_0^\infty (\mathscr{L}(x,D)u^+, \mathscr{L}(x,D)(u^+)')e^{-\gamma t}\,dt \geqslant \frac{\gamma}{2}\mu_1 \int_0^\infty \|M^+u^+\|_0^2 e^{-\gamma t}\,dt,$$

where $M^+u^+ = \sqrt{\mathscr{L}^+u^+}$, and $\mu_1 > 0$ is the first positive eigenvalue of the operator $\mathscr{L}$.

PROOF. We have

$$u(x,t) = \sum_{i=0}^\infty c_i(t)\,\omega_i(x),$$

where $c_i(t) = (h(x,t), \omega_i(x))$. Consequently

$$\int_0^\infty (\mathscr{L}(x,D)u, u')e^{-\gamma t}dt = \sum_{i=0}^\infty \lambda_i \int_0^\infty c_i(t)\,c_i'(t)\,e^{-\gamma t}\,dt$$

$$= \sum_{i=0}^\infty \frac{\gamma}{2}\lambda_i \int_0^\infty c_i^2(t)\,e^{-\gamma t}\,dt \geqslant \frac{\gamma}{2}\lambda_0 \int_0^\infty \|u\|_0^2 e^{-\gamma t}dt.$$

Inequality b) is proved in exactly the same way. The lemma is proved.

We proceed to derive inequality (4.3). As in Theorem 1.1, we first estimate $u^+(x,t)$. Projecting equation (4.1) onto the subspace H_γ^+, we obtain

$$\mathscr{P}_s\left(\frac{\partial}{\partial t}\right)u^+ + \mathscr{L}^+(x,D)u^+ = h(x,t), \tag{4.4}$$

where $\mathscr{L}^+$ is the restriction of $\mathscr{L}$ to H_γ^+. We multiply equation (4.4) by $\mathscr{L}^+(u^+)'e^{-\gamma t}$ and integrate over the cylinder Q:

$$\int_0^\infty \left(\mathscr{P}_s\left(\frac{\partial}{\partial t}\right)u^+, \mathscr{L}^+(u^+)'\right)e^{-\gamma t}dt + \int_0^\infty (\mathscr{L}^+u^+, \mathscr{L}^+(u^+)')\,e^{-\gamma t}dt$$

$$= \int_0^\infty (h^+, \mathscr{L}^+(u^+)')\,e^{-\gamma t}dt. \tag{4.5}$$

We consider the first term. We have

$$\int_0^\infty \left(\mathscr{P}_s\left(\frac{\partial}{\partial t}\right)u^+,\ \mathscr{L}^+(u^+)'\right)e^{-\gamma t}dt = \int_0^\infty \left(\mathscr{P}_s\left(\frac{\partial}{\partial t}\right)v^+,\ (v^+)'\right)e^{-\gamma t}dt,$$

where $v^+(x,t) = M^+u^+$. Further, we easily see that

$$\sum_{q=0}^{s} a_q\,(v^+)^{(q)}\,(v^+)'\,e^{-\gamma t} \equiv \sum_{q=1}^{s} a_q\,(v^+)^{(q)}\,(v^+)'\,e^{-\gamma t} + a_0 v^+\,(v^+)'\,e^{-\gamma t}$$

$$= \sum_{q=0}^{s-1} a_{q+1} \sum_{p=0}^{q} \beta_{pq}\,((v^+)'\,e^{-\frac{\gamma}{2}t})^{(p)}\,((v^+)'\,e^{-\frac{\gamma}{2}t}) + \frac{a_0}{2}\,\frac{d}{dt}\,(v^+)^2\,e^{-\gamma t}.$$

Hence, integrating by parts and using Parseval's equality, we obtain that

$$\int_0^\infty \left(\mathscr{P}_s\left(\frac{\partial}{\partial t}\right)v^+,\ (v^+)'\right)e^{-\gamma t}dt = \int_0^\infty \bar{A}(\gamma;\tau)\,\|\widetilde{v^+e^{-\frac{\gamma}{2}t}}\|_0^2\,d\tau,$$

where $\bar{A}(\gamma;\tau)$ is the polynomial in Definition 4.1. In addition, by the lemma

$$\int_0^\infty (\mathscr{L}^+u^+,\ \mathscr{L}^+(u^+)')\,e^{-\gamma t}dt \geqslant \frac{\gamma}{2}\lambda_0 \int_0^\infty \|v^+\|_0^2 e^{-\gamma t}dt = \frac{\gamma}{2}\lambda_0 \int_0^\infty \|\widetilde{v^+e^{-\frac{\gamma}{2}t}}\|_0^2\,d\tau.$$

From equation (4.4) we then derive the inequality

$$\varepsilon\int_0^\infty (\mathscr{L}^+u^+,\ \mathscr{L}^+(u^+)')\,e^{-\gamma t}dt + \int_0^\infty \left(\bar{A}(\gamma;\tau) + (1-\varepsilon)\,\frac{\gamma}{2}\lambda_0\right)\|\widetilde{v^+e^{-\frac{\gamma}{2}t}}\|_0^2\,d\tau$$

$$\leqslant \int_0^\infty (h^+,\ \mathscr{L}^+(u^+)')\,e^{-\gamma t}dt = -\int_0^\infty ((h^+e^{-\gamma t})',\ \mathscr{L}^+u^+)\,dt.$$

By Definition 4.1, for any fixed $\gamma\in\Gamma_0$

$$\bar{A}(\gamma;\tau) + (1-\varepsilon)\,\frac{\gamma}{2}\lambda_0 > 0,$$

if $\epsilon > 0$ is sufficiently small.

Discarding this term, we find from the last inequality that

$$\int_0^\infty (\mathscr{L}^+u^+,\ \mathscr{L}^+(u^+)')\,e^{-\gamma t}dt \leqslant C\left(\int_0^\infty \|(h^+e^{-\gamma t})'\|_0^2 e^{\gamma t}dt\right)^{\frac{1}{2}}\left(\int_0^\infty \|\mathscr{L}^+u^+\|_0^2\,e^{-\gamma t}dt\right)^{\frac{1}{2}},$$

from which we derive the estimate

$$\int_0^\infty \|u^+\|_{2m}^2 e^{-\gamma t}dt \leqslant C\int_0^\infty \|(h^+e^{-\gamma t})'\|_0^2\,e^{\gamma t}dt.$$

Now equation (4.4) implies the desired estimate of the theorem for $u^+(x,t)$. To estimate the function $u^-(x,t)$, we must multiply the equation

$$\mathscr{P}_s\left(\frac{\partial}{\partial t}\right)u^- + \mathscr{L}(x, D)\,u^- = h^-(x, t) \tag{4.6}$$

by $(u^-)'e^{-\gamma t}$ and repeat the previous reasoning. We obtain the inequality

$$\int_0^\infty \|u^-\|_0^2\, e^{-\gamma t}dt \leqslant C\int_0^\infty \|(h^- e^{-\gamma t})'\|_0^2 e^{\gamma t}dt,$$

which implies that

$$\int_0^\infty \|u^-\|_{2m}^2 e^{-\gamma t}dt \leqslant \tilde{C}\int_0^\infty \|(h^- e^{-\gamma t})'\|_0^2\, e^{\gamma t}dt,$$

since the space H_γ^- is generated by a finite number of generators $\omega_i(x)$, $\lambda_i \leqq 0$. Equation (4.6) now implies estimate (4.3) for $u^-(x,t)$ as well. Consequently the estimate is also true for $u(x,t) = u^+(x,t) + u^-(x,t)$. The theorem is proved.

REMARK. As in §1, for subsequent considerations it is important to note that, to derive estimate (4.3), it is sufficient that

$$u(x, 0) = 0, \ldots, u^{(s-1-k)}(x, 0) = 0,$$

where k is defined below in Definition 5.1.

§5. Generalized mixed problem for elliptico-hyperbolic equations. The regularizer. Normal solvability

1. We begin with some definitions.

DEFINITION 5.1. By the *generalized mixed problem* for the elliptico-hyperbolic equation

$$\mathfrak{A}(u) \equiv \mathscr{P}_s\left(\frac{\partial}{\partial t}\right)u + \mathscr{L}(x, D)\,u = h(x, t) \tag{5.1}$$

we mean the problem of finding a function $u(x,t)$ which satisfies the conditions

$$u(x, 0) = 0, \ldots, u^{(s-1)}(x, 0) = 0,\ B_j(x, D)\,u|_S = 0, \quad j = 1, \ldots, m, \tag{5.2}$$

and a system of functions $\rho_1(x), \rho_2(x), \cdots, \rho_k(x)$, such that

$$\mathscr{P}_s\left(\frac{\partial}{\partial t}\right)u + \mathscr{L}(x, D)\,u = h(x, t) + \mathscr{P}_{k-1}(t),$$

where $\mathscr{P}_{k-1}(t) = \rho_1(x) + \rho_2(x)t + \cdots + \rho_k(x)t^{k-1}$. Here the number k of new unknown functions $\rho_1(x), \cdots, \rho_k(x)$ is defined as follows:

a) $k = [s/2]$ is s is odd,

b) $k = s/2$, if s is even and $(-1)^{s/2}a_s > 0$,

c) $k = s/2 - 1$ if s is even and $(-1)^{s/2}a_s < 0$.

DEFINITION 5.2. By *the boundary value problem* for the elliptico-hyperbolic equation

$$\mathfrak{A}(u) \equiv \mathscr{P}_s\left(\frac{\partial}{\partial t}\right) u + \mathscr{L}(x, D) u = h(x, t) \tag{5.3}$$

we mean the problem of finding a solution $u(x,t)$ which satisfies the conditions

$$u(x, 0) = 0, \ldots, u^{(s-1-k)}(x, 0) = 0, \quad B_j(x, D) u|_S = 0, \quad j = 1, \ldots, m. \tag{5.4}$$

As in the elliptico-parabolic case, the following two theorems are equivalent.

THEOREM 5.1. *If the operator* $\mathfrak{A}(u)$ *is strictly elliptico-hyperbolic on the set* Γ_0, *then for any function* $h(x,t) \in H_\gamma^{k+1,0}(Q)$, $\gamma \in \Gamma_0$, *there is a unique solution* $u(x,t), \rho_1(x), \cdots, \rho_k(x)$ *of the generalized mixed problem* (5.1), (5.2), *where*

$$\int_0^\infty \| u^{(s+k)} \|_0^2 e^{-\gamma t}\, dt + \int_0^\infty \| u^{(k)} \|_{2m}^2 e^{-\gamma t}\, dt + \sum_{i=1}^k \| \rho_i \|_0^2 \leqslant C \int_0^\infty \| (h^{(k)} e^{-\gamma t})' \|_0^2 e^{\gamma t}\, dt.$$

THEOREM 5.2. *Under the same conditions there is a solution* $u(x,t)$ *of the boundary value problem* (5.4) *for the equation*

$$\mathscr{P}_s\left(\frac{\partial}{\partial t}\right) u + \mathscr{L}(x, D) u = h^{(k)}(x, t),$$

where

$$\int_0^\infty \| u^{(s)} \|_0^2 e^{-\gamma t}\, dt + \int_0^\infty \| u \|_{2m}^2 e^{-\gamma t}\, dt \leqslant C \int_0^\infty \| (h^{(k)} e^{-\gamma t})' \|_0^2 e^{\gamma t}\, dt.$$

Equivalence is proved exactly as in §2.

We now prove a theorem on solvability of the boundary value problem.

THEOREM 5.3. *For any function* $h(x,t) \in H_\gamma^{1,0}(Q)$ *there is a unique solution of the boundary value problem* (5.3), (5.4). *Here the following inequality holds*:

$$\int_0^\infty \| u^{(s)} \|_0^2 e^{-\gamma t}\, dt + \int_0^\infty \| u \|_{2m}^2 e^{-\gamma t}\, dt \leqslant C \int_0^\infty \| (h e^{-\gamma t})' \|_0^2 e^{\gamma t}\, dt.$$

PROOF. Let $\omega_0(x), \omega_1(x), \cdots$ be an orthonormal system of eigenfunctions of the operator $\mathscr{L}(x, D)u$. We look for a solution in the form of a series

$$u(x, t) = \sum_{i=0}^\infty c_i(t)\, \omega_i(x), \tag{5.5}$$

where the functions $c_i(t)$ are defined as the solutions of the ordinary differential equations

$$\mathscr{P}_s\left(\frac{d}{dt}\right) c_i(t) + \lambda_i c_i(t) = h_i(t), \tag{5.6}$$

$$c_i(0) = 0, \ldots, c_i^{(s-1-k)}(0) = 0, \tag{5.7}$$

where $h_i(t) = (h, \omega_i)$.

Lemma. *The ordinary differential equations* (5.6) *have unique solutions* $c_i(t) \in H^s_\gamma(0, \infty)$ *which satisfy the initial conditions* (5.7).

This lemma is proved just like Lemma 2.1, except that the approximate solution $c_{in}(t)$ must be determined from the relation

$$\frac{dc_{in}(t)}{dt} e^{-\gamma t} = \sum_{\nu=1}^{n} \alpha_{in}^{\nu} v_{\nu}(t), \quad c_{in}(0) = 0,$$

where $\{v_\nu(t), \nu = 1, 2, \cdots\}$ is a basis system of functions with compact support.

We now show that the series (5.5) converges in $H^{s,2m}_\gamma(Q)$ and gives the desired solution. We let $u_n(x,t)$ designate the nth partial sum of the series:

$$u_n(x, t) = \sum_{i=0}^{n} c_i(t)\,\omega_i(x).$$

Then

$$\mathscr{P}_s\left(\frac{\partial}{\partial t}\right) u_n + \mathscr{L}(x, D)\, u_n = P_n h(x, t),$$

where $P_n h(x,t) = \sum_{i=0}^{n} h_i(t)\,\omega_i(x)$. We have the estimate

$$\int_0^\infty \| u_n^{(s)} \|_0^2 e^{-\gamma t}\, dt + \int_0^\infty \| u_n \|_{2m}^2 e^{-\gamma t}\, dt \leqslant C \int_0^\infty \| (he^{-\gamma t})' \|_0^2 e^{\gamma t}\, dt.$$

This estimate is derived in exactly the same way as in §4 (see the Remark at the end of §4).

This estimate implies that the sequence of partial sums of the series (5.5) is fundamental in $H^{s,2m}_\gamma(Q)$, i.e. this series converges and obviously gives a solution of the boundary value problem. The theorem is proved.

2. Regularizer of the mixed problem. It follows from the theorem we have proved that the generalized mixed problem (5.1), (5.2) is also uniquely solvable. We show that it is the regularizer of the incorrect mixed problem (4.1), (4.2).

Theorem 5.2. *Suppose that* $h(x,t) \in H^{k+1,0}_\gamma(Q) \cap K$ *and* $u(x,t) \in H^{s+k,2m}_\gamma(Q)$ *is a solution of the mixed problem*

$$\mathscr{P}_s\left(\frac{\partial}{\partial t}\right) u + \mathscr{L}(x, D)\, u = h(x, t), \tag{5.8}$$

$$u(x, 0) = 0, \ldots, u^{(s-1)}(x, 0) = 0, \quad B_j(x, D)\, u|_S = 0, \tag{5.9}$$

$$j = 1, \ldots, m.$$

Further, let $h_\epsilon(x,t) \in H^{k+1,0}_\gamma(Q)$, *and let the system of functions* $u_\epsilon(x,t), \rho_{1\epsilon}(x), \cdots, \rho_{k\epsilon}(x)$ *be a solution of the generalized mixed problem*

$$\mathscr{P}_s\left(\frac{\partial}{\partial t}\right) u_\varepsilon + \mathscr{L}(x, D)\, u_\varepsilon = h_\varepsilon(x, t) + \mathscr{P}_{k-1,\varepsilon}(t), \tag{5.10}$$

$$u(x, 0) = 0, \ldots, u^{(s-1)}(x, 0) = 0, \quad B_j(x, D)\, u|_S = 0, \tag{5.11}$$

where $\mathscr{P}_{k+1,\varepsilon}(t) = \rho_{1\varepsilon}(x) + \rho_{2\varepsilon}(x)t + \cdots + \rho_{k\varepsilon}(x)t^{k-1}$, *with the number* k *given by Definition* 5.1.

Then the following inequality holds:

$$\int_0^\infty \|(u-u_\varepsilon)^{(s+k)}\|_0^2 e^{-\nu t}\,dt + \int_0^\infty \|(u-u_\varepsilon)^{(k)}\|_{2m}^2 e^{-\nu t}\,dt \leqslant C\int_0^\infty \|((h^{(k)}-h_\varepsilon^{(k)})e^{-\nu t})'\|_0^2 e^{\nu t}\,dt. \tag{5.12}$$

PROOF. Differentiating equations (5.8) and (5.10) k times with respect to t, we find that the difference $v_\varepsilon(x,t) = u^{(k)}(x,t) - u_\varepsilon^{(k)}(x,t)$ is a solution of the boundary value problem

$$\mathscr{P}_s\left(\frac{\partial}{\partial t}\right)v_\varepsilon + \mathscr{L}(x, D)v_\varepsilon = h^{(k)} - h_\varepsilon^{(k)},$$

$$v_\varepsilon(x, 0) = 0, \ldots, v_\varepsilon^{(s-1-k)}(x, 0) = 0, \ B_j(x, D)v_\varepsilon|_S = 0.$$

It follows from Theorem 5.1 that

$$\int_0^\infty \|v_\varepsilon^{(s)}\|_0^2 e^{-\nu t}\,dt + \int_0^\infty \|v_\varepsilon\|_{2m}^2 e^{-\nu t}\,dt \leqslant C\int_0^\infty \|((h^{(k)}-h_\varepsilon^{(k)})e^{-\nu t})'\|_0^2 e^{\nu t}\,dt,$$

but this is the desired inequality (5.12). The theorem is proved.

3. Normal solvability of elliptico-hyperbolic equations.

DEFINITION 5.3. The operator

$$\mathfrak{A}(u) \equiv \mathscr{P}_s\left(\frac{\partial}{\partial t}\right)u + \mathscr{L}(x, D)u$$

is called *elliptico-hyperbolic* on the set Γ if for any $\gamma \in \Gamma$ there is a constant $\alpha(\gamma)$ such that the polynomial

$$\overline{A}(\gamma; \tau) + \alpha(\gamma) > 0,$$

where

$$\overline{A}(\gamma; \tau) \equiv \left(\tau^2 + \frac{\gamma^2}{4}\right)\sum_{q=0}^{s-1} a_{q+1}\sum_{p=0}^{\left[\frac{q}{2}\right]}(-1)^p\beta_{2p,q}\tau^{2p} + \frac{a_0\gamma}{2},$$

$$\beta_{2p,q} = \frac{q!}{(q-2p)!\,(2p)!}\left(\frac{\gamma}{2}\right)^{q-2p}.$$

THEOREM 5.4. *If the operator* $\mathfrak{A}(u)$ *is elliptico-hyperbolic, then for any* $\gamma \in \Gamma$ *the generalized mixed problem* (5.1), (5.2), *along with the boundary value problem* (5.3), (5.4), *is normally solvable. This means that, for any right side* $h(x,t) \in H_\gamma^{1,0}(Q)$ *which satisfies a finite number of conditions, these problems are solvable in the space* $H_\gamma^{s,2m}(Q)$. *In addition, the kernel of the operator* $\mathfrak{A}(u)$ *is finite dimensional.*

The proof of this theorem is completely analogous to that of Theorem 3.1 in §3.

§6. Structure of the subspace of right sides of the mixed problem

1. Elliptico-parabolic case. In the first section of this article it was shown that the mixed problem

$$\mathfrak{A}(u) \equiv \mathscr{P}_s\left(\frac{\partial}{\partial t}\right)u + \mathscr{L}(x, D)\,u = h(x, t), \tag{6.1}$$

$$u(x, 0) = 0, \dots, u^{(s-1)}(x, 0) = 0, \quad B_j(x, D)\,u|_S = 0, \quad j = 1, \dots, m, \tag{6.2}$$

is Tihonov correct in the spaces $H_\gamma^{s,2m}(Q)$ for any $\gamma \in \Gamma_0$ (Γ_0 is the correctness set of the operator $\mathfrak{A}(u)$ in Definition 1.1).

In particular, this means that problem (6.1), (6.2) is solvable not for an arbitrary right side $h(x,t) \in \mathscr{L}_{2,\gamma}(Q)$, but only when $h(x,t) \in K$, where $K \subset \mathscr{L}_{2,\gamma}(Q)$ is a certain subspace. This subspace is completely described by the theorem below.

We consider the roots μ_{ij}, $i = 0, 1, \cdots, j = 1, \cdots, s$, of the characteristic polynomials

$$\mathscr{P}_s(\mu) + \lambda_i,$$

where the λ_i are the eigenvalues of the operator $\mathscr{L}(x, D)u$. We suppose that the roots μ_{ij} are not multiple (the case of multiple roots is treated analogously).

THEOREM 6.1. *In order that the mixed problem* (6.1), (6.2) *be solvable in the space* $H_\gamma^{s,2m}(Q)$ *for* $h(x,t) \in \mathscr{L}_{2,\gamma}(Q)$, *it is necessary and sufficient that the following conditions hold for any* $i = 0, 1, \cdots$:

$$\sum_{\operatorname{Re}\mu_{ij} \geq \gamma} \frac{W_j}{W} e^{\mu_{ij}t} \int_0^t e^{-\mu_{ij}\tau} h_i(\tau)\, d\tau \in \mathscr{L}_{2,\gamma}(0, \infty), \tag{*}$$

where $h_i(\tau) = (h, \omega_i)$, W *is the Vandermonde determinant, and the* W_j *are the cofactors of the last row of* W.

PROOF. NECESSITY. For $h(x,t) \in \mathscr{L}_{2,\gamma}(Q)$ let there exist a solution $u(x,t) \in H_\gamma^{s,2m}(Q)$ of the mixed problem. We then have

$$u(x, t) = \sum_{i=0}^{\infty} c_i(t)\, \omega_i(x), \tag{6.3}$$

where $c_i(t) = (u, \omega_i)$ and the series (6.3) converges in $H_\gamma^{s,2m}(Q)$. In particular, this means that

$$c_i(t) \in \mathscr{L}_{2,\gamma}(0, \infty).$$

Substituting the series (6.3) in the equation

$$\mathscr{P}_s\left(\frac{\partial}{\partial t}\right)u + \mathscr{L}(x, D)\,u = h(x, t),$$

we find that the projections $c_i(t)$ are a solution of the system of ordinary differential equations

$$\mathscr{P}_s\left(\frac{d}{dt}\right)c_i(t) + \lambda_i c_i(t) = h_i(t), \tag{6.4}$$

$$c_i(0) = 0, \ldots, c_i^{(s-1)}(0) = 0. \tag{6.5}$$

It is then easy to see that

$$c_i(t) = \frac{1}{W}\sum_{j=1}^{s} W_j e^{\mu_{ij}t}\int_0^t e^{-\mu_{ij}\tau} h_i(\tau)\,d\tau. \tag{6.6}$$

Formula (6.6) shows that $c_i(t) \in \mathscr{L}_{2,\gamma}(0, \infty)$ (with, we note, all derivatives through the sth order) if and only if conditions (*) are fulfilled. Necessity is proved.

SUFFICIENCY. We set

$$u(x, t) = \sum_{i=0}^{\infty} c_i(t)\,\omega_i(x), \tag{6.7}$$

where the $c_i(t)$ are defined by formulas (6.6). It follows from conditions (*) that $c(t) \in H_\gamma^s(0, \infty)$. We show that the series (6.7) converges in $H_\gamma^{s,2m}(Q)$ and gives a solution of the mixed problem.

To do this, we multiply each equation (6.4) by $\omega_i(x)$ and sum over i from 0 to n. We then obtain that

$$\mathscr{P}_s\left(\frac{\partial}{\partial t}\right)u_n + \mathscr{L}(x, D)\,u_n = P_n h(x, t), \tag{6.8}$$

where

$$u_n(x, t) = \sum_{i=0}^{n} c_i(t)\,\omega_i(x), \quad P_n h(x, t) = \sum_{i=0}^{n} h_i(t)\,\omega_i(x).$$

In addition,

$$u_n(x, 0) = 0, \ldots, u_n^{(s-1)}(x, 0) = 0.$$

In this case (see §1) for $u_n(x,t)$ we have the inequality

$$\int_0^\infty \| u_n^{(s)} \|_0^2 e^{-\nu t}dt + \int_0^\infty \| u_n \|_{2m}^2 e^{-\nu t}dt \leqslant C\int_0^\infty \| P_n h \|_0^2 e^{-\nu t}dt,$$

which means that the sequence of partial sums $u_n(x,t)$ is fundamental in $H_\gamma^{s,2m}(Q)$, i.e. the series (6.7) converges. Passing to the limit as $n \to \infty$ in equation (6.8), we obtain that $u(x,t)$ is the desired solution. The theorem is completely proved.

2. Elliptico-hyperbolic case. Now let the operator

$$\mathfrak{A}(u) \equiv \mathcal{P}_s\left(\frac{\partial}{\partial t}\right) v + \mathcal{L}(x, D) u$$

be elliptico-hyperbolic for $\gamma \in \Gamma_0$, and let μ_{ij}, $i = 0, 1, \cdots$, $j = 1, \cdots, s$, be the roots of the characteristic equations

$$\mathcal{P}_s(\mu) + \lambda_i = 0,$$

where the λ_i are the eigenvalues of the operator $\mathcal{L}(x, D)u$.

In this case we have the analogous

THEOREM 6.2. *In order that the mixed problem* (6.1), (6.2) *be solvable in the space* $H_\gamma^{s,2m}(Q)$ *for* $h(x,t) \in H_\gamma^{1,0}(Q)$, *it is necessary and sufficient that for any* $i = 0, 1, \cdots$

$$\sum_{\operatorname{Re}\mu_{ij} \geqslant \gamma} e^{\mu_{ij}t} \int_0^t e^{-\mu_{ij}\tau} h_i(\tau)\, d\tau \in \mathcal{L}_{2,\gamma}(0, \infty).$$

The proof is just like that of Theorem 6.1.

§7. Smoothness theorem

1. Smoothness of the solutions of the mixed problem. It was shown in the previous section that if $h(x,t) \in K \subset \mathcal{L}_{2,\gamma}$, then the mixed problem for the equation

$$\mathfrak{A}(u) \equiv \mathcal{P}_s\left(\frac{\partial}{\partial t}\right) u + \mathcal{L}(x, D) u$$

has a unique solution $u(x,t) \in H_\gamma^{s,2m}(Q)$. Here we prove smoothness theorems for this solution which are analogous to well-known theorems in the theory of elliptic and parabolic equations.

To do this, we first note that conditions (*) of the previous section are invariant relative to differentiation with respect to t. Namely, if $h(x,t) \in K \cap H_\gamma^{r,0}(Q)$ and $h(x,0) = 0, \cdots, h^{(r-1)}(x,0) = 0$, then $h^{(i)}(x,t) \in H_\gamma^{r-i,0}(Q) \cap K$. In other words, if $h \in K$, $h(x,0) = 0$ and $h(x,t)$ is differentiable with respect to t, then we also have $h'(x,t) \in K$.

In fact, integrating by parts, we have

$$e^{\mu_{ij}t} \int_0^t e^{-\mu_{ij}\tau} h_i'(\tau)\, d\tau = h_i(t) + \frac{e^{\mu_{ij}t}}{\mu_{ij}} \int_0^t e^{-\mu_{ij}\tau} h_i(\tau)\, d\tau,$$

so that if $h(x,t) \in \mathcal{L}_{2,\gamma}(Q)$ and

$$\sum_{\operatorname{Re}\mu_{ij} \geqslant \gamma} \frac{W_j}{W} e^{\mu_{ij}t} \int_0^t e^{-\mu_{ij}\tau} h_i(\tau)\, d\tau \in \mathcal{L}_{2,\gamma}(0, \infty),$$

then also

$$\sum_{\operatorname{Re}\mu_{ij} \geqslant \gamma} e^{\mu_{ij}t} \int_0^\infty e^{-\mu_{ij}\tau} h_i'(\tau)\, d\tau \in \mathcal{L}_{2,\gamma}(0, \infty).$$

In this subsection we adopt the following notation:

a) $\overset{\circ}{H}(r,s,2m;\gamma)$, $r = ls + s_0$, $l \geqq 0$, $0 \leqq s_0 < s$, is the space of functions $u(x,t)$ which take zero initial values at $t=0$, $u(x,0) = 0, \cdots, u^{(r-1)}(x,0) = 0$, and have a finite norm

$$|||u|||^2_{r,\gamma} = \sum_{i=0}^{l} \int_0^\infty \| u^{(is+s_0)} \|^2_{2m(l-i)} e^{-\gamma t} dt.$$

b) $\overset{\circ}{H}(r,s,2m;\gamma)\{B_j\}$ is the subspace of $\overset{\circ}{H}(r,s,2m;\gamma)$ of functions which on the lateral surface of the cylinder S satisfy the boundary conditions

$$B_j(x,D)u|_S = 0, \quad j = 1, \dots, m.$$

THEOREM 7.1. *Let the operator* $\mathfrak{A}(u)$ *be strictly elliptico-parabolic on* Γ_0. *Then, if* $h(x,t) \in K \cap \overset{\circ}{H}(r,s,2m;\gamma)$, *it follows that the solution* $u(x,t)$ *of the mixed problem found in Theorem* 6.1 *belongs to the space* $\overset{\circ}{H}(r+s,s,2m;\gamma)\{B_j\}$, *where*

$$|||u|||_{r+s,\gamma} \leqslant C|||h|||_{r,\gamma}, \tag{7.1}$$

where $C > 0$ *is a constant.*

PROOF. We consider the mixed problem for the equation

$$\mathcal{P}_s\left(\frac{\partial}{\partial t}\right)v + \mathcal{L}(x,D)v = h^{(r)}(x,t), \tag{7.2}$$

$$v(x,0) = 0, \dots, v^{(s-1)}(x,0) = 0, \quad B_j(x,D)u|_S = 0, \tag{7.3}$$

$$j = 1, \dots, m.$$

By the property of invariance relative to differentiation with respect to t noted above for the subspace K, we have $h^{(r)}(x,t) \in K$. Hence there is a unique solution $v(x,t) \in \overset{\circ}{H}(s,s,2m;\gamma)\{B_j\}$ of problem (7.2), (7.3) with

$$|||v|||_{s,\gamma} \leqslant C|||h^{(r)}|||_{0,\gamma}.$$

We now set

$$u(x,t) = \frac{1}{(r-1)!}\int_0^t (t-\tau)^{r-1} v(x,\tau)\,d\tau.$$

Obviously $u(x,t)$ is a solution of the mixed problem for the equation

$$\mathcal{P}_s\left(\frac{\partial}{\partial t}\right)u + \mathcal{L}(x,D)u = h(x,t),$$

where

$$|||u^{(r)}|||_{s,\gamma} \leqslant C|||h^{(r)}|||_{0,\gamma} \leqslant C|||h|||_{r,\gamma}.$$

Hence, successively taking into account the definition of the norm $|||u|||_{r,\gamma}$ and the sharp inequalities for the elliptic operator $\mathcal{L}(x,D)u$ (see, for example, [**3**]), we arrive at inequality (7.1). In exactly the same way we prove

THEOREM 7.2. *Let the operator* $\mathfrak{A}(u)$ *be strictly elliptico-hyperbolic on the set* Γ_0. *Then, if* $h \in K \cap \overset{\circ}{H}(r, s, 2m; \gamma)$ *and* $h^{(r+1)}(x,t) \in \mathscr{L}_{2,\gamma}(Q)$, *it follows that the solution of the mixed problem* $u(x,t) \in \overset{\circ}{H}(r+s, s, 2m; \gamma)\{B_j\}$, *where*

$$|||u|||_{r+s,\gamma} \leqslant C\,|||h|||_{r+1,\gamma},$$

and $C > 0$ *is a constant.*

2. Smoothness of the solutions of the boundary value problem. Let $u(x,t)$ be a solution of the boundary value problem

$$\mathscr{P}_s\left(\frac{\partial}{\partial t}\right) u + \mathscr{L}(x, D)\, u = h(x, t), \tag{7.4}$$

$$u(x, 0) = 0, \ldots, u^{(s-1-k)}(x, 0) = 0, \ B_j(x, D)\, u|_S = 0, \quad j = 1, \ldots, m. \tag{7.5}$$

We show that a smoothness theorem analogous to the previous theorem holds for problem (7.4), (7.5). To do this, we introduce the necessary spaces. We let L designate the linear span of the system of eigenfunctions $\omega_0(x), \omega_1(x), \cdots$ of the operator $\mathscr{L}(x, D)u$:

$$L = \left\{u(x, t) \mid u(x, t) = \sum_{i=0}^{N} c_i(t)\, \omega_i(x), \ \forall N\right\}.$$

Then the space $H(r, s, 2m; \gamma)\{B_j\}$, where $r = 2ml$, l being an integer, is defined as the closure of L in the norm

$$|||u|||_{r,\gamma}^2 = \sum_{i=0}^{l} \int_0^\infty \|u^{(is)}\|_{2m(l-i)}^2 e^{-\gamma t} dt. \tag{7.6}$$

Further, $\overset{\circ}{H}(r, s, 2m; \gamma)\{B_j\}$ is the subspace of $H(r, s, 2m; \gamma)\{B_j\}$ of functions $u(x,t)$ such that $u(x,0) = 0, \cdots, u^{(s-1-k)}(x,0) = 0$.

THEOREM 7.3. *If* $h(x,t) \in H(r, s, 2m; \gamma)\{B_j\}$, *then the solution of the boundary value problem* (7.4), (7.5) *found in Theorem* 2.3 *belongs to the space* $\overset{\circ}{H}(r+2m, s, 2m; \gamma)\{B_j\}$, *where*

$$|||u|||_{r+2m,\gamma} \leqslant C\,|||h|||_{r,\gamma}. \tag{7.7}$$

PROOF. We recall that

$$u(x, t) = \lim_{n\to\infty} u_n(x, t),$$

where the $u_n(x,t) = \sum_{i=0}^{n} c_i(t)\,\omega_i(x)$ are solutions of the equation

$$\mathscr{P}_s\left(\frac{\partial}{\partial t}\right) u_n + \mathscr{L}(x, D)\, u_n = P_n h(x, t), \tag{7.8}$$

where $P_n h(x,t) = \sum_{i=0}^{n} h_i(t)\,\omega_i(x)$. Hence we obtain that the function $v_n(x,t) = \mathscr{L}^l(x, D)\, u_n(x,t)$ is a solution of the equation

$$\mathscr{P}_s\left(\frac{\partial}{\partial t}\right)v_n + \mathscr{L}(x, D)\, v_n = P_n(\mathscr{L}^l h(x, t)),$$

where obviously

$$v_n(x, 0) = 0, \dots, v_n^{(s-1-k)}(x, 0) = 0, \quad B_j(x, D)\, v_n|_S = 0, \quad j = 1, \dots, m.$$

Consequently we have the inequality

$$\int_0^\infty (\|v_n^{(s)}\|_0^2 + \|v_n\|_{2m}^2)\, e^{-\gamma t} dt \leqslant C \int_0^\infty \|P_n(\mathscr{L}^l h)\|_0^2\, e^{-\gamma t} dt \leqslant C\, |||P_n h|||_{r,\gamma}^2,$$

or, equivalently, the inequality

$$\int_0^\infty \|u_n^{(s)}\|_{2ml}^2 e^{-\gamma t}\, dt + \int_0^\infty \|u_n\|_{2m(l+1)}^2 e^{-\gamma t} dt \leqslant C\, |||P_n h|||_{r,\gamma}^2.$$

Taking this inequality into account, from equation (7.8) we find that

$$|||u_n|||_{r+2m,\gamma} \leqslant C\, |||P_n h|||_{r,\gamma},$$

where $|||\cdot|||_{r,\gamma}$ is defined by (7.6). Letting $n \to \infty$, we obtain (7.7). The theorem is proved.

REMARK. Theorem 7.3 holds for any $l \geqq 0$. In the case when the operator $\mathscr{L}(x, D)u$ is positive definite, the proof is exactly as before. In the general case we must estimate $u_n^+(x,t)$ and $u_n^-(x,t)$ separately, as was done, for example, in the proof of Theorem 1.1.

Similar theorems hold for elliptico-hyperbolic equations, except that the smoothness of the right side must be one greater.

§8. Examples. Remarks

1. We consider the parabolic equation

$$\mathfrak{A}(u) \equiv \frac{\partial u}{\partial t} + \mathscr{L}(x, D)\, u = h(x, t).$$

In this case the strict elliptico-parabolicity condition has the form

$$A(\gamma; \tau) + \lambda_0 \equiv \frac{\gamma}{2} + \lambda_0.$$

Hence for $\gamma > -2\lambda_0$ the boundary value problem, which in this case coincides with the mixed problem, is uniquely solvable, i.e. the correctness set is $\Gamma_0 = (-2\lambda_0, \infty)$. For $\gamma \leqq -2\lambda_0$ the mixed problem is normally solvable. Since the solution is unique in this case, it follows that the index $\kappa = \dim \operatorname{Ker} \mathfrak{A}(u) - l$, where l is the number of conditions on the right side, is nonpositive. A simple example shows that the index may be negative.

2. Inversely parabolic equation. Let

$$\mathfrak{A}(u) \equiv -\frac{\partial u}{\partial t} + \mathscr{L}(x, D)\, u = h(x, t). \tag{8.1}$$

The boundary value problem consists of finding a solution of equation (8.1) which satisfies the conditions

$$B_j(x, D)\, u|_S = 0, \quad j = 1, \dots, m,$$

and is free from any conditions at $t = 0$. In this case we have

$$A(\gamma; \tau) + \lambda_0 \equiv -\frac{\gamma}{2} + \lambda_0.$$

Obviously the problem is correct for $\gamma \in (-\infty, 2\lambda_0)$; for $\gamma \in [2\lambda_0, \infty)$ it is normally solvable.

3. Elliptic equation. Let

$$\mathfrak{A}(u) \equiv -\frac{\partial^2 u}{\partial t^2} + \mathscr{L}(x, D)\, u = h(x, t).$$

The boundary value problem is the Dirichlet problem (in t)

$$u(x, 0) = 0, \quad B_j(x, D)\, u|_S = 0, \quad j = 1, \dots, m.$$

We have

$$A(\gamma; \tau) + \lambda_0 \equiv \tau^2 - \frac{\gamma^2}{4} + \lambda_0,$$

which implies that the Dirichlet problem is correct for $|\gamma| < 2\sqrt{\lambda_0}$, $\lambda_0 > 0$. If $\lambda_0 \leqq 0$, then the entire axis $\mathbf{R}^1$ is the normal solvability set.

4. Hyperbolic equation. We set

$$\mathfrak{A}(u) \equiv \frac{\partial^2 u}{\partial t^2} + \mathscr{L}(x, D)\, u = h(x, t).$$

In this case the boundary value problem coincides with the mixed problem:

$$u(x, 0) = 0, \quad u'(x, 0) = 0, \quad B_j(x, D)\, u|_S = 0, \quad j = 1, \dots, m.$$

Further (see Definition 4.1),

$$\bar{A}(\gamma; \tau) + \frac{\gamma}{2}\lambda_0 \equiv \frac{\gamma}{2}\left(\tau^2 + \frac{\gamma^2}{4} + \lambda_0\right).$$

Consequently, $\Gamma_0 = \{\gamma \mid \gamma > 0, \gamma^2 > -4\lambda_0\}$ (see [**10**]). In the case $\lambda_0 \leqq 0$ the mixed problem is normally solvable for $\gamma \in (0, 2\sqrt{|\lambda_0|})$.

We proceed to consider examples for $s \geqq 3$.

5. For $k \geqq 1$ we set

$$\mathfrak{A}(u) \equiv (-1)^k u^{(2k+1)} + \mathscr{L}(x, D)\, u = h(x, t). \tag{8.2}$$

The boundary value problem reduces to finding a solution of equation (8.2) under the conditions

$$u(x, 0) = 0, \dots, u^{(k)}(x, 0) = 0, \quad B_j(x, D)\, u|_S = 0, \quad j = 1, \dots, m. \tag{8.3}$$

In this case the polynomial $A(\gamma; \tau)$ has the form

$$A(\gamma;\tau)+\lambda_0 \equiv \frac{\gamma}{2}\tau^{2k}+a_1\gamma^3\tau^{2k-2}+\ldots+\lambda_0.$$

Consequently for $\lambda_0>0$ we have $\Gamma_0=(0,\gamma_0)$, where $\gamma_0>0$ is the least upper bound of all $\gamma>0$ for which

$$A(\gamma;\tau)+\lambda_0>0.$$

We obtain that in this case the boundary value problem is correct in the sense of Hadamard and Petrovskiĭ for $\gamma\in\Gamma_0$; for $\gamma\geqq\gamma_0$ the boundary value problem is normally solvable. But if $\lambda_0\leqq 0$, then for any $\gamma>0$ problem (8.2), (8.3) generally has finite defect.

6. For $k\geqq 1$ and $a>0$ let

$$(-1)^{k+1}u^{(2k+1)}+(-1)^k a u^{(2k)}+\mathscr{L}(x,D)u=h(x,t). \tag{8.4}$$

Then to equation (8.4) we must add the conditions

$$u(x,0)=0,\ \ldots,\ u^{(k-1)}(x,0)=0,\ B_j(x,D)u|_S=0,\ j=1,\ldots,m. \tag{8.5}$$

Computation shows that

$$A(\gamma;\tau)+\lambda_0\equiv\left(-\frac{\gamma}{2}+a\right)\tau^{2k}+\ldots+\lambda_0.$$

This implies that the correctness set is $\Gamma_0=(-\infty,\gamma_0)$, where $\gamma_0<2a$ is the least upper bound of all γ for which $A(\gamma;\tau)+\lambda_0>0$. If $\lambda_0\leqq 0$, this set may be absent. On the other hand, the semiaxis $(-\infty,2a)$ is always a set where problem (8.4), (8.5) is normally solvable.

7. We consider the equation of even order

$$(-1)^k u^{(2k)}+\mathscr{L}(x,D)u=h(x,t)$$

with initial and boundary conditions

$$u(x,0)=0,\ \ldots,\ u^{(k-1)}(x,0)=0,\ B_j(x,D)u|_S=0,\ j=1,\ldots,m.$$

We have

$$A(\gamma;\tau)+\lambda_0\equiv\tau^{2k}+\ldots+\lambda_0,$$

so that for $\lambda_0>0$ we have the correctness interval $\Gamma_0=(-\gamma_0,\gamma_0)$; the problem is normally solvable for any γ.

8. We conclude by considering the boundary value problem for the equation of elliptico-hyperbolic type

$$(-1)^{k-1}u^{(2k)}+\mathscr{L}(x,D)u=h(x,t),$$

$$u(x,0)=0,\ \ldots,\ u^{(k)}(x,0)=0,\ B_j(x,D)u|_S=0,\ j=1,\ldots,m.$$

In this example

$$\bar{A}(\gamma;\tau)+\frac{\gamma}{2}\lambda_0\equiv\frac{\gamma}{2}\tau^{2k}+\ldots+\frac{\gamma}{2}\lambda_0,$$

so that for $\lambda_0 > 0$ we obtain the correctness interval $\Gamma_0 = (0, \gamma_0)$. For any $\gamma > 0$ and any sign of λ_0 we have normal solvability.

After these examples it is appropriate to make the following two remarks.

REMARK 1. It is easily seen that the same equation

$$\mathfrak{A}(u) \equiv \mathcal{P}_s\left(\frac{\partial}{\partial t}\right)u + \mathcal{L}(x, D)u = h(x, t)$$

can be both elliptico-parabolic and elliptico-hyperbolic (for example, the heat conduction equation). But it is important to emphasize here that the boundary value problem, and hence the generalized mixed problem, will be the same.

Further, we easily see that if $s = 2k$ the polynomials $A(\gamma; \tau)$ and $\overline{A}(\gamma; \tau)$ in Definitions 1.1 and 4.1 have the form

$$A(\gamma; \tau) \equiv (-1)^k a_{2k}\tau^{2k} + \dots,$$
$$\overline{A}(\gamma; \tau) \equiv (-1)^{k-1}\left(\frac{\gamma}{2}a_{2k} + a_{2k-1}\right)\tau^{2k} + \dots .$$

In exactly the same way, for $s = 2k + 1$

$$A(\gamma; \tau) \equiv (-1)^k\left(\frac{\gamma}{2}a_{2k+1} + a_{2k}\right)\tau^{2k} + \dots,$$
$$\overline{A}(\gamma; \tau) \equiv (-1)^k a_{2k+1}\tau^{2k+2} + \dots .$$

It is clear from this that for $(-1)^k a_{2k} < 0$ the operator $\mathfrak{A}(u)$ must be elliptico-hyperbolic, and for $(-1)^k a_{2k+1} < 0$ it must be elliptico-parabolic.

Because of this, it is natural to consider the operator $\mathfrak{A}(u)$ to be elliptico-parabolic for $(-1)^k a_{2k} > 0$ and $(-1)^k a_{2k+1} > 0$. We thereby eliminate the intersection of classes noted at the beginning of the remark. In addition, we find that the definitions of the classes of elliptico-parabolic and elliptico-hyperbolic equations actually do not depend on γ.

REMARK 2. All these considerations have been examined in the domain of real numbers, but the results can be carried over to the complex case without difficulty. We shall only give the definitions of elliptico-parabolic and elliptico-hyperbolic operators for this case.

Let $\mathfrak{A}(u) \equiv \mathcal{P}_s(\partial/\partial t)u + \mathcal{L}(x, D)u$ be an operator with complex coefficients.

DEFINITION 8.1. The operator $\mathfrak{A}(u)$ is called *strictly elliptico-parabolic* for $\gamma \in \Gamma_0$ if the polynomial

$$A(\gamma; \tau) + \lambda_0 \equiv \operatorname{Re}\sum_{p \leqslant q = 0}^{s} a_q \beta_{pq}(-i\tau)^p + \lambda_0 > 0,$$

where

$$\beta_{pq} = \frac{q!}{(q-p)!\,p!}\left(\frac{\gamma}{2}\right)^{q-p}.$$

DEFINITION 8.2. The operator $\mathfrak{A}(u)$ is called *elliptico-parabolic* on the set $\Gamma \in \mathbf{R}^1$ if for any $\gamma \in \Gamma$ there is a constant $\alpha(\gamma) \geqq 0$ such that

$$A(\gamma;\tau)+\lambda_0>0.$$

DEFINITION 8.3. The operator $\mathfrak{A}(u)$ is called *strictly elliptico-hyperbolic* for $\gamma\in\Gamma_0$ if the polynomial

$$A^1(\gamma;\tau)+\frac{\gamma}{2}\lambda_0\equiv \operatorname{Re}\sum a_q\beta_{pq}\beta_{l1}(-i\tau)^{p+l}+\frac{\gamma}{2}\lambda_0>0$$

$$(0\leqslant p\leqslant q\leqslant s;\ 0\leqslant l\leqslant 1).$$

DEFINITION 8.4. The operator $\mathfrak{A}(u)$ is called *elliptico-hyperbolic* on the set $\Gamma\in\mathbf{R}^1$ if for any $\gamma\in\Gamma$ there is a constant $\alpha(\gamma)\geqq 0$ such that

$$\overline{A}(\gamma;\tau)+\alpha(\gamma)>0.$$

All of the results of this paper, without exception, remain valid under these new definitions.

Received 20 NOV 1967

BIBLIOGRAPHY

[1] M. S. Agranovič and M. I. Višik, *Elliptic problems with a parameter and parabolic problems of general type,* Uspehi Mat. Nauk **19** (1964), no. 3 (117), 53-161 = Russian Math. Surveys **19** (1964), no. 3, 53-157. MR **33** #415.

[2] S. Agmon, *On the eigenfunctions and on the eigenvalues of general elliptic boundary value problems,* Comm. Pure Appl. Math. **15** (1962), 119-147. MR **26** #5288.

[3] ———, *The coerciveness problem for integro-differential forms,* J. Analyse Math. **6** (1958), 183-223. MR **24** #A2748.

[4] S. Agmon, A. Douglis and L. Nirenberg, *Estimates near the boundary for solutions of elliptic partial differential equations satisfying general boundary conditions.* I, Comm. Pure Appl. Math. **12** (1959), 623-727; Russian transl., IL, Moscow, 1962. MR **23** #A2610.

[5] N. Aronszajn and A. N. Milgram, *Differential operators on Riemannian manifolds,* Rend. Circ. Mat. Palermo (2) **2** (1953), 266-325. MR **16**, 252.

[6] N. I. Briš and I. N. Valeškevič, *The Fourier method for nonstationary equations with general boundary conditions,* Differencial'nye Uravnenija **1** (1965), 393-399 = J. Differential Equations **1** (1965), 300-305. MR **34** #3115.

[7] M. I. Višik, *Solution of a system of quasilinear equations having divergence form, under periodic boundary conditions,* Dokl. Akad. Nauk SSSR **137** (1961), 502-505 = Soviet Math. Dokl. **2** (1961), 293-296. MR **27** #451.

[8] M. I. Višik and G. I. Èskin, *Convolution equations in a bounded region,* Uspehi Mat. Nauk **20** (1965), no. 3 (123), 89-152 = Russian Math. Surveys **20** (1965), no. 3, 85-151. MR **32** #2741.

[9] V. K. Ivanov, *On linear problems which are not well-posed,* Dokl. Akad. Nauk SSSR **145** (1962), 270-272 = Soviet Math. Dokl. **3** (1962), 981-983. MR **25** #4357.

[10] O. A. Ladyženskaja, *The mixed problem for a hyperbolic equation,* GITTL, Moscow, 1953. (Russian) MR **17**, 160.

[11] M. M. Lavrent'ev, *Some ill-posed problems of mathematical physics,* Izdat. Sibirsk. Otdel. Akad. Nauk SSSR, Novosibirsk, 1962; English transl., Springer Tracts in Natural Philosophy, vol. 11, Springer-Verlag, Berlin, 1967. MR **36** #5487; 7363.

[12] J.-L. Lions, *Équations différentielles opérationnelles et problèmes aux limites,* Die Grundlehren der math. Wissenschaften, Band 111, Springer-Verlag, Berlin, 1961. MR **27** #3935.

[13] S. L. Sobolev, *Applications of functional analysis in mathematical physics,* Izdat. Leningrad. Gos. Univ., Leningrad, 1950; English transl., Transl. Math. Monographs, vol. 7, Amer. Math. Soc., Providence, R. I., 1963. MR **14**, 565; **29** #2624.

[14] A. N. Tihonov, *On the regularization of ill-posed problems,* Dokl. Akad. Nauk SSSR **153** (1963), 49-52 = Soviet Math. Dokl. **4** (1963), 1624-1627. MR **28** #5577.

[15] ———, *On the solution of ill-posed problems and the method of regularization,* Dokl. Akad. Nauk SSSR **151** (1963), 501-504 = Soviet Math. Dokl. **4** (1963), 1035-1038. MR **28** # 5576.

[16] E. Hille, *Une généralisation du problème de Cauchy,* Ann. Inst. Fourier Grenoble **4** (1952), 31-48. MR **15**, 718.

[17] M. Schecter, *General boundary value problems for elliptic partial differential equations,* Comm. Pure Appl. Math. **12** (1959), 457-486; Russian transl., Matematika **4** (1960), no. 5, 93-122. MR **23** # A2626.

[18] ———, *Remarks on elliptic boundary value problems,* Comm. Pure Appl. Math. **12** (1959), 561-578; Russian transl., Matematika **4** (1960), no. 6, 3-21. MR **23** # A2633.

[19] G. E. Šilov, *Mathematical analysis: Second special course,* "Nauka", Moscow, 1965. (Russian) MR **36** # 2943.

Translated by:
N. Koblitz

Trudy Moskov. Mat. Obšč.
Tom 20 (1969)

Trans. Moscow Math. Soc.
Vol. 20 (1969)

HYPOELLIPTIC PSEUDODIFFERENTIAL OPERATORS IN THE THEORY OF FUNCTIONAL SPACES

*L. R. VOLEVIČ and V. M. KAGAN**

Contents

INTRODUCTION 243
§1. Prerequisites on pseudodifferential operators 244
§2. The definition of the spaces $H^{\mu(x)}$. The isomorphism of $H^{\mu(x)}$ and L_2 . 253
§3. The embedding theorems 262
§4. Localization in the spaces $H^{\mu(x)}$ 268
§5. The space $W^{s(x)}$ 272
APPENDIX 280
BIBLIOGRAPHY 282

Introduction

The present article is devoted to the study of the function spaces $H^{\mu(x)}$ consisting of those tempered distributions u for which

$$\mu(x, D)\, u \in L_2, \tag{1}$$

where $\mu(x, D)$ is a hypoelliptic pseudodifferential operator (see [**3**], [**7**], [**12**]). If $\mu(x, D) = (1 + |D|^2)^{s(x)}$, then the spaces $H^{\mu(x)}$ become the spaces $H^{(s(x))}$ of variable order. These spaces arose in the studies of M. I. Višik and G. I. Èskin [**1**], [**2**] on discontinuous boundary problems; they have been studied independently by A. Unterberger and J. Bokobza [**15**] (see also [**12**]). When $\mu(x,\xi)$ is independent of x, $\mu(x,\xi) \equiv \mu(\xi)$, the spaces $H^{\mu(x)}$ are the spaces H^{μ} studied in [**5**], [**6**] and [**11**].

The present article consists of five sections.

In §1 we present the commutation formulas and theorems on the continuity and complete continuity of pseudodifferential operators. The majority of the statements in this section are modifications of the corresponding theorems in [**8**], [**12**] and [**3**].

In §2 we introduce the hypoelliptic pseudodifferential operators studied in [**10**], [**7**], [**3**] and [**12**], and with their help we define the spaces $H^{\mu(x)}$. The most important property of hypoelliptic pseudodifferential operators is that they have a regularizer (parametrix); that is, a pseudodifferential operator $r(x, D)$ such that $T = 1 - r(x, D)\mu(x, D)$ is a smoothing operator of arbitrarily high order. The regularizer plays an important role in the study of the spaces $H^{\mu(x)}$; however, in many problems it is necessary to have the exact inverse

* The present translation incorporates suggestions made by the authors.

operator. The main result in §2 is that, if $\mu(x,\xi)$ is stabilized at infinity, then the equation

$$\mu(x, D)u + Hu = v \in L_2,$$

with a special choice of the subordinate operator H, is uniquely solvable. By solving this equation we can obtain an "integral representation" for the function $u \in H^{\mu(x)}$:

$$u = r(x, D)v + Tv, \quad v \in L_2, \tag{2}$$

where $r(x,D)$ is a pseudodifferential operator with an explicitly written symbol and T is a smoothing operator of arbitrarily high order. The representation (2) determines an isomorphism between L_2 and $H^{\mu(x)}$ and is an analog of the representation of functions belonging to the Sobolev spaces as Bessel potentials (see, for example, [**13**]).

In §3 we establish precise embedding theorems for the spaces $H^{\mu(x)}$ that generalize theorems from [**5**] and [**11**].

In §5 we introduce the spaces $W^{s(x)}$ that generalize the Aronszajn-Slobodeckiĭ spaces to the case of variable order. The main result of this section is that the spaces $W^{s(x)}$ and $H^{(s(x))}$ are isomorphic. In proving this result we essentially use Hörmander's localization technique [**12**] which is developed in §4 in connection with the spaces $H^{\mu(x)}$.

A preliminary version of the present article was presented by L. R. Volevič at the Symposium on Embedding Theorems in Baku in May, 1966, and at the International Congress of Mathematicians in Moscow [**4**]. In this article a crude (for the given problems) technique ([**8**], [**3**]) was used to estimate the norms of the pseudodifferential operators; this led to gaps between the necessary and sufficient conditions.

In a preprint of an article by Hörmander [**12**] a new, remarkable technique was put forward for the estimation of the norms of pseudodifferential operators; this technique is reminiscent of the technique of lacunary partitions in the theory of Fourier series. By using this technique we easily succeeded in removing the gaps in [**4**] and in putting the embedding theorems for $H^{\mu(x)}$ in a final form.

§1. Prerequisites on pseudodifferential operators

1. Notation. $x = (x_1, \cdots, x_n)$ is a point of R^n, $\xi = (\xi_1, \cdots, \xi_n)$ are dual variables; that is, $x \cdot \xi = x_1\xi_1 + \cdots + x_n\xi_n$. We put $D = (D_1, \cdots, D_n)$ and $\partial = (\partial_1, \cdots, \partial_n)$ where $D_j = -i\partial/\partial x_j$ and $\partial_j = \partial/\partial\xi_j$, $j = 1, \cdots, n$. If $\alpha = (\alpha_1, \cdots, \alpha_n)$ is a set of n integers, then $|\alpha| = \alpha_1 + \cdots + \alpha_n$ and $x^\alpha = x_1^{\alpha_1} \cdots x_n^{\alpha_n}$. The symbols ξ^α, D^α and ∂^α are defined similarly.

We denote the Fourier transform of the function $u(x)$ by $\tilde{u}(\xi)$:

$$\tilde{u}(\xi) = (2\pi)^{-n/2} \int e^{-ix\xi} u(x)\, dx, \tag{1.1}$$

and let $H^{(s)}$ denote the space of all distributions u for which $\tilde{u}(\xi)$ is locally summable and

$$\|u\|_{(s)}^2 = \int (1+|\xi|^2)^s |\tilde{u}(\xi)|^2 d\xi < \infty. \tag{1.2}$$

We denote by $H^{(\pm\infty)}$ the intersection (union) of the spaces $H^{(s)}$, endowed with the natural topology of the projective (inductive) limit.

2. Pseudodifferential operators. We follow [**12**] and let $S^m_{\rho,\delta}$, $0 \leq \delta < \rho \leq 1$, denote the class of functions $a(x,\xi) \in C^\infty(R^n \times R^n)$ that are stabilizing for large x:

$$a(x, \xi) = a(\xi) + a'(x, \xi), \quad a'(x, \xi) = 0 \text{ for } |x| > X, \tag{1.3}$$

and that satisfy

$$|\partial^\alpha D^\beta a(x, \xi)| < C_{\alpha\beta}(1+|\xi|)^{m-\rho|\alpha|+\delta|\beta|}. \tag{1.4}$$

We denote the union of all the classes $S^m_{\rho,\delta}$ by $S^{-\infty}_{\rho,\delta}$.

Following [**8**], with each $a(x,\xi) \in S^m_{\rho,\delta}$ we associate the pseudodifferential operator

$$a(x, D)u = (2\pi)^{-n/2} \int e^{ix\xi} a(x, \xi) \tilde{u}(\xi)\, d\xi. \tag{1.5}$$

The function $a(x,\xi)$ is said to be the *symbol* of the operator (1.5).

It was noted in [**3**] that when $\delta < 1$ the operator (1.5) is a continuous operator from $H^{(\infty)}$ into $H^{(\infty)}$ and can be extended to a continuous operator from $H^{(-\infty)}$ into $H^{(-\infty)}$.

PROPOSITION 1.1. *Let* $a(x,\xi) \in S^{m_1}_{\rho_1,\delta_1}$ *and* $b(x,\xi) \in S^{m_2}_{\rho_2,\delta_2}$, *where* $\delta_2 < \rho_1$ $(\delta_1, \delta_2 < 1)$. *We put*

$$a(x, D)\, b(x, D) = \sum_{|\alpha|<N} c_\alpha(x, D) + T_N, \tag{1.6}$$

$$c_\alpha(x, \xi) = \partial^\alpha a(x, \xi)\, D^\alpha b(x, \xi)/\alpha!. \tag{1.6'}$$

Then for any real s and l we can find an $N_0(s,l)$ *such that for* $N > N_0(s,l)$ *the following operator is continuous*:

$$T_N : H^{(s)} \to H^{(s+l)}. \tag{1.7}$$

PROOF. When $\delta_2 < \rho_1/(1+\rho_1)$ the proof is completely elementary (in the embedding theorems in which we are interested this restriction will be fulfilled). For the general case we refer the reader to the Appendix, or to [**12**].

First let us note that on the Fourier transforms the operator (1.5) has the form

$$\widetilde{a(x, D)u}(\xi) = \int \tilde{a}(\xi-\eta, \eta)\, \tilde{u}(\eta)\, d\eta, \tag{1.8}$$

where $\tilde{a}(\xi,\eta)$ denotes the Fourier transform of $a(x,\eta)$ with respect to x. But then the operator $a(x, D)b(x, D) - c(x, D)$ on the Fourier transforms is an integral operator with kernel

$$(2\pi)^{-n/2} \int [\tilde{a}(\xi - \tau, \tau) - \tilde{a}(\xi - \tau, \eta)] \tilde{b}(\tau - \eta, \eta)\, d\tau.$$

By expanding the integrand by Taylor's formula and representing the symbols a, b, c in the form (1.3), we find that, on the Fourier transforms, T_N is an integral operator with kernel

$$K_N(\xi, \eta) = K_N'(\xi, \eta) + K_N''(\xi, \eta) = \sum_{|\alpha| = N} \frac{(\xi - \eta)^\alpha}{\alpha!} \partial^\alpha a(\zeta) \tilde{b}'(\xi - \eta, \eta)$$
$$+ (2\pi)^{-n/2} \sum_{|\alpha| = N} \int \frac{(\tau - \eta)^\alpha}{\alpha!} \partial^\alpha \tilde{a}'(\xi - \tau, \theta) \tilde{b}'(\tau - \eta, \eta)\, d\tau, \quad (1.9)$$

where ζ and θ are points of the segments $[\xi, \eta]$ and $[\tau, \eta]$. We will estimate the kernel $K_N''(\xi, \eta)$ (see [3]); the kernel K_N' is estimated even more simply. In view of (1.4), for large $|\alpha| = N$ we have

$$|\partial^\alpha \tilde{a}'(\xi - \tau, \theta)| < \text{const}\, (1 + |\xi - \tau|)^{-M_1} (1 + |\theta|)^{m_1 - \rho_1 N + \delta_1 M_1}$$
$$\leqslant \text{const}\, (1 + |\xi - \eta|)^{-M_1} (1 + |\eta|)^{-M_2} (1 + |\tau - \eta|)^{M_1 + M_2},$$

where

$$0 < M_2 < \rho_1 N - \delta_1 M_1 - m_1.$$

Next,

$$|(\tau - \eta)^\alpha \tilde{b}'(\tau - \eta, \eta)| < \text{const}\, (1 + |\tau - \eta|)^{-M_3} (1 + |\eta|)^{\delta_2(M_3 + N) + m_2}.$$

Setting

$$M_3 = M_1 + M_2 + n + 1,$$

we obtain the following estimate for the integral (1.9):

$$|K_N''(\xi, \eta)| < \text{const}\, (1 + |\xi - \eta|)^{-M_1} (1 + |\eta|)^{-M_4},$$

where

$$M_4 = M_2 - \delta_2(M_3 + N) - m_2 = (1 - \delta_2) M_2 - \delta_2 N - \delta_2 (M_1 + n + 1) - m_2.$$

If $(1 - \delta_2)\rho_1 > \delta_2$, for large N we can make $M_4 \geqq l$, where l is arbitrary. By estimating K_N' analogously we find that

$$|K_N(\xi, \eta)| < \text{const}\, (1 + |\xi - \eta|)^{-M_1} (1 + |\eta|)^{-l}.$$

Our assertion is easily proved by taking $M_1 > |s + l| + n + 1$.

REMARK. Our proof also goes through when $\rho_1 > \delta_2$, but the symbol $D^\alpha(x, \xi)$ cannot grow more rapidly than a fixed (possibly large) power; that is (cf. [7]),

$$|D^\alpha b(x, \xi)| < C_\alpha (1 + |\xi|)^{m_\alpha}, \quad m_\alpha = \min(M, \delta |\alpha|).$$

In what follows we shall require two refinements of Proposition 1.1 to the case when the symbols depend on a parameter.

PROPOSITION 1.2. *Suppose that, in the hypotheses of Proposition 1.1, the symbol $b(x,\xi)$ depends on a parameter $t\in R^m$ and that*

$$\left(\int |\partial^{\alpha} D^{\beta} b(x,\xi,t)|^2\,dt\right)^{1/2} \leqslant C_{\alpha\beta}(1+|\xi|)^{m_2-\rho_2|\alpha|-\delta_2|\beta|}. \tag{1.10}$$

Then, if $N > N_0(s,l)$, we have

$$\left\|\int T_N(t)\,u(\ ,t)\,dt\right\|^2 \leqslant C\int \|u(\ ,t)\|^2\,dt, \tag{1.11}$$

where the constant in (1.11) *depends only on the constants in* (1.4) *and* (1.10).

PROOF. Let $K_N(\xi,\eta,t)$ denote the kernel of the operator T_N on the Fourier transforms. We have

$$\left|\int \widetilde{T_N u(\xi,t)}\,dt\right| \leqslant \int\int |K_N(\xi,\eta,t)|\,|\tilde u(\eta,t)|\,dt\,d\eta$$

$$\leqslant \int\left[\int |K_N(\xi,\eta,t)|^2\,dt\right]^{1/2}\left[\int |\tilde u(\eta,t)|^2\,dt\right]^{1/2} d\eta \equiv \int K_N(\xi,\eta)\,v(\eta)\,d\eta.$$

In view of (1.10), the estimates that hold for the kernel (1.9) also hold for $K_N(\xi,\eta)$. Hence when $\delta_2 < \rho_1/(1+\rho_1)$ we can repeat literally the arguments in Proposition 1.1. In the general case the proof can be extracted from results of Hörmander [**12**].

PROPOSITION 1.3. *Suppose that for each t, $1\leqq t<\infty$, the symbols $a(\xi,t)$ and $b(x,t)$ belong to C^∞, $b(x,t)\equiv$ const for $|x|>X$ and that the following estimates hold:*

$$|\partial^{\alpha} a(\xi,t)| < C_\alpha t^{m_1-\rho|\alpha|},\quad |D^{\beta} b(x,t)| < C_\beta t^{m_2+\delta|\beta|}, \tag{1.12}$$

where the constants C_α and C_β are independent of t, and $\delta<\rho$. Then for any $l>0$ we can find an $N_0(l)$ such that for $N>N_0(l)$

$$\|T_N u\| \leqslant C_N t^{-l}\|u\|, \tag{1.13}$$

where C_N is independent of t and T_N is the operator in (1.6).

PROOF. By Proposition 1.1, on the Fourier transforms T_N is an integral operator with kernel

$$K_N(\xi,\eta) = \sum_{|\alpha|=N} \frac{(\xi-\eta)^{\alpha}}{\alpha!}\,\partial^{\alpha} a(\zeta)\,\tilde b'(\xi-\eta),$$

where $\zeta\in[\xi,\eta]$. In view of (1.12)

$$|\partial^{\alpha} a(\zeta)| < \text{const}\ t^{m_1-\rho N}$$

and

$$|\tilde b'(\xi-\eta)| < \text{const}\,(1+|\xi-\eta|)^{-M}\,t^{\delta M+m_2}.$$

By multiplying out these estimates and putting $M = N + n + 1$, we show that the L_2 norm of T_N does not exceed

$$\text{const}\ t^{-(\rho-\delta)N+\delta(n+1)+m_1+m_2}.$$

Our statement is proved by choosing N so that $(\rho-\delta)N > \delta(n+1) + m_1 + m_2 + l$.

We follow Kohn and Nirenberg [8] and, together with the operator (1.5), consider the operator

$$^t a(x, D)\, u = (2\pi)^{-n/2} \int e^{ix\cdot\xi} (2\pi)^{-n/2} \int e^{-ix\cdot\xi} a(x, \xi)\, u(x)\, dx. \tag{1.14}$$

The operators (1.5) and (1.14) are adjoints relative to the usual scalar product in L_2; that is,

$$(a(x, D)\, u, v) = (u, {}^t\bar{a}(x, D)\, v), \quad u, v \in H^{(\infty)}. \tag{1.15}$$

Hence if $a(x,\xi) \in S^{-\infty}_{\rho,\delta}$, $\delta < 1$, the operator (1.14) sends $H^{(\infty)}$ and $H^{(-\infty)}$ into themselves.

On the Fourier transforms the operator (1.14) takes the form

$$\widetilde{a(x, D)\, u}(\xi) = (2\pi)^{-n/2} \int \tilde{a}(\xi - \eta, \xi)\, \tilde{u}(\eta)\, d\eta. \tag{1.16}$$

It follows from (1.8) and (1.16) that, on the Fourier transforms, the operator $a(x, D) - {}^t a(x, D)$ is an integral operator with kernel

$$\tilde{a}(\xi - \eta, \eta) - \tilde{a}(\xi - \eta, \xi).$$

Expanding this kernel by Taylor's formula, we can (by following Proposition 1.1) easily prove

PROPOSITION 1.4. *If $a(x,\xi) \in S^m_{\rho,\delta}$, then*

$$a(x, D) - {}^t a(x, D) = \sum_{|\alpha|<N} \partial^\alpha D^\alpha a(x, D)/\alpha! + \Phi_N, \tag{1.17}$$

and for $N > N_0(s, l)$ the following operator is continuous:

$$\Phi_N : H^{(s)} \to H^{(s+l)}. \tag{1.18}$$

3. Theorems on boundedness. In the present article we make substantial use of the results in §3 of Hörmander's paper [12]. Since, for our purposes, not only the statements of these theorems but also the methods of proving them are important, we briefly outline proofs of Hörmander's theorems.

THEOREM 1.1. *Let $a(x,\xi) \in S^{-\infty}_{\rho,\delta}$, $0 \leqq \delta < \rho \leqq 1$. The inequality*

$$\| a(x, D)\, u \| \leqslant C \| u \|, \quad u \in H^{(\infty)}, \tag{1.19}$$

holds if and only if $a(x,\xi) \in S^0_{\rho,\delta}$.

We can prove the necessity of the inclusion $a(x,\xi) \in S^0_{\rho,\delta}$ by modifying the arguments of Theorem 3.3 of [12]. Let

$$|a(x^j, \xi^j)| \to \infty, \quad x^j \to x^0, \quad |\xi^j| \to \infty. \tag{1.20}$$

We consider the sequence

$$u_j(x) = |\xi^j|^{n\theta/2} u(|\xi^j|^{\theta}(x - x^j)) e^{ix\cdot\xi^j}, \quad u \in C_0^\infty, \quad \theta = \frac{\rho+\delta}{2}, \tag{1.21}$$

and can verify that

$$\|u_j\| = \|u\|, \quad \|a(x, D)u_j\| \to \infty, \quad j \to \infty;$$

which contradicts (1.19).

In his proof of the sufficiency of the condition Hörmander constructed a special partition of unity

$$\sum \varphi_k^2(\xi) \equiv 1, \tag{1.22}$$

with the properties:

1) $\phi_k(\xi) \geqq 0$, $\phi_k(\xi) \in C_0^\infty$, and, at each point $\xi \in R^n$, only a finite number J (independent of ξ) of the functions ϕ_j are nonzero;

2) for any multi-index α

$$|\partial^\alpha \varphi_k(\xi)| < C_\alpha (1 + |\xi|)^{-\theta|\alpha|}, \quad \delta < \theta < \rho; \tag{1.23}$$

3) for any $\xi, \eta \in \operatorname{supp} \phi_k$, $k \neq 0$,

$$|\xi - \eta| < C|\xi|^\theta. \tag{1.24}$$

In view of (1.22)

$$\|a(x, D)u\|^2 = \sum_j \|\varphi_j(D)a(x, D)u\|^2.$$

We put

$$\varphi_j(D)a(x, D)u = \sum_{|\alpha+\beta|<N} \frac{1}{\alpha!\beta!} a_{(\beta)}^{(\alpha)}(x, \xi^j)(D - \xi^j)^\alpha \varphi_j^{(\beta)}(D)u + R_j^N u, \tag{1.25}$$

where

$$a_{(\beta)}^{(\alpha)}(x, \xi) = \partial^\alpha D^\beta a(x, \xi), \quad \varphi_j^{(\beta)}(\xi) = \partial^\beta \varphi_j(\xi); \quad \xi^j \in \operatorname{supp} \varphi_j.$$

Following this, Hörmander verified that

$$\sum_j \|a_{(\beta)}^{(\alpha)}(x, \xi^j)(D - \xi^j)^\alpha \varphi_j^{(\beta)}(D)u\|^2 < C_1 \|u\|^2_{(-\lambda|\alpha+\beta|)}, \tag{1.26'}$$

$$\sum_j \|R_j^N u\|^2 \leqslant C_2 \|u\|^2_{(\delta(n+1)-N\lambda)}, \tag{1.26''}$$

where $\delta + \lambda \leqq \theta \leqq \rho - \lambda$, $\lambda > 0$. If N is sufficiently large, the right-hand side of (1.26″) does not exceed $\text{const}\|u\|$, which leads to the proof of the theorem.

COROLLARY 1. *If $a(x, \xi) \in S^m_{\rho,\delta}$, then for any s*

$$\|a(x, D)u\|_{(s)} \leqslant \text{const}\|u\|_{(m+s)}. \tag{1.19'}$$

PROOF. The estimate (1.19′) is equivalent to the boundedness in L_2 of the operator

$$(1+|D|^2)^{s/2}a(x,D)(1+|D|^2)^{-(s+m)/2} = (1+|D|^2)^{s/2}a_1(x,D).$$

By Proposition 1.1 this operator is equal to

$$\sum c_\alpha(x,D) + T_N,$$

where

$$\alpha!\, c_\alpha(x,\xi) = \partial^\alpha (1+|\xi|^2)^{s/2} D^\alpha a_1(x,D) \in S_{\rho,\delta}^{-(\rho-\delta)|\alpha|} \subset S_{\rho,\delta}^0,$$

and the estimate (1.19′) holds for T_N. By applying Theorem 1.1 to the operators $c_\alpha(x,D)$ we prove (1.19′).

As is usual we let $\mathscr{E}'(\bar\Omega)$ denote the set of distributions whose supports lie in the closure of the bounded domain Ω and, following [**5**], we write

$$H_{\bar\Omega}^{(s)} = H^{(s)} \cap \mathscr{E}'(\bar\Omega).$$

COROLLARY 2. *If $a(x,\xi)\in S_{\rho,\delta}^{-\infty}$ and $\psi(x)a(x,\xi)\in S_{\rho,\delta}^m$ for any $\psi(x)\in C_0^\infty(\Omega_1)$, $\bar\Omega\subset\Omega_1$, then the operator $a(x,D): H_{\bar\Omega}^{(s+m)}\to H_{\bar\Omega}^{(s)}$ is continuous.*

PROOF. Let $\psi(x)\in C_0^\infty(\Omega_1)$ and let $\psi(x)=1$ when $x\in\bar\Omega$. If $u\in H_{\bar\Omega}^{(s+m)}$, then $u=\psi u$. Hence

$$a(x,D)u = a(x,D)(\psi u) = \sum D^\alpha\psi a^{(\alpha)}(x,D)u/\alpha! + T_N u = a_1(x,D)u + T_N u.$$

If N is large then the operator $T_N: H^{(s+m)}\to H^{(s)}$ is continuous and we can apply Corollary 1 to the operator $a_1(x,D)$.

Theorem 1.1 can be carried over without difficulty to the case when the symbols depend on a parameter.

THEOREM 1.2. *Suppose that the symbol $a(x,\xi,t)\in S^{-\infty}$ for each $t\in R^m$ and that it satisfies the conditions*

$$\left(\int |\partial^\alpha D^\beta a(x,\xi,t)|^2 dt\right)^{1/2} < C_{\alpha\beta}(1+|\xi|)^{-\rho|\alpha|+\delta|\beta|}. \tag{1.27}$$

Then

$$\left\|\int a(x,D,t)u(x,t)\,dt\right\|^2 \leqslant C\iint |u(x,t)|^2 dx\,dt, \tag{1.28}$$

where the constant C depends only on the constants $C_{\alpha\beta}$ in (1.27).

PROOF. By using (1.25) and Schwarz' inequality we can estimate the right-hand side of (1.28) by

$$\sum_j \sum_{|\alpha+\beta|<N} \left\| \left(\int |a_{(\beta)}^{(\alpha)}(x,\xi^j,t)|^2 dt\right)^{1/2} \left(\int |(D-\xi^j)^\alpha \psi_j^{(\beta)}(D)u(x,t)|^2 dt\right)^{1/2} \right\|$$

$$+ \sum_j \int d\eta \left[\int \left(\int |r_j^N(\xi,\eta,t)|^2 dt\right)^{1/2} \left(\int |\tilde u(\xi,t)|^2 dt\right)^{1/2} d\xi\right]^2. \tag{1.29}$$

The estimates (1.27) allow us to repeat literally the arguments of Theorem 1.1 and to estimate the sums in (1.29) in terms of the right-hand sides of (1.26′) and (1.26″).

4. Theorems on complete continuity. We pause to consider questions connected with the complete continuity of the operators referred to in Propositions 1.1 and 1.4. We follow [8] and start from the simple

LEMMA 1.1. *Let Σ be the set of functions $u \in H^{(s)}$ with the properties:*

1) $\| (1+|D|)^\epsilon u \|_{(s)} \leqq C \| u \|_{(s)} < C_1, \quad \epsilon > 0$;

2) *in each ball $S_R = \{\xi \in R^n : |\xi| < R\}$ the family of Fourier transforms $\tilde{u}(\xi)$ $(u \in \Sigma)$ is compact.*

Then Σ is precompact in $H^{(s)}$; that is, from any sequence $u \in \Sigma$ we can choose a subsequence that converges in $H^{(s)}$.

LEMMA 1.2. *Suppose that $a(x,\xi) \in S^m_{\rho,\delta}$ and $a(x,\xi) \equiv 0$ for $|x| > X$. Let Σ denote the set of functions $w = a(x,D)u(x)$ where $\| u \|_t <$ const. Then this family satisfies condition 2) of Lemma 1.1.*

PROOF. We will follow [8] and show that the family $\tilde{w}(\xi)$ is uniformly bounded and equicontinuous. Then the assertion of the lemma will follow from Arzelà's Theorem. In view of (1.8) we have

$$\tilde{w}(\xi) = (2\pi)^{-n/2} \int \tilde{a}'(\xi - \eta, \eta)(1+|\eta|)^{-t}(1+|\eta|)^t \tilde{u}(\eta)\, d\eta. \qquad (1.30)$$

Since $a(x,\xi) \in S^m_{\rho,\delta}$ and $a(x,\xi) \equiv 0$ when $|x| > X$, for any integer $p > 0$ we obtain that

$$|\tilde{a}'(\xi-\eta,\eta)| < C_p(1+|\xi-\eta|)^{-p}(1+|\eta|)^{m+p\delta}$$
$$< C_p(1+|\xi|)^p(1+|\eta|)^{m-(1-\delta)p}.$$

By choosing p so that $(1-\delta)p > m-t-n$ and applying the Schwarz inequality to the right-hand side of (1.30), we find that

$$|\tilde{w}(\xi)| < C(R)\|u\|_{(t)} < C'(R), \quad |\xi| < R.$$

To prove the equicontinuity let us note that

$$\tilde{w}(\xi+t) - \tilde{w}(\xi) = (2\pi)^{-n/2} \int [\tilde{a}(\xi+\tau-\eta,\eta) - \tilde{a}(\xi-\eta,\eta)]\, \tilde{u}(\eta)\, d\eta.$$

It is not difficult to check that, for any integer $p > 0$,

$$|\tilde{a}(\xi+\tau-\eta,\eta) - \tilde{a}(\xi-\eta,\eta)| < C_p|\tau|(1+|\xi-\eta|)^{-p}(1+|\eta|)^{m+p\delta}.$$

In fact,

$$(\xi-\eta)^\alpha[\tilde{a}(\xi+\tau-\eta,\eta) - \tilde{a}(\xi-\eta,\eta)$$
$$= (2\pi)^{-n/2} \int_{|x|<X} [D_x^\alpha e^{ix\cdot(\xi-\eta)}][e^{ix\tau}-1]\, a(x,\eta)\, dx.$$

By integrating the right-hand side of the last equation by parts and using (1.4) we obtain the required estimate of (1.30). By repeating the arguments of the first part of the lemma we can show that

$$|\tilde{w}(\xi+\tau)-\tilde{w}(\xi)| \leqslant C''(R)\,|\tau|\,\|u\|_{(t)} \leqslant C'''(R)\,|\tau|$$

and the lemma is proved.

Lemma 1.2 and Theorem 1.1 allow us to sharpen Propositions 1.1 and 1.4.

PROPOSITION 1.1′. *Under the conditions of Proposition* 1.1 *the operators*

$$c_\alpha(x, D) : H^{(s)} \to H^{(s-m_1-m_2+\rho_1|\alpha|-\delta_2|\alpha|-\lambda)}, \quad \lambda > 0, \quad \alpha > 0, \tag{1.31}$$

are completely continuous. The operator (1.7) *is also completely continuous.*

PROPOSITION 1.4′. *Under the conditions of Proposition* 1.4 *the operators*

$$\partial^\alpha D^\alpha a(x, D) : H^{(s)} \to H^{(s-m-\rho|\alpha|+\delta|\alpha|+\lambda)}, \quad \lambda > 0, \quad \alpha > 0, \tag{1.32}$$

are completely continuous. The operator (1.18) *is also completely continuous.*

We prove Proposition 1.1′ (Proposition 1.4′ is proved even more simply). In view of Theorem 1.1 the operators (1.31) are continuous at $\lambda = 0$. Hence the family $w = c_\alpha(x, D)u$, where $\|u\|_{(s)} <$ const, satisfies the first condition of Lemma 1.1 (with s replaced by $s - m_1 - m_2 + (\rho_1 - \delta_2)|\alpha| - \lambda$). Because $D^\alpha b(x, \xi) \equiv 0$ when $|x| > X$, in view of Lemma 1.2 the second condition of Lemma 1.1 holds, whence it follows that the operators (1.31) are compact.

We next check that the operator T_N in (1.7) satisfies the second condition of Lemma 1.1. Since the operators $c_\alpha(x, D)$ satisfy this condition when $\alpha > 0$, it is sufficient to consider the operator

$$a(x, D)\, b(x, D) - c(x, D). \tag{1.33}$$

Using (1.4) we can rewrite the operator (1.33) in the form

$$a'(x, D)\, b(D) + a'(x, D)\, b'(x, D) - c'(x, D) + a(D)\, b'(x, D). \tag{1.34}$$

Lemma 1.2 can be directly applied to all the operators in (1.34) (except for the last one). Since

$$\widetilde{a(D)b'(x, D)u}\,(\xi) = a(\xi)\, \widetilde{b'(x, D)u}\,(\xi),$$

it also follows from Lemma 1.2 that condition 2) of Lemma 1.1 holds for the operator $a(D)b'(x, D)$, and our assertion is proved.

In conclusion let us note a theorem of Hörmander [**12**].

THEOREM 1.3. *Suppose that* $a(x, \xi) \equiv 0$ *for* $|x| > X$ *and that*

$$\overline{\lim_{|\xi|\to\infty}} \sup_x a(x, \xi) = 0.$$

Then the operator $a(x, D) : L_2 \to L_2$ *is completely continuous.*

§2. The definition of the spaces $H^{\mu(x)}$. The isomorphism of $H^{\mu(x)}$ and L_2

1. The class $\mathfrak{w}$. We denote by $\mathfrak{w}_{\rho,\delta}$ the class of real symbols $\mu(x,\xi)\in S^{-\infty}_{\rho,\delta}$ that satisfy

$$C^{-1}(1+|\xi|)^{\underline{\mu}} < \mu(x,\xi) < C(1+|\xi|)^{\bar{\mu}}, \tag{2.1}$$

$$|\partial^{\alpha}D^{\beta}\mu(x,\xi)| < C_{\alpha\beta}(1+|\xi|)^{-\rho|\alpha|+\delta|\beta|}\mu(x,\xi). \tag{2.2}$$

This class was introduced by Hörmander [**10**] and has been studied in [**7**], [**3**] and [**12**]. Let us note certain elementary properties of the class $\mathfrak{w}=\bigcup\mathfrak{w}_{\rho,\delta}$.

1) The symbols belonging to $\mathfrak{w}$ form a group under multiplication.

2) If $l(x)\in C^{\infty}$ is a real function and if $l(x)=l$ for large $|x|$, then $\mu^{l(x)}(x,\xi)\in\mathfrak{w}_{\rho,\delta}$ when $\mu(x,\xi)\in\mathfrak{w}_{\rho,\delta}$.

3) If $\mu(x,\xi)\in\mathfrak{w}_{\rho,\delta}$, then

$$\mu(x,\xi)\in S^{\bar{\mu}}_{\rho,\delta},\quad \mu^{-1}(x,\xi)\in S^{-\underline{\mu}}_{\rho,\delta}.$$

4) If $\mu(x,\xi)\in\mathfrak{w}_{\rho,\delta}$, then we can find an N such that

$$\mu(x,\xi)\mu^{-1}(x,\eta) < C(1+|\xi-\eta|)^{N}\quad\text{for all}\quad \xi,\eta\in R^{n}. \tag{2.3}$$

Let us verify property 4) (the verification of properties 1)—3) is left to the reader). By expanding $\mu(x,\xi)=\mu(x,\eta+(\xi-\eta))$ by Taylor's Theorem, we find that

$$\frac{\mu(x,\xi)}{\mu(x,\eta)}=\sum_{|\alpha|<N}\frac{(\xi-\eta)^{\alpha}}{\alpha!}\frac{\partial^{\alpha}\mu(x,\eta)}{\mu(x,\eta)}+\sum_{|\alpha|=N}\frac{(\xi-\eta)^{\alpha}}{\alpha!}\frac{\partial^{\alpha}\mu(x,\zeta)}{\mu(x,\eta)},$$

where $\xi=\eta+t(\xi-\eta)$, $0\leqq t\leqq 1$. If N is large then

$$\mu^{-1}(x,\eta)|\partial^{\alpha}\mu(x,\zeta)| < \text{const}\,(1+|\eta|)^{-\underline{\mu}}(1+|\zeta|)^{\bar{\mu}-\rho N}$$

$$< \text{const}\,(1+|\xi-\eta|)^{-\bar{\mu}+\rho N}(1+|\eta|)^{\bar{\mu}-\underline{\mu}-\rho N} < \text{const}\,(1+|\xi-\eta|)^{N_1},$$

and we arrive at the estimate (2.3) without difficulty.

Note that the simplest (and, in the present article, fundamental) example of a symbol belonging to $\mathfrak{w}$ is $(1+|\xi|^{2})^{s(x)}$.

We will say that $b(x,\xi)\dashv\mu(x,\xi)$ if $b(x,\xi)\in S^{-\infty}_{\rho,\delta}$, $\mu(x,\xi)\in\mathfrak{w}_{\rho,\delta}$ and

$$|\partial^{\alpha}D^{\alpha}b(x,\xi)| < C_{\alpha\beta}(1+|\xi|)^{-\rho|\alpha|+\delta|\beta|}\mu(x,\xi). \tag{2.4}$$

If we can find a $\lambda>0$ such that $b(x,\xi)(1+|\xi|^{2})^{\lambda}\dashv\mu(x,\xi)$, then we shall write $b(x,\xi)\dashv\dashv\mu(x,\xi)$.

LEMMA 2.1. *If $a(x,\xi)\dashv\mu(x,\xi)$, $b(x,\xi)\dashv\mu^{-1}(x,\xi)$ (where $\mu\in\mathfrak{w}_{\rho,\delta}$), then the operator*

$$a(x,D)\,b(x,D): H^{(s)}\to H^{(s)} \tag{2.5}$$

is continuous and the operator

$$a(x, D)\, b(x, D) - c(x, D) : H^{(s)} \to H^{(s)},\ c(x, \xi) = a(x, \xi)\, b(x, \xi) \quad (2.5')$$

is completely continuous. Analogous statements hold for the operator $b(x, D)a(x, D)$.

PROOF. According to Proposition 1.1

$$a(x, D)\, b(x, D) = \sum c_\alpha(x, D) + T_N,$$

where

$$c_\alpha(x, \xi) = \partial^\alpha a(x, \xi)\, D^\alpha b(x, \xi)/\alpha!,$$

and the operator

$$T_N : H^{(s)} \to H^{(s+l_N)}, \quad l_N \to +\infty,\ N \to \infty$$

is completely continuous (in view of Proposition 1.1′). It is easy to verify that if $a \dashv \mu$ and $b \dashv \mu^{-1}$, then $c(x, \xi) \in S^0_{\rho,\delta}$ and, in view of Proposition 1.1′, the operators

$$c_\alpha(x, D) : H^{(s)} \to H^{(s+(\rho-\delta)|\alpha|-\lambda)} \subset H^{(s)}$$

are completely continuous when $\alpha > 0$. The lemma is proved.

LEMMA 2.2. *Let* $a(x, \xi) \dashv \mu(x, \xi)$ *and*

$$|a(x, \xi)| > c\mu(x, \xi) \text{ when } |\xi| > \Xi. \quad (2.6)$$

Then for any N we can find a symbol $r(x, \xi) \dashv \mu^{-1}(x, \xi)$ *such that*

$$r(x, D)\, a(x, D) = E + T_N, \quad (2.7)$$

where E is the identity operator and T_N *maps* $H^{(s)}$ *into* $H^{(s+N)}$.

This lemma is proved with the help of an explicit construction [**10**] (for a detailed proof see [**3**] or [**12**]).

If $|a(x, \xi)| > 0$, then the symbol $r(x, \xi)$ is constructed in the form of the finite sum

$$r(x, \xi) = r_0(x, \xi) + \ldots + r_J(x, \xi), \quad (2.8)$$

where

$$r_0(x, \xi) = a^{-1}(x, \xi),\ r_{j+1}(x, \xi) = -a^{-1}(x, \xi) \sum \frac{1}{\alpha!} \partial^\alpha r_j(x, \xi)\, D^\alpha a(x, \xi).$$

If the symbol $|a(x, \xi)|$ is zero for $|\xi| < \Xi$ we must multiply all the symbols in (2.8) by a function $\psi(\xi) \in C^\infty$ which is zero when $|\xi| < \Xi$ and unity when $|\xi| > \Xi + 1$.

REMARK. If $|a(x, \xi)| > 0$, then, as is clear from the proof of Lemma 2.2, $r(x, \xi)$ is equal to

$$a^{-1}(x, \xi)(1 + p(x, \xi)), \quad p(x, \xi) \in S^{-\lambda}_{\rho,\delta}, \quad \lambda > 0,$$

where the symbol $p(x,\xi)$ is a linear combination of expressions of the form

$$\frac{\partial^{\alpha_1}a}{a}\cdots\frac{\partial^{\alpha_k}a}{a}\,\frac{D^{\beta_1}a}{a}\cdots\frac{D^{\beta_l}a}{a},$$

where $\alpha_1,\cdots,\beta_l$ are integral multi-indices and

$$|\alpha_1+\ldots+\alpha_k|>0.$$

The operator T_N is completely continuous when $|a(x,\xi)|>0$ (see Proposition 1.1′).

2. The definition of the spaces $H^{\mu(x)}$. Let $\mu(x,\xi)\in\mathfrak{m}_{\rho,\delta}$. We denote by $H^{\mu(x)}$ the set of distributions $u\in H^{(\underline{\mu})}$ (the numbers $\underline{\mu}$ and $\bar{\mu}$ are related to $\mu(x,\xi)$ as in (2.1)) such that $\mu(x,D)u\in H^{(0)}=L_2$. $H^{\mu(x)}$ becomes a Banach space when we introduce in it the norm

$$\|u\|_\mu^2=\|\mu(x,D)u\|^2+\|u\|_{(\underline{\mu})}. \tag{2.9}$$

Since $\mu(x,\xi)\in S^{\mu}_{\rho,\delta}$, in view of Corollary 1 to Theorem 1.1 we have

$$\|\mu(x,D)u\|<\text{const}\,\|u\|_{(\bar{\mu})}. \tag{2.10}$$

Hence

$$C^{-1}\|u\|_{(\underline{\mu})}\leqslant\|u\|_\mu\leqslant C\|u\|_{(\bar{\mu})},$$

from which follow the embeddings (together with the topology)

$$H^{(\bar{\mu})}\subset H^{\mu(x)}\subset H^{(\underline{\mu})}. \tag{2.11}$$

We will show later that each of the spaces (2.11) is dense in the enveloping space.

We can prove without difficulty that $H^{\mu(x)}$ is complete. In fact if $\{u_j\}$ is a fundamental sequence in $H^{\mu(x)}$, then it is a fundamental sequence in $H^{(\underline{\mu})}$; that is, $u_j\to u$ in $H^{(\underline{\mu})}$. But then $\mu(x,D)u_j\to\mu(x,D)u$ in the topology of $H^{(\underline{\mu}-\mu)}$. Since the sequence $\{\mu(x,D)u_j\}$ is fundamental in $H^{(0)}$, we have that $\mu(x,D)u\in H^{(0)}$; that is, $u\in H^{\mu(x)}$, which proves that $H^{\mu(x)}$ is complete.

LEMMA 2.3. *If* $b(x,\xi)\prec\mu(x,\xi)$, *then*

$$\|b(x,D)u\|\leqslant C_m(\|\mu(x,D)u\|+\|u\|_{(m)}), \tag{2.12}$$

where $m\leqq\underline{\mu}$ *is arbitrary and* $u\in H^{\mu(x)}$.

PROOF. The proof is based on Lemma 2.2. Let $\mu(x,D)u=v$. Then $u=r(x,D)v+T_Nu$, whence

$$b(x,D)u=b(x,D)r(x,D)v+b(x,D)T_Nu.$$

In view of Lemma 2.1 the operator $b(x,D)r(x,D)$ is bounded. If N is sufficiently large, then

$$\|b(x,D)T_Nu\|\leqslant C\|T_Nu\|_{(\bar{\mu})}<C_1\|u\|_{(\bar{\mu}-N)}\leqslant C_2\|u\|_{(m)}$$

and the lemma is proved.

COROLLARY 1. *The distribution* $u \in H^{(-\infty)}$ *belongs to* $H^{\mu(x)}$ *if and only if* $\mu(x, D)u \in L_2$.

PROOF. It is sufficient to verify that $u \in H^{(\underline{\mu})}$. Since $u \in H^{(-\infty)}$, $u \in H^{(m)}$ for some m. Because $(1+|\xi|^2)^{\underline{\mu}/2} \twoheadrightarrow \mu(x,\xi)$, in view of (2.12) we have that $(1+|D|^2)^{\underline{\mu}/2}u \in L_2$; that is, $u \in H^{(\underline{\mu})}$.

COROLLARY 2. *Let* $m < \underline{\mu}$. *Then the norm* (2.9) *is equivalent to the norm*

$$\|u\|_\mu'^2 = \|\mu(x, D)u\|^2 + \|u\|^2_{(m)}.$$

COROLLARY 3. *Let* $a(x,\xi) = \mu(x,\xi) + b(x,\xi)$, *where* $b(x,\xi) \twoheadrightarrow \twoheadrightarrow \mu(x,\xi)$. *Then the norm* (2.9) *is equivalent to the norm*

$$\|u\|_\mu''^2 = \|a(x, D)u\|^2 + \|u\|^2_{(m)}. \tag{2.13}$$

PROOF. The inequality $\| \ \|''_\mu \leqq \text{const} \| \ \|_\mu$ follows trivially from the lemma. To prove the reverse inequality note that the symbol $a(x,\xi)$ satisfies the conditions of Lemma 2.2. By repeating literally the arguments of Lemma 2.3 we arrive at the inequality

$$\|\mu(x, D)u\| \leqslant \text{const}\,(\|a(x, D)u\| + \|u\|_{(m)}), \quad m \leqslant \underline{\mu}.$$

REMARK. It is obvious that $\| \ \|''_\mu$ is equivalent to the norm

$$\|u\|_\mu''' = [\|a(x, D)u + Tu\|^2 + \|u\|^2_{(m)}]^{1/2}, \tag{2.14}$$

where T is a smoothing operator of sufficiently high order.

COROLLARY 4. *The distribution* $u \in H^{(-\infty)}$ *belongs to* $H^{\mu(x)}$ *if and only if* ${}'\mu(x, D)u \in L_2$, *where* ${}'\mu(x, D)$ *is the operator in* (1.14) *corresponding to the symbol* $\mu(x,\xi)$.

To prove this it is sufficient to compare Proposition 1.4 and the Remark to Corollary 3.

COROLLARY 5. *The space* $H^{\mu(x)}$ *is a module over* C_0^∞; *that is*,

$$\|u\|_\mu''' = [\|a(x, D)u + Tu\|^2 + \|u\|^2_{(m)}]^{1/2}.$$

PROOF. As is well known (see, for example, [**5**]),

$$\|\varphi u\|_{(\underline{\mu})} \leqslant C_\varphi \|u\|_{(\underline{\mu})}.$$

Next, if $\phi \in C_0^\infty$, then $b(x,\xi) = \phi(x)\mu(x,\xi) \twoheadrightarrow \mu(x,\xi)$. Hence, in view of the lemma,

$$\|\varphi(x)\mu(x, D)u\| \leqslant \text{const}\,\|u\|_\mu.$$

3. The isomorphism of the spaces $H^{\mu(x)}$ **and** L_2. If the symbol $\mu(x,\xi) \in \mathfrak{w}$ does not depend on x, then, as is easily verified, the mapping

$$\mu(D) : H^\mu \ni u \to \mu(D)u \in L_2$$

is an isomorphism (that is, it is a continuous mapping with a trivial kernel and such that the image of H^μ is all of L_2). With this isomorphism is associated the "integral" representation

$$u = \mu^{-1}(D)\, v, \quad v \in L_2, \tag{2.15}$$

on which, in fact, the theory of the spaces H^μ in L_2 is based. The mapping

$$\mu(x, D) : H^{\mu(x)} \ni u \to \mu(x, D)\, u \in L_2,$$

in general, is not an isomorphism. However, we now show that if we add the operator $b(x, D) + T$ to $\mu(x, D)$, where $b \prec \mu$ and T is a smoothing operator of high order, then the operator $\mathfrak{M} = \mu + b + T$ effects an isomorphism of $H^{\mu(x)}$ and L_2. This isomorphism leads to an analog of the representation (2.15) for distributions from $H^{\mu(x)}$.

Theorem 2.1. *Let* $\mu_1(x, \xi) = \sqrt{\mu(x, \xi)}$ *and*

$$\mathfrak{M} = {}^t\mu_1(x, D)\, \mu_1(x, D) + (1 + |D|^2)^{m/2}, \quad m \leq \underline{\mu}. \tag{2.16}$$

Then the mapping

$$\mathfrak{M} : H^{\mu(x)} \ni u \to \mathfrak{M}u \in L_2 \tag{2.17}$$

is an isomorphism.

The proof of the theorem can be split into the proof of a number of simple assertions which we state in the form of lemmas.

Lemma 2.4. *The operator* (2.16) *can be represented as*

$$\mathfrak{M} = \mu(x, D) + (1 + |D|^2)^{m/2} + b_N(x, D) + T_N, \tag{2.18}$$

where $b_N(x, \xi) \prec\prec \mu(x, \xi)$, $b_N(x, \xi) \equiv 0$ *for* $|x| > X$ *and the operator* $T_N : H^{(s)} \to H^{(s+N)}$ *is completely continuous (when* $|s| < s_0$, s_0 *being fixed).*

Proof. In view of Proposition 1.4′ we have

$${}^t\mu_1(x, D) = \mu_1(x, D) + \mu_2(x, D) + T_N', \tag{2.19}$$

where $\mu_2 \prec\prec \mu_1$, $\mu_2(x, \xi) \equiv 0$ for large $|x|$, and T_N is a compact smoothing operator. We obtain (2.18) without difficulty by using (2.19) and Proposition 1.1′.

Corollary. *The operator* (2.17) *is continuous.*

This statement follows trivially from (2.18) and Lemma 2.3.

Lemma 2.5. *The operator* (2.16) *is formally selfadjoint and positive.*

Proof. By (2.16) and the duality (1.15), for $u, v \in H^{(\infty)}$ we have

$$\begin{aligned}(\mathfrak{M}u, v) &= (\mu_1(x, D)\, u, \mu_1(x, D)\, v) \\ &+ ((1 + |D|^2)^{m/4} u, (1 + |D|^2)^{m/4} v) = (u, \mathfrak{M}v).\end{aligned} \tag{2.20}$$

By putting $u = v$ we obtain that

$$(\mathfrak{M}u, u) = \| \mu_1(x, D) u \|^2 + \| u \|^2_{(m/2)}.$$

COROLLARY. *The equation* $\mathfrak{M}u = 0$ *does not have nontrivial solutions* $u \in H^{(\infty)}$.

LEMMA 2.6 (CF. [**12**], [**7**] AND [**3**]). *The operator* (2.16) *is hypoelliptic; that is,*

$$\{u \in H^{(-\infty)},\ \mathfrak{M}u = v \in H^{(\infty)}\} \Rightarrow \{u \in H^{(\infty)}\},$$

where the equation $\mathfrak{M}u = v$ *is to be understood in the sense of distribution theory.*

PROOF. In accordance with Lemma 2.4 we can write $\mathfrak{M}$ as

$$\mathfrak{M} = \mu(x, D) + b_N(x, D) + (1 + |D|^2)^{m/2} + T_N = \mu_N(x, D) + T_N.$$

It is not difficult to check that the symbol $\mu_N(x, \xi)$ satisfies the conditions of Lemma 2.2. Hence there is an operator $r_N(x, D)$ such that

$$r_N(x, D)\mu_N(x, D) = E + T'_N,$$

and the operator T'_N is smoothing by N. By applying the operator $r_N(x, D)$ to the equation $\mathfrak{M}u = v$ we find that

$$u = -T'_N u - r_N(x, D) T_N u + r_N(x, D) v.$$

If $u \in H^{(s)}$, then when N is sufficiently large the right-hand side of the last equation belongs to $H^{(s+1)}$, etc. The lemma is proved.

Thus we have proved that, if $u \in H^{(-\infty)}$, $v \in H^{(\infty)}$ and

$$(u, \mathfrak{M}\varphi) = (v, \varphi) \quad \text{for all} \quad \varphi \in H^{(\infty)}, \tag{2.21}$$

then $u \in H^{(\infty)}$ (the left-hand side is the value of the functional $u \in H^{(-\infty)}$ at $\mathfrak{M}\phi \in H^{(\infty)}$; the right-hand side is the scalar product in L_2).

COROLLARY 1. *The kernel of the operator* (2.17) *is trivial.*

In fact, if $\mathfrak{M}u = 0$ in the sense of distribution theory, then in view of (2.20) we have

$$(u, \mathfrak{M}^*\varphi) = (u, \mathfrak{M}\varphi) = 0,$$

whence it follows that $u \in H^{(\infty)}$ and $\mathfrak{M}u = 0$. Then, from what has been proved, $u = 0$.

COROLLARY 2. *The set* $\mathfrak{M}H^{(\infty)}$ *is dense in* L_2.

In fact, if $\mathfrak{M}H^{(\infty)}$ is not dense in L_2, then there is a $u \in L_2$ such that

$$(u, \mathfrak{M}\varphi) = 0 \quad \text{for all} \quad \varphi \in H^{(\infty)}.$$

But then $u \in H^{(\infty)}$ and $\mathfrak{M}u = 0$; that is, $u = 0$.

Thus to prove Theorem 2.1 we need only establish that the set $\mathfrak{M}H^{\mu(x)}$ is closed in L_2. It is easy to see that this is equivalent to the inequality

$$\| u \|_\mu \leqslant \text{const} \| \mathfrak{M}u \|, \quad u \in H^{\mu(x)}. \tag{2.22}$$

In fact, if $\mathfrak{M}u_j = f_j \to f$ in L_2, then by (2.22) the sequence $\{u_j\}$ is fundamental in $H^{\mu(x)}$ and so, since $H^{\mu(x)}$ is complete, $u_j \to u_0$ in $H^{\mu(x)}$, and $f = \mu(x, D)u_0$. Conversely, if we could establish that $\mathfrak{M}H^{\mu(x)}$ is closed, because this set is dense in L_2 it would coincide with L_2 and so (2.22) would follow from a theorem of Banach.

In view of the Remark to Corollary 3 of Lemma 2.3,

$$\|u\|_{\mu} \leqslant \text{const}\,(\|\mathfrak{M}u\| + \|u\|_{(\underline{\mu})}), \quad u \in H^{(\infty)},$$

and so (2.22) reduces to

$$\|u\|_{(\underline{\mu})} \leqslant \text{const}\,\|\mathfrak{M}u\|, \quad u \in H^{(\infty)}. \tag{2.23}$$

For our purposes we will prove a more general bound from which will follow (2.23) and, together with it, Theorem 2.1.

LEMMA 2.7. *Let $\mathfrak{M}$ be the operator in* (2.16). *Then for any real s we have*

$$\|u\|_{(\underline{\mu}+s)} \leqslant \text{const}\,\|\mathfrak{M}u\|_{(s)}, \quad u \in H^{(\infty)}. \tag{2.24}$$

PROOF. Let us assume that (2.24) does not hold. Then there is a sequence $u_j \in H^{(\underline{\mu}+s)}$ such that

$$\|u_j\|_{(s+\underline{\mu})} = 1, \quad \|\mathfrak{M}u_j\|_{(s)} \to 0, \ j \to \infty. \tag{2.25}$$

If this sequence (or a subsequence of it) has a limit we are led to a contradiction. In fact, if $\|u_j - u_0\|_{(s+\underline{\mu})} \to 0$ then $\|u_0\|_{(s+\underline{\mu})} = 1$ and $\|\mathfrak{M}u_j - \mathfrak{M}u_0\|_{(\sigma)} \to 0$, where $\sigma = s + \underline{\mu} + \bar{\mu}$. Because the sequence $\mathfrak{M}u_j$ converges to zero in the stronger topology of $H^{(s)}$, then $\mathfrak{M}u_0 = 0$ and so, from what we have proved, $u_0 \equiv 0$; that is, we obtain a contradiction.

To prove the compactness we represent u_j in the form

$$u_j = u_j' + u_j'' = r(x, D)\,\mathfrak{M}u_j + (E - r(x, D))\,\mathfrak{M}u_j,$$

where

$$r(x, \xi) = (\mu(x, \xi) + (1 + |\xi|^2)^{m/2})^{-1}.$$

Since $r(x, \xi) \in S_{\rho,\delta}^{(-\underline{\mu})}$ (see (2.21)), we have

$$\|u_j'\|_{(s+\underline{\mu})} \leqslant \text{const}\,\|\mathfrak{M}u_j\|_{(s)},$$

and, in view of (2.25), the sequence $\{u_j'\}$ converges to zero. Let us prove that the sequence $\{u_j''\}$ is compact. By Lemma 2.4,

$$\begin{aligned} E - r(x, D)\,\mathfrak{M} &= \{E - r(x, D)\,[\mu(x, D) + (1 + |D|^2)^{m/2}]\} \\ &\quad + r(x, D)\,b_N(x, D) + r(x, D)\,T_N = T_1 + T_2 + T_3. \end{aligned}$$

The operator $T_1 : H^{(s+\underline{\mu})} \to H^{(s+\underline{\mu})}$ is completely continuous in view of Lemma 2.1; for large N the complete continuity of T_3 follows from Lemma 2.4. By using Lemma 2.4 it is not difficult also to prove that the operator $T_2 : H^{(s+\underline{\mu})} \to H^{(s+\underline{\mu})}$ is completely continuous. In fact, if $h_N(x, \xi) = r(x, \xi)\,b_N(x, \xi)$, then

$h_N(x,\xi) \in S^{-\lambda}_{\rho,\delta}$ and $h_N(x,\xi) \equiv 0$ for large $|x|$. Hence the operator $h_N(x,D)$: $H^{(s+\mu)} \to H^{(s+\mu)}$ is completely continuous. By representing T_2 as

$$T_2 = h_N(x, D) + [r(x, D)\, b_N(x, D) - h_N(x, D)]$$

and using Proposition 1.1′, we prove that T_2 is a compact operator, and hence the lemma is proved.

4. The operator $\mathfrak{M}^{-1}$. By Theorem 2.1 the operator (2.17) has the inverse

$$\mathfrak{M}^{-1} : L_2 \to H^{\mu(x)};$$

that is, any $u \in H^{\mu(x)}$ can be written as

$$u = \mathfrak{M}^{-1} v, \quad v \in L_2, \quad \|v\| \leqslant \text{const}\, \|u\|_\mu. \tag{2.26}$$

The structure of the operator $\mathfrak{M}^{-1}$ will be of importance when we prove the embedding theorems. With the help of Lemma 2.2 we can easily prove

THEOREM 2.2. *For arbitrary $N > 0$, any $u \in H^{\mu(x)}$ can be represented as*

$$u = r_N(x, D)\, v_N + T_N v_N, \tag{2.27}$$

where

$$v_N \in L_2, \quad \|v_N\| \leqslant C_N \|u\|_\mu, \tag{2.28}$$

the symbol $r_N(x,\xi)$ has the form

$$r_N(x, \xi) = \mu_N^{-1}(x, \xi)\,(1 + p_N(x, \xi)), \tag{2.29}$$

where

$$\mu_N(x, \xi) \in \mathfrak{w}_{\rho,\delta}, \quad C^{-1}\mu(x, \xi) < \mu_N(x, \xi) < C\mu(x, \xi), \tag{2.30}$$

$$p_N(x, \xi) \in S^{-\lambda}_{\rho,\delta}, \ \lambda > 0, \ p_N(x, \xi) \equiv 0 \quad npu \quad |x| > X, \tag{2.31}$$

and the operator $T_N\colon H^{(s)} \to H^{(s+N)}$ is completely continuous for $|s| < s_0$.

PROOF. We put

$$\mathfrak{M}_N = {}^t\mu_1(x, D)\, \mu_1(x, D) + K_N (1 + |D|^2)^{m/2}, \tag{2.32}$$

where K_N is a large positive constant to be chosen later. In accordance with Lemma 2.4 we represent the operator (2.32) in the form

$$\mathfrak{M}_N = \mu(x, D) + K_N (1 + |D|^2)^{m/2} + b_N(x, D) + T'_N \equiv \mu_N(x, D) + T'_N. \tag{2.32′}$$

Because $b_N(x,\xi) \to\!\!\to \mu(x,\xi)$, at the expense of choosing the constant K_N we can assure that conditions (2.30) are satisfied. According to Lemma 2.2 (see the Remark to this lemma) there exists a symbol $r_N(x,\xi)$ satisfying (2.29)—(2.31) such that

$$r_N(x, D)\, \mu_N(x, D) = E + T''_N, \qquad T''_N : H^{(s)} \to H^{(s-\mu+N)}, \ |s| < s_0,$$

is a completely continuous operator. By putting $\mathfrak{M} u_N = v_N$ and applying the operator $r_N(x, D)$ to this equation we obtain that

$$u = r_N(x, D) v_N - (T''_N + r_N(x, D) T'_N) u = r_N(x, D) v_N + T_N v_N,$$

where

$$T_N = -(T''_N + r_N(x, D) T'_N) \mathfrak{M}_N^{-1}$$

According to Lemma 2.7 $\mathfrak{M}_N^{-1}$ is a continuous operator from $H^{(s)}$ into $H^{(s+\underline{\mu})}$. Taking the properties of the operators T'_N and T''_N into account, we see that the operator T_N satisfies the hypothesis of the theorem; hence the theorem is proved.

Let $H^{\mu(x)}_{\overline{\Omega}}$ denote the set of distributions whose supports lie in the closure $\overline{\Omega}$ of the bounded domain Ω; that is,

$$H^{\mu(x)}_{\overline{\Omega}} = H^{\mu(x)} \cap \mathcal{E}'(\overline{\Omega}).$$

For the spaces $H^{\mu(x)}_{\overline{\Omega}}$ we can modify Theorem 2.2 as follows.

THEOREM 2.2′. *Let $N > 0$ and $\Omega_1 \supset \Omega$. Any element $u \in H^{\mu(x)}_{\overline{\Omega}}$ can be represented as*

$$u = r_N(x, D) v_N + T_N u, \tag{2.33}$$

where $v_N \in L_2$, $\operatorname{supp} v_N \subset \Omega_1$ *and*

$$\|v_N\| \leqslant C(N, \Omega_1, \Omega) \|u\|_\mu;$$

the operators r_N and T_N have the same properties as the corresponding operators in Theorem 2.2.

PROOF. Let $\psi(x) \in C_0^\infty(\Omega_1)$ and $\psi(x) = 1$ when $x \in \Omega$. If $u \in H^{\mu(x)}_{\overline{\Omega}}$, then obviously $u = \psi u$. Let $\mathfrak{M}_N$ be the operator in Theorem 2.2. Then

$$\begin{aligned} \mathfrak{M}_N u &= \mu_N(x, D)(\psi u) + T'_N u \\ &= \mu'_N(x, D) u + T''_N u + T'_N u \equiv v_N + T'''_N u, \end{aligned} \tag{2.34}$$

where

$$\mu'_N(x, \xi) = \sum_{|\alpha| \leqslant N_1} \frac{D^\alpha \psi}{\alpha!} \partial^\alpha \mu_N(x, \xi), \quad \mu'_N(x, \xi) \equiv 0, \quad x \notin \Omega_1. \tag{2.34′}$$

We rewrite (2.34) as

$$\mu_N(x, D) u = v_N + T^{\mathrm{IV}}_N u.$$

Our statement is proved by applying the operator $r_N(x, D)$ in Theorem 2.2 to the last equation.

Let $\mathfrak{M}_\infty$ denote the restriction of the operator (2.17) to $H^{(\infty)}$. It follows easily from the above arguments that the mapping

$$\mathfrak{M}_\infty : H^{(\infty)} \to H^{(\infty)} \tag{2.35}$$

is an isomorphism. In fact, the continuity of $\mathfrak{M}_\infty$ follows from the continuity of operators of the type $\mu_1(x, D)$ and ${}^t\mu_1(x, D)$, which was pointed out in §1. The operator (2.35) does not have a kernel, in view of the Corollary to Lemma 2.5. If $v \in H^{(\infty)}$, then, according to Theorem 2.1, the equation $\mathfrak{M}u = v$ has a solution $u \in H^{\mu(x)}$; since $\mathfrak{M}$ is hypoelliptic (Lemma 2.6), $u \in H^{(\infty)}$.

From duality considerations we see that the adjoint operator

$$\mathfrak{M}_{-\infty} = \mathfrak{M}_{\infty}^{*} : H^{(-\infty)} \to H^{(-\infty)} \tag{2.36}$$

is also an isomorphism. By using the formal selfadjointness of $\mathfrak{M}_{\infty}$ (Lemma 2.5) it is not difficult to check that the restriction of $\mathfrak{M}_{-\infty}$ to $H^{\mu(x)}$ is the same as the operator (2.17). In what follows we shall not distinguish between the operator (2.36) and its restriction and will denote this operator by $\mathfrak{M}$.

Since $H^{(\infty)}$ is dense in L_2 (in the topology of L_2), it follows trivially from the isomorphism (2.35) that $H^{(\infty)}$ is dense in $H^{\mu(x)}$. Because $H^{(\infty)}$ is dense in $H^{(-\infty)}$, $H^{\mu(x)}$ has this property.

We show that $H^{1/\mu(x)}$ can be identified with the Banach space conjugate to $H^{\mu(x)}$. In fact, in view of the isomorphism (2.17), we can identify $(H^{\mu(x)})^*$ in a natural way with the image of L_2 under the mapping (2.36). Let us show that $\mathfrak{M}H^{(0)} = H^{1/\mu(x)}$. This equation is a particular case of the more general

LEMMA 2.8. *Let $\mu(x, \xi)$, $\nu(x, \xi) \in \mathfrak{w}_{\rho,\delta}$ and let $\mathfrak{M}$ be an operator of the form (2.32). Then the mapping*

$$\mathfrak{M} : H^{\mu(x)\nu(x)} \to H^{\nu(x)} \tag{2.37}$$

is an isomorphism.

PROOF. Let $\lambda(x, \xi) = \mu(x, \xi)\nu(x, \xi)$. The continuity of the operator (2.37) is equivalent to the inequality

$$\| \nu(x, D)\, \mathfrak{M}u \| \leqslant \text{const} \| u \|_{\lambda}, \quad u \in H^{(\infty)}. \tag{2.38}$$

By using (2.18) and Proposition 1.1 it is not difficult to verify that $\nu(x, D)\mathfrak{M} = a(x, D) + T$, where $a(x, \xi) \dashv \lambda(x, \xi)$ and T is a smoothing operator of high order N. With the help of Lemma 2.3 we arrive at (2.38). Because the mapping (2.36) does not have a kernel, the mapping (2.37) also has this property. If $v \in H^{\nu(x)}$, then in view of the isomorphism (2.37) we can find a distribution $u \in H^{(-\infty)}$ such that $\mathfrak{M}u = v$. By applying the operator $\nu(x, D)$ to this equation we find that

$$\lambda(x, D) + \lambda_1(x, D) + T_N u = \nu(x, D)\, v \in L_2.$$

By using Corollary 3 to Lemma 2.3 we prove that $u \in H^{\mu\nu(x)}$, and so our lemma is proved.

§3. The embedding theorems

1. The embedding of $H^{\mu(x)}$ in $H^{\nu(x)}$.

THEOREM 3.1. *The space $H^{\mu(x)}$ is embedded in $H^{\nu(x)}$ if and only if*

$$\sup_x \sup_\xi \nu(x, \xi)\, \mu^{-1}(x, \xi) < \infty. \tag{3.1}$$

PROOF. According to the isomorphism (2.17) the embedding of $H^{\mu(x)}$ in $H^{\nu(x)}$ is equivalent to the boundedness of the operator $\nu(x, D)\mathfrak{M}^{-1}$ (as an operator from L_2 into L_2). By Theorem 2.2 this operator coincides (up to a bounded operator) with the operator

$$\lambda_N(x, D) + \lambda'_N(x, D), \tag{3.2}$$

where $\lambda_N(x,\xi) = \nu(x,\xi)\mu_N(x,\xi)$, $\mu_N(x,\xi)$ is the symbol in (2.30), and $\lambda'_N(x,\xi) \prec\!\!\prec \lambda_N(x,\xi)$. But then, by Corollary 3 to Lemma 2.3, the boundedness of the operator (3.2) is equivalent to the boundedness of the operator $\lambda_N(x, D)$. By applying Theorem 1.1 we arrive at (3.1), and our theorem is proved.

THEOREM 3.2. *Let* $\lambda(x,\xi) = \nu(x,\xi)\mu^{-1}(x,\xi)$ *and suppose that, for some domain* $\Omega_1 \supset \Omega$,

$$\sup_{x\in\Omega_1}\sup_{\xi}\lambda(x, \xi) < \infty. \tag{3.3}$$

Then $H^{\mu(x)}_{\overline{\Omega}}$ *is embedded in* $H^{\nu(x)}_{\overline{\Omega}}$. *If*

$$\sup_{x\in\Omega_1}\overline{\lim_{|\xi|\to\infty}}\lambda(x, \xi) = 0, \tag{3.4}$$

then the operator that embeds $H^{\mu(x)}_{\overline{\Omega}}$ *in* $H^{\nu(x)}_{\overline{\Omega}}$ *is completely continuous.*

PROOF. The first part of the theorem follows trivially from Theorem 2.2′ and Corollary 2 to Theorem 1.1.

We now show that if (3.4) holds then the family $\Sigma = \{u \in H^{\mu(x)}_{\overline{\Omega}}; \|u\|_\mu \leqq 1\}$ is compact in $H^{\nu(x)}$. To do this we need to check that the set $\{v = \nu(x, D)u;\ u \in \Sigma\}$ is compact in L_2. According to Theorem 2.2′ we have

$$\nu(x, D)u = \nu(x, D)r_N(x, D)v_N + \nu(x, D)T_N u = w_1 + w_2.$$

If N is large then T_N is a compact operator from $H^{(\nu)}$ into $H^{(\nu)}$, so that the family $\{w_2\}$ is compact in L_2. By applying Proposition 1.1′ we can write

$$w_1 = \lambda_N(x, D)v_N + \lambda'_N(x, D)v_N + T'_N v_N = w_3 + w_4 + w_5;$$
$$\lambda_N(x, \xi) = \nu(x, \xi)\mu_N(x, \xi).$$

Because $\lambda_N(x,\xi) \in S^0_{\rho,\delta}$, the families $\{w_4\}$ and $\{w_5\}$ are compact in L_2 in view of Proposition 1.1′. To prove that the family $\{w_3\}$ is compact we choose a function $\psi(x) \in C_0^\infty(\Omega_1)$ which is equal to unity when $x \in \overline{\Omega}$. Then

$$\begin{aligned} w_3 &= \lambda_N(x, D)v_N = \lambda_N(x, D)(\psi v_N) \\ &= \tilde{\lambda}_N(x, D)v_N + Tv_N = w_6 + w_7, \end{aligned}$$

where

$$\tilde{\lambda}_N(x, \xi) = \sum_{|\alpha|<\tilde{N}} D^\alpha\psi\partial^\alpha\lambda_N(x, \xi)/\alpha!.$$

The family $\{w_7\}$ is compact in view of Proposition 1.1′. If (3.4) holds, then the symbol $\tilde{\lambda}_N(x,\xi)$ satisfies the conditions of Theorem 1.3, whence it follows that the family $\{w_6\}$ is compact and our theorem is proved.

REMARKS. 1) By using the sequence (1.21) we can show that, if we replace Ω_1 by Ω in (3.3) and (3.4), the results are necessary conditions for the (complete) continuity of the corresponding embeddings.

2) If the symbol $a(x,\xi) \rightarrow \omega(x,\xi) \in \mathfrak{w}$ and

$$\omega\,|\,a\,(x,\ \xi)\,|\,\nu\,(x,\ \xi)\,\mu^{-1}\,(x,\ \xi) < \text{const},$$

then $a(x,D)$ is a continuous operator from $H^{\mu(x)}$ into $H^{\nu(x)}$.

2. The embedding of $H^{\mu(x)}$ in spaces of a different character.

THEOREM 3.2. *The space $H^{\mu(x)}$ is embedded in C if and only if*

$$\sup_x \int \mu^{-2}\,(x,\ \xi)\,d\xi < \infty. \tag{3.5}$$

PROOF. In view of Theorem 2.2 the question concerning the embedding of $H^{\mu(x)}$ in C reduces to the following: under what conditions on the symbol $\mu(x,\xi)$ is the following operator continuous:

$$r\,(x,\ D) : L_2 \rightarrow C, \tag{3.6}$$

where $r(x,\xi) = r_N(x,\xi)$ is the symbol given by (2.29)? Let $R(x,z)$ denote the inverse Fourier transform (with respect to ξ) of $r(x,\xi)$. Then

$$r\,(x,\ D)\,v\,(x) = \int R\,(x,\ x - y)\,v\,(y)\,dy$$

and the continuity of the operator (3.6) means that at any point $x_0 \in R^n$

$$|\,(r\,(x_0,\ D)\,v)\,(x_0)\,| = \left|\int R\,(x_0,\ x_0 - y)\,v\,(y)\,dy\right| \leqslant K\,\|v\|. \tag{3.7}$$

But then, by the reverse of Schwarz' inequality the last estimate is equivalent to the inequality

$$\int |\,R\,(x_0,\ x_0 - y)\,|^2\,dy = \int |\,R\,(x_0,\ y)\,|^2\,dy \leqslant K^2. \tag{3.8}$$

By applying Parseval's equality to (3.8) and by using (2.30) we are lead to (3.5).

REMARKS. 1) It is easy to show by differentiating (3.7) that $H^{\mu(x)}$ is embedded in C^k (k is an integer) if and only if

$$\sup_x \int (1 + |\,\xi\,|)^{2k}\,\mu^{-2}\,(x,\ \xi)\,d\xi < \infty. \tag{3.5′}$$

2) With the help of (3.7) we can show (cf. [**5**], Theorem 9.1) that, if in (3.5′) k is not an integer, then $H^{\mu(x)}$ is embedded in the Hölder space C^k.

THEOREM 3.3. *If*

$$\sup_{x\in\Omega} \int \mu^{-2}\,(x,\ \xi)\,d\xi < \infty, \tag{3.9}$$

then the operator of the embedding $H^{\mu(x)}_{\overline{\Omega}} \rightarrow C(\overline{\Omega})$ is completely continuous.

PROOF. In our proof of Theorem 3.2 we have, in fact, shown that if the family $\{v(x)\}$ is uniformly bounded in L_2, then the family (3.7) is uniformly bounded in $C(\overline{\Omega})$. It is not difficult to prove that this family is equicontinuous. We have

$$(rv)(x+z)-(rv)(x)=\int[R(x+z,x-y)-R(x,x-y)]\,v(y+z)\,dy$$

$$+\int R(x,x-y)[v(y+z)-v(y)]\,dy=w_1(x,z)+w_2(x,z).$$

In view of Schwarz' inequality and the integral continuity of functions from L_2 we have that

$$|w_2(x,z)|^2<\int r^2(x,\xi)\,d\xi\int|v(y+z)-v(y)|^2\,dy\to 0 \text{ when } |z|\to 0.$$

For $w_1(x,z)$ we have the estimate

$$|w_1(x,z)|^2<\int|r(x+z,\xi)-r(x,\xi)|^2\,d\xi\,\|v\|^2.$$

Since $\mu(x,\xi)$, and consequently $r(x,\xi)$, is a continuous function of x, in view of (3.9) the right-hand side of the last inequality can be made arbitrarily small when $|z|\to 0$.

Because, according to Theorem 2.2,

$$u=r_N(x,D)\,v_N+T_Nv_N,\quad u\in H^{\mu(x)},\quad v_N\in L_2,$$

and the operator T_N is completely continuous from L_2 into $H^{(l)}$, where $l=l_N$ is large, for $l>n/2$ this operator sends a sequence bounded in L_2 into a sequence in C that is uniformly bounded and equicontinuous; hence the family $\{u\in H^{\mu(x)}_{\overline{\Omega}};\ \|u\|_\mu<\text{const}\}$ satisfies the hypotheses of Arzelà's Theorem. Thus our theorem is proved.

THEOREM 3.4. *If*

$$\mu^{-1}(x,\xi)\in S^{-\lambda}_{\rho,\delta},\quad \lambda\geqslant n\left(\frac{1}{2}-\frac{1}{p}\right)>0, \tag{3.10}$$

then the operator of the embedding $H^{\mu(x)}\to L_p$ *is continuous.*

PROOF. If (3.10) holds then we can write the operator $r(x,D)$ as

$$r(x,D)=(1+|D|^2)^{-\lambda/2}(1+|D|^2)^{\lambda/2}r(x,D)$$

$$=(1+|D|^2)^{-\lambda/2}r_1(x,D)+(1+|D|^2)^{-\lambda/2}T_N,$$

where $r_1(x,\xi)\in S^0_{\rho,\delta}$ and the operator T_N maps L_2 into itself. If (3.10) holds, then, by a well-known lemma of S. L. Sobolev, the operator $(1+|D|^2)^{-\lambda/2}$ maps L_2 into L_p. Our theorem follows from these results.

3. The restrictions of functions belonging to $H^{\mu(x)}$. Suppose that the space R^n is decomposed into the direct sum of two subspaces R^m and R^{n-m}. We write $x=(x',x'')$, where $x'=(x_1,\cdots,x_m)$ and $x''=(x_{m+1},\cdots,x_n)$. We introduce a similar decomposition for the dual variables: $\xi = (\xi', \xi'')$. Let ∂' and D' denote the sets of operators $\partial' = (\partial_1, \cdots, \partial_m)$ and $D' = (D_1, \cdots, D_m)$. We retain the notation $\mathfrak{w}_{\rho,\delta}$ for the class of symbols $\nu(x', \xi')$ satisfying the estimates (2.1) and (2.2) (with respect to x' and ξ').

We consider the following operator, which acts on smooth functions in R^n:

$$\tau : u(x', x'') \to u(x', 0). \tag{3.11}$$

If this operator is continuous as an operator from $H^{\mu(x)}(R^n)$ into $H^{-\infty}(R^m)$, we shall say that the elements of $H^{\mu(x)}$ have a restriction to the plane $x'' = 0$.

For the spaces H^{μ} considered in [5], a necessary and sufficient condition for the restriction to exist is the convergence of

$$\int \mu^{-2}(\xi', \xi'')\, d\xi''; \tag{3.12}$$

here the restrictions of functions from H^{μ} fill out the space $H^{\mu'}$, where

$$\mu'(\xi') = \left(\int \mu^{-2}(\xi', \xi'')\, d\xi''\right)^{-1/2}. \tag{3.13}$$

Next we establish an analog of this result for the variable symbols $\mu(x, \xi)$. Here we shall impose conditions that will ensure that the integral (3.12) converges and that functions of the type (3.13) belong to the class $\mathfrak{w}$.

THEOREM 3.5. *Suppose that the symbol $\mu(x, \xi) \in \mathfrak{w}$ satisfies the conditions*

1)
$$\int \mu^{-2}(x', 0, \xi', \xi'')\, d\xi'' = M(x', \xi') < \infty \tag{3.14}$$

for each $x \in R^n, \xi \in R^m$,[1)]

2) *for any set of multi-indices $\alpha^1, \cdots, \alpha^k, \beta^1, \cdots, \beta^k$*

$$\begin{aligned}&\int |\partial'^{\alpha^1} D'^{\beta^1}\mu(x', \xi) \ldots \partial'^{\alpha^k} D'^{\beta^k}\mu(x', \xi)|\, \mu^{-2-k}(x, \xi)\, d\xi'' \\ &\leqslant C(\alpha^1, \ldots, \beta^k)\, M(x', \xi')\,(1 + |\xi'|)^{-\rho|\alpha^1+\ldots+\alpha^k| + \delta|\beta^1+\ldots+\beta^k|}.\end{aligned} \tag{3.15}$$

Suppose that $\nu(x', \xi') \in \mathfrak{w}$ and that

$$\nu(x', \xi')\, M^{1/2}(x', \xi') < \text{const}. \tag{3.16}$$

Then the operator (3.11) *can be extended to a continuous operator from $H^{\mu(x)}(R^n)$ into $H^{\nu(x')}(R^m)$.*

REMARK. We put

$$\mu'(x', \xi') = M^{-1/2}(x', \xi'). \tag{3.17}$$

As is not difficult to verify, it follows from (3.15) that $M(x', \xi) \in \mathfrak{w}_{\rho,\delta}$, and consequently $\mu'(x', \xi') \in \mathfrak{w}_{\rho,\delta}$. Since the function (3.17) satisfies (3.16), for

---

1) In view of (2.3) it is sufficient to require that the integral (3.14) converges for one value of ξ', say $\xi' = 0$.

$\nu(x', \xi')$ we can take the function (3.17); that is, the restrictions of functions from $H^{\mu(x)}$ belong to $H^{\mu'(x')}$. In the case of constant symbols μ the mapping $\tau: H^{\mu} \to H^{\mu'}$ is an epimorphism; unfortunately we have not succeeded in proving this result for variable symbols.

PROOF OF THEOREM 3.5. According to Theorem 2.2 any $u \in H^{\mu(x)}$ has the form

$$u = u_1 + u_2 = r_N(x, D)\, v + T_N v, \quad v \in L_2,$$

where the operator T_N maps L_2 into $H^{(N)}$, $N > 0$; for large N the restrictions of the functions $T_N v$ belong to $H^{(\nu)}$ and so to $H^{\nu(x')}$. Hence we must prove that the operator $\nu(x', D')\tau r_N(x, D): L_2(R^n) \to L_2(R^m)$ is continuous. We have

$$\nu(x\ , D')\, u_1(x', 0) = \nu(x', D') \int r_N(x', 0, D', \xi'')\hat{v}(x', \xi'')\, d\xi'', \quad (3.18)$$

where $\hat{v}(x', \xi'')$ denotes the Fourier transform of v with respect to x''. By carrying the operator $\nu(x', D')$ in (3.18) under the integral sign and using Proposition 1.1, we find that

$$\nu(x', D')\, u_1(x', 0) = \int \lambda_{NM}(x', D', \xi'')\hat{v}(\ , \xi'')\, d\xi'' + \int T_M(\xi'')\, \hat{v}(\ , \xi'')\, d\xi'', \quad (3.19)$$

where

$$\lambda_{NM}(x', \xi) = \sum_{|\alpha| \leqslant M} \frac{1}{\alpha!} \partial'^{\alpha}\nu(x', \xi')\, D'^{\alpha} r_N(x', 0, \xi). \quad (3.20)$$

If we show that Theorem 1.2 (Proposition 1.2) can be applied to the operator $\lambda_{NM}(x', D', \xi'')$ $(T_M(\xi''))$, we will have proved our assertion. This, in turn, follows easily from

LEMMA 3.1. *Suppose that the symbol $\mu(x, \xi) \in \mathfrak{w}$ satisfies the conditions of Theorem 3.5, and let $r_N(x, \xi)$ be the symbol in Theorem 2.2 that corresponds to the symbol $\mu(x, \xi)$. Then*

$$\left(\int \left| \partial'^{\alpha} D'^{\beta} r_N(x', 0, \xi', \xi'') \right|^2 d\xi'' \right)^{1/2} \leqslant C_{\alpha\beta} (1 + |\xi'|)^{-\rho|\alpha| + \delta|\beta|} (M(x', \xi'))^{1/2}. \quad (3.21)$$

PROOF. To prove (3.21) we must recall the algorithm for the construction of $r_N(x, \xi)$.

We took $\mu(x, \xi)$ as the initial symbol and first constructed the symbol $\mu_N(x, \xi)$ from (2.32′). It follows from (2.32) that the symbol $b_N(x, \xi)$ from (2.32′) is a linear combination of expressions of the form

$$\mu^{-k-1}(x, \xi)\, \partial^{\alpha^1} D^{\beta^1} \mu(x, \xi) \ldots \partial^{\alpha^k} D^{\beta^k} \mu(x, \xi),$$

where

$$|\alpha^1 + \ldots + \alpha^k| \geqslant |\beta^1 + \ldots + \beta^k|.$$

By using this result and (2.30) it is easy to verify that the symbol $\mu_N(x, \xi)$ satisfies conditions of the type (3.14), (3.15).

The symbol $r_N(x, \xi)$ was constructed in terms of $\mu_N(x, \xi)$ with the help of the algorithm in Lemma 2.2. It is not difficult to prove (3.21) by using the Remark to Lemma 2.2.

4. Examples. Let $\overline{m} = (m_1, \cdots, m_n)$ be a set of n nonnegative integers. We put

$$\mu(x, \xi) = (1 + \xi_1^{2m_1} + \ldots + \xi_n^{2m_n})^{s(x)},$$

where $s(x) \in C^\infty$, $s(x) = s$ for large $|x|$. The space corresponding to the symbol (3.22) will be denoted by $H^{(\overline{m}, s(x))}$. It can easily be verified that the symbol (3.22) satisfies both (2.1), (2.2) and conditions of the type (3.14), (3.15) (with the appropriate values of $s(x)$). From the general theorems of the present section we easily deduce embedding theorems for the spaces $H^{(\overline{m}, s(x))}$ that generalize the theorems of Sobolev, Aronszajn and Slobodeckiĭ; to do this it is only necessary to carry out uncomplicated calculations analogous to those performed in [**5**], §9.

When $m_1 = m_2 = \cdots = m_n = 1$ we denote the corresponding space by $H^{(s(x))}$. As we remarked in the Introduction, these spaces have been studied in quite a number of articles. When s is a constant the space $H^{(s)}$ becomes the space $L_2^{(s)}$ of Bessel potentials introduced by Aronszajn and Smith [**13**].

§4. Localization in the spaces $H^{\mu(x)}$

Let $1 = \sum \phi_k^2(\xi)$ be the partition of unity referred to in Theorem 1.1. We define a new norm in $H^{\mu(x)}$ by

$$|[u]|_\mu^2 = \sum_j \|\mu(x, \xi^j)\varphi_j(D)u\|^2 + \|u\|_{(m)}, \tag{4.1}$$

where $\xi^j \in \operatorname{supp} \phi_j$. The main result of this section is

THEOREM 4.1. *The norm* (4.1) *is equivalent to the norm* (2.9); *that is, the following inequality holds*:

$$C^{-1}\|u\|_\mu \leqslant |[u]|_\mu \leqslant C\|u\|_\mu. \tag{4.2}$$

PROOF. By (1.25) we have

$$\|\mu(x, D)u\|^2 = \sum \|\varphi_j(D)\mu(x, D)u\|^2 = \sum_j \Big\| \mu(x, \xi^j)\varphi_j(D)u + \sum_{|\alpha+\beta|<N} \frac{1}{\alpha!\beta!}\mu_{(\beta)}^{(\alpha)}(x, \xi^j)(D - \xi^j)^\alpha \varphi_j^{(\beta)}(D)u + R_j^N u\Big\|^2, \tag{4.3}$$

where

$$\mu_{(\beta)}^{(\alpha)}(x, \xi) = \partial^\alpha D^\beta \mu(x, \xi), \quad \varphi_j^{(\beta)}(\xi) = \partial^\beta \varphi_j(\xi).$$

If N is sufficiently large, then, according to (1.26″),

$$\sum_j \|R_j^N u\|^2 \leqslant C\|u\|_{(m)}. \tag{4.4}$$

We show that for any small $\epsilon > 0$ and for $|\alpha + \beta| > 0$

$$\sum_j \| \mu_{(\beta)}^{(\alpha)} (x, \xi^j) (D - \xi^j) \varphi_j^{(\beta)} (D) u \|^2 \leqslant \varepsilon | [u] |_\mu + C_\varepsilon \| u \|_{(m)}. \tag{4.5}$$

The desired inequality follows trivially from the relations (4.3)—(4.5). We preface the derivation of (4.5) with two lemmas.

LEMMA 4.1. *Let ξ^j and η^j be arbitrary points belonging to the support of ϕ_j. Then*

$$| \mu (x, \xi^j) \ \mu^{-1} (x, \eta^j) - 1 | < C (1 + | \eta^j |)^{-\lambda}, \tag{4.6}$$

where $\lambda > 0$ and the constant C is independent of j.

COROLLARY. *For any ξ^j, $\eta^j \in \operatorname{supp} \phi_j$*

$$C_1 \mu (x, \xi^j) < \mu (x, \eta^j) < C_2 \mu (x, \xi^j). \tag{4.6'}$$

PROOF. By Taylor's formula the left-hand side of (4.6) does not exceed

$$\begin{aligned} &\sum_{0<|\alpha|<N} | (\xi^j - \eta^j)^\alpha | \, | \partial^\alpha \mu (x, \eta^j) | \, \mu^{-1} (x, \eta^j)/\alpha! \\ &+ \sum_{|\alpha|=N} | (\xi^j - \eta^j)^\alpha | \, | \partial^\alpha \mu (x, \zeta^j) | \, \mu^{-1} (x, \eta^j)/\alpha!, \end{aligned} \tag{4.7}$$

where ζ^j is a point of the segment $[\xi^j, \eta^j]$. In view of (1.24) and (2.2) the first sum in (4.7) does not exceed

$$\text{const} \sum_{|\alpha|>0} (1 + | \eta^j |)^{(\theta-\rho) |\alpha|} < \text{const} (1 + | \eta^j |)^{-\lambda}, \quad \lambda < \rho - \theta.$$

Let us note that, by (1.24),

$$C_1 (1 + | \eta^j |) < (1 + | \zeta^j |) < C_2 (1 + | \eta^j |). \tag{4.8}$$

Hence the second sum in (4.7) does not exceed

$$\text{const} (1 + | \eta^j |)^{(\theta-\rho)N + \bar{\mu} - \underline{\mu}}.$$

If N is sufficiently large (that is, $N(\rho - \theta) > \bar{\mu} - \underline{\mu}$), we obtain (4.6).

LEMMA 4.2. *Suppose that the functions $\psi_j(\xi)$ have all the properties of the functions $\phi_j(\xi)$ in Theorem 1.1 except the one given by equation (1.22). Then*

$$\sum \| \mu (x, \eta^k) \psi_k (D) u \|^2 \leqslant C | [u] |_\mu, \tag{4.9}$$

where $\eta^k \in \operatorname{supp} \psi_k(\xi)$.

COROLLARY. *The norm (4.1) goes into an equivalent norm when the partition of unity $\{ \phi_j(\xi) \}$ is replaced by a partition of unity $\{ \psi_j(\xi) \}$ satisfying the properties (1.22)—(1.24).*

PROOF. We have

$$\psi_k(D)\,u=\psi_k(D)\sum\varphi_j^2(D)\,u=\sum_{j\in U(k)}c_{jk}(D)\,\varphi_j(D)\,u, \tag{4.10}$$

where $U(k)$ is a finite set, the number of points of which is independent of k, and

$$c_{jk}(D)=\psi_k(D)\,\varphi_j(D). \tag{4.11}$$

Using (4.10) we write

$$\|\mu(x,\eta^k)\,\psi_k(D)\,u\|=\Big\|\mu(x,\eta^k)\sum_{j\in U(k)}c_{jk}(D)\,\varphi_j(D)\,u\Big\|$$

$$\leqslant\sum_{j\in U(k)}\|\mu(x,\eta^k)\,c_{jk}(D)\,\mu^{-1}(x,\eta^j)\,\mu(x,\eta^j)\,\varphi_j(D)\,u\|;$$

the inequality (4.9) easily follows from the inequality

$$\|\mu(x,\eta^k)\,c_{jk}(D)\,\mu^{-1}(x,\eta^j)\,v\|\leqslant C\|v\|, \tag{4.12}$$

where the constant C is independent of j and k.

By (4.11) we have

$$|\partial^\alpha c_{jk}(\xi)|<C_\alpha(1+|\xi^j|)^{-\theta|\alpha|}. \tag{4.13}$$

Hence we can apply Proposition 1.3 (taking t to be $1+|\xi^j|$) to the operators $c_{jk}(D)$ and $\mu^{-1}(x,\xi^j)$. Then

$$\mu(x,\eta^k)\,c_{jk}(D)\,\mu^{-1}(x,\xi^j)\,v=\sum_{|\alpha|<N}\frac{1}{\alpha!}\mu(x,\eta^k)\,D^\alpha\mu^{-1}(x,\xi^j)\,\partial^\alpha c_{jk}(D)$$

$$+\mu(x,\eta^k)\,T_N v.$$

The estimate (4.12) holds for the sum on the right-hand side of the last equation in view of Lemma 4.1 and the estimates (4.13). If N is sufficiently large, then by Proposition 1.3 we can make the norm of T_N smaller than $\mathrm{const}(1+|\eta^j|)^{-\bar\mu}<\mathrm{const}(1+|\eta^k|)^{-\bar\mu}$; that is, the estimate (4.12) holds for $\mu(x,\eta^k)\,T_N v$. Our lemma is proved.

Proof of (4.5). Suppose that the functions $\psi_j(\xi)$ satisfy all the conditions of Lemma 4.2 and that

$$\psi(\xi)=1,\quad \xi\in\operatorname{supp}\varphi,\ \text{т. е.}\ \psi(\xi)\,\varphi(\xi)=\varphi(\xi).$$

Then we can rewrite the left-hand side of (4.5) in the form

$$\sum_j\|T_{\alpha\beta j}\mu(x,\xi^j)\,\psi_j(D)\,u\|^2, \tag{4.14}$$

where

$$T_{\alpha\beta j}=\mu^{(\alpha)}_{(\beta)}(x,\xi^j)\,(D-\xi^j)^\alpha\,\varphi_j^{(\beta)}(D)\,\mu^{-1}(x,\xi^j). \tag{4.15}$$

First of all let us note that, since the support of $\phi_j^{(\beta)}(\xi)$ is compact,

$$\|T_{\alpha\beta j}v\|\leqslant C_{\alpha\beta,m}\|v\|_{(m)}, \tag{4.16}$$

which yields an estimate for the norm of the operator (4.15) for small j. For large j we use Proposition 1.3, repeating the arguments of Lemma 4.2. We have

$$T_{\alpha\beta j}v = \sum_{|\gamma|<N} \frac{1}{\gamma!}\mu_{(\beta)}^{(\alpha)}(x, \xi^j)\, D^\gamma \mu^{-1}(x, \xi^j)\, \partial^\gamma [(D-\xi^j)^\alpha \varphi_j^{(\beta)}(D)]\, v$$

$$+ \mu_{(\beta)}^{(\alpha)}(x, \xi^j)\, T_{\alpha\beta j N} v = \sum w_\gamma + w_N.$$

If N is large, then, by Proposition 1.3,

$$\|w_N\| \leqslant \text{const}\,(1+|\xi^j|)^{-l}\|v\|. \tag{4.17}$$

Since

$$|\partial^\gamma [(\xi-\xi^j)^\alpha \varphi_j^{(\beta)}(\xi)]| \leqslant C\,(1+|\xi^j|)^{\theta|\alpha-\beta-\gamma|},$$

we have

$$\begin{aligned}\|w_\gamma\| &\leqslant \text{const}\,(1+|\xi^j|)^{(\theta-\rho)|\alpha|-(\theta-\delta)(|\beta|+|\gamma|)}\|v\| \\ &\leqslant \text{const}\,(1+|\xi^j|)^{-\lambda}\|v\|, \quad \delta+\lambda<\theta<\rho-\lambda.\end{aligned} \tag{4.18}$$

By combining the estimates (4.16)—(4.18) we prove that for any $\epsilon>0$ the sum (4.14) does not exceed

$$\varepsilon^2 \sum \|\mu(x, \xi^j)\,\psi_j(D)\,u\|^2 + C_\varepsilon \|u\|_{(m)}.$$

By applying Lemma 4.2 we prove (4.5) and hence also Theorem 4.1.

Together with localization with respect to ξ we can develop localization with respect to x. The following holds.

THEOREM 4.2. *Let* $\{\phi_j^2(x)\}$ *and* $\{\psi_j^2(x)\}$ *be two locally finite partitions of unity in* R^n *with*

$$\operatorname{supp}\psi_j \subset \operatorname{supp}\varphi_j, \quad \varphi_j(x)\,\psi_j(x) = \psi_j(x).$$

Then the norm (2.9) *is equivalent to the norm*

$$\sum \|\psi_j(x)\,\mu(x, D)\,(\varphi_j u)\|^2 + \|u\|_{(m)}^2. \tag{4.19}$$

PROOF. Because $\sum \psi_j^2 = 1$ we have that

$$\|\mu(x, D)\,u\|^2 = \sum \|\psi_j(x)\,\mu(x, D)\,u\|^2 = \sum \|\psi_j(x)\,\mu(x, D)\,(\varphi_j u) + T_j u\|^2, \tag{4.20}$$

where

$$T_j u = \psi_j(x)\,\mu(x, D)\,(1-\varphi_j)\,u. \tag{4.21}$$

We are going to prove that for any m

$$\sum \|T_j u\|^2 \leqslant \text{const}\,\|u\|_{(m)}^2. \tag{4.22}$$

The proof of the theorem will follow trivially from (4.20) and (4.22). In deriving (4.22) we use a method of Hörmander (see [**12**], Theorem 3.5). If we denote the inverse Fourier transform of $\mu(x,\xi)$ (with respect to ξ) by $M(x,z)$, we can rewrite (4.21) in the form

$$T_j u(x) = \psi_j(x) \int M(x, x-z)(1-\varphi_j(z)) u(z)\, dz$$
$$= \psi_j(x) \int_{|z-x^j|>\delta} M(x, x-z)(1-\varphi_j(z)) u(z)\, dz, \tag{4.23}$$

where $x^j \in \operatorname{supp} \psi_j$ and $\delta > 0$. If $\mu(x,\xi) \in S^{-\infty}_{\rho,\delta}$, then, for $|z| > \delta$, $M(x,z)$ decreases (with respect to z) more rapidly than any power of $|z|$; in particular,

$$|M(x, z)| < C_\delta (1+|z|^2)^{-n-1} = K(z).$$

Hence

$$|T_j u(x)| < \psi_j(x) |(K * u)(x)|,$$

whence

$$\sum \| T_j u(x) \|^2 \leqslant \| K * u \|^2 = \int |\widetilde{K}(\xi) \widetilde{u}(\xi)|^2 d\xi \leqslant \sup (1+|\xi|^2)^{-m} |\widetilde{K}(\xi)|^2 \| u \|^2_{(m)},$$

where $\widetilde{K}(\xi)$ is the Fourier transform of $K(x)$. Since the function $\widetilde{K}(\xi)$ is bounded and decreases exponentially at infinity, the right-hand side of the last inequality does not exceed $\mathrm{const} \| u \|^2_{(m)}$. Thus our theorem is proved.

§5. The space $W^{s(x)}$

1. Let x and t denote points of R^n and let $\Delta_t^l u(x)$ denote the difference of order l; that is,

$$\Delta_t^l u(x) = \sum_{k=0}^{l} \binom{l}{k} (-1)^k u(x+kt).$$

Let $s(x) \in C^\infty$ be a real valued function that is stabilized for large $|x|$, and let

$$0 < \underline{s} \leqslant s(x) \leqslant \bar{s} < \widetilde{s},$$

where $\widetilde{s}$ stands for $[s]+1$.

We denote by $W^{s(x)}$ the set of functions $u \in L_2$ for which the following norm is finite:

$$||| u |||^2_{s(x),l} = \iint |\Delta_t^l u(x)|^2 |t|^{-n-2s(x)} dt dx + \| u \|^2, \tag{5.2}$$

where $l \geqq \widetilde{s}$.

First of all we consider the elementary properties of the norm (5.2). It is not difficult to see that this norm is equivalent to the norm

$$\iint_{|t| \leqq a} |\Delta_t^l u(x)|^2 |t|^{-n-2s(x)} dx dt + \| u \|^2. \tag{5.2'}$$

In fact the norm (5.2′) differs from (5.2) by

$$\mathscr{I} = \iint_{|t| \geqq a} |\Delta_t^l u(x)|^2 |t|^{-n-2s(x)} dx dt.$$

It is not difficult to observe that this integral does not exceed

$$\mathrm{const} \| u \|^2.$$

In fact, if $c(a) = \max|t|^{2\underline{s}-2s(x)}$, $x \in R^n$, $|t| \geqq a$, then

$$\mathscr{I} \leqq c(a) \int_{|t| \geqq a} \int \sum \binom{l}{k} |u(x+kt)|^2 dx \frac{dt}{|t|^{n+2\underline{s}}}$$
$$\leqq c(a) 2^l \|u\|^2 \int_{|t| \geqq a} |t|^{-n-2\underline{s}} dt \leqq \text{const } \|u\|^2.$$

PROPOSITION 5.1. *If* $s(x) \leqq k(x)$ *then*

$$|||u|||_{s(x),l} \leqq \text{const } |||u|||_{k(x),l}.$$

If $k(x) - s(x) > \delta > 0$, *then for any* $\epsilon > 0$

$$|||u|||_{s(x),l} \leqq \epsilon |||u|||_{k(x),l} + C_\epsilon \|u\|,$$

where $l \geqq [\max k(x)] + 1$.

PROOF. Since the norms (5.2) and (5.2′) differ by const $\|u\|^2$, we can prove the inequality for the norm (5.2′) by a judicious choice of the number a. If we take $a \leqq 1$ we trivially obtain the first assertion of the proposition. Since

$$|\Delta_t^l u| \, |t|^{n-2s(x)} \leqq a^\delta |\Delta_t^l u| \, |t|^{-n-2k(x)},$$

we can prove the second assertion by choosing a so that $a^\delta < \epsilon$. Thus our proposition is proved.

PROPOSITION 5.2. *If* $0 < s(x) \leqq \bar{s} < l$, *then the norm* (5.2) *is equivalent to the norm* $|||\cdot|||_{s(x),k+l}$, *where* $k > 0$ *is an arbitrary integer.*

PROOF. Without loss of generality we can assume that $k = 1$. We have

$$|||u|||_{s(x),l+1} = \left[\iint |\Delta_t^l u(x+t) - \Delta_t^l u(x)|^2 |t|^{-n-2s(x)} dx dt \right]^{1/2}$$
$$\leqq |||u|||_{s(x),l} + \left[\iint |\Delta_t^l u(x)|^2 |t|^{-n-2s(x+t)} dx dt \right]^{1/2}$$
$$\leqq 2|||u|||_{s(x),l} + \left[\iint |\Delta_t^l u(x)|^2 \right.$$
$$\left. \times |t|^{-n-2s(x)} |1 - |t|^{2s(x)-2s(x+t)}| dx dt \right]^{1/2}.$$

Since

$$|1 - |t|^{2s(x)-2s(x+t)}| < \text{const } |t| \, |\ln|t|\,|, \quad |t| \leqq 1,$$

the last integral does not exceed

$$\text{const } |||u|||_{s(x)-\epsilon,l}, \quad 0 < \epsilon < \underline{s}.$$

When we use Proposition 5.1 we obtain that

$$|||u|||_{s(x),l+1} \leqq \text{const } |||u|||_{s(x),l}.$$

To prove the reverse inequality we use the identity ([**14**], p. 228)

$$\Delta_t^l u(x) - \frac{1}{2^l}\Delta_{2t}^l u(x) = -\frac{1}{2}\sum_{k=0}^{l-1}\sum_{m=0}^{k} 2^{-k}\binom{k}{m}\Delta_t^{l+1}u(x+mt).$$

By dividing by $|t|^{s(x)}$ we find that

$$\frac{|\Delta_t^l u(x)|}{|t|^{s(x)}} - \frac{1}{2^{l-\bar{s}}}\frac{|\Delta_{2t}^l u(x)|}{|2t|^{s(x)}}$$

$$\leqq \frac{1}{2}\sum_{k=0}^{l-1}\sum_{m=0}^{k} 2^{-k}\binom{k}{m}\left[\frac{|\Delta_t^{l+1}u(x+mt)|}{|t|^{s(x+mt)}} + \frac{|\Delta_t^{l+1}u(x+mt)|}{|t|^{s(x+mt)}}\left|1-|t|^{s(x+mt)-s(x)}\right|\right].$$

Next we calculate the L_2-norm of both sides of the inequality with respect to the measure $|t|^{-n}dxdt$. The left-hand side is bounded from below by

$$(1-2^{-l+\bar{s}})\, |||u|||_{s(x),l}.$$

The right-hand side is bounded from above by

$$\frac{l}{2}\, |||u|||_{s(x),l+1} + \text{const}\, |||u|||_{s(x)-\epsilon,l+1}.$$

When we use Proposition 5.1 we find that

$$|||u|||_{s(x),l} \leqq \text{const}\, |||u|||_{s(x),l+1}$$

and so our proposition is proved.

In view of Proposition 5.2, in what is to follow we shall omit the number l when writing the norm (5.2).

PROPOSITION 5.3. *If* $s(x) = \text{const}$, *then the space* W^s *coincides with* $H^{(s)}$ *and the norm* (5.2) *is equivalent to the norm*

$$\int (1+|\xi|^2)^s |\tilde{u}(\xi)|^2 d\xi = \|u\|_{(s)}^2 .$$

PROOF. By Parseval's equality we have

$$\iint |\Delta_t^l u(x)|^2 |t|^{-n-2s} dxdt = \int |\tilde{u}(\xi)|^2 d\xi \int |1-e^{i\xi\cdot t}|^{2l} |t|^{-n-2s} dt. \quad (5.4)$$

Because

$$\int |1-e^{i\xi\cdot t}|^{2l} |t|^{-n-2s} dt = c(n,s,l)\, |\xi|^{2s}, \quad (5.5)$$

the norms (5.2) and (5.3) are equivalent.

The result we have proved is due to Aronszajn and Slobodeckiĭ. Our problem is to extend this result to the case of a variable $s(x)$. The following holds.

THEOREM 5.1. *If* (5.1) *is fulfilled, then the spaces* $H^{(s(x))}$ *and* $W^{s(x)}$ *coincide and the norms in these spaces are equivalent*:

$$C^{-1} |||u|||_{s(x)} \leqslant \|u\|_{(s(x))} \leqslant C\, |||u|||_{s(x)}. \quad (5.6)$$

We prove Theorem 5.1 in two stages. First we consider functions $s(x)$ whose oscillation is small; that is,

$$0 < \bar{s} - \underline{s} < \theta_0 < \min\left\{\frac{1}{2}, \underline{s}\right\}. \tag{5.7}$$

In this case we can establish an analog of Theorem 4.1 for the norm (5.2) and so reduce Theorem 5.1 to the case of functions with small (with respect to ξ) support. For this case our theorem is proved quite simply. Next, with the help of localization with respect to x, the case of arbitrary functions $s(x)$ can be reduced to the case of functions with small support.

2. Proof of Theorem 5.1 for $s(x)$ with small support.

LEMMA 5.1. *Let $v(x) \in H^{(\infty)}$ and*

$$\tilde{v}(\xi) = 0 \text{ when } |\xi - \xi^0| > C|\xi^0|^\theta, \ \theta < \frac{1}{2}, \ |\xi^0| > 1. \tag{5.8}$$

Suppose that the oscillation of $s(x)$ is small; that is, that

$$0 \leqslant \bar{s} - \underline{s} < \underline{s} \leqslant s(x) \leqslant \bar{s}. \tag{5.9}$$

Then we can find a constant C, independent of $|\xi^0|$, such that

$$C^{-1} |||v|||_{s(x)} \leqslant \int |\xi^0|^{2s(x)} |v(x)|^2 dx + \|v\|^2 \leqslant C |||v|||^2_{s(x)}. \tag{5.10}$$

PROOF. According to the identity (5.4) we have

$$|\xi^0|^{2s(x)} = c(n, s(x), l) \int |1 - e^{i\xi^0 \cdot t}|^{2l} |t|^{-n-2s(x)} dt.$$

Because the function $c(n, s(x), l)$ is uniformly bounded from above and from below (the explicit formulas for this function can be found in [**13**]), the integral in (5.10) can be replaced by

$$\iint |1 - e^{i\xi^0 \cdot t}|^{2l} |v(x)|^2 |t|^{-n-2s(x)} dx dt. \tag{5.11}$$

By rewriting the integrand in (5.11) as

$$[1 - e^{i\xi^0 \cdot t}]^l v(x) = \Delta_t^l v(x) + [(1 - e^{i\xi^0 \cdot t})^l - (1 - e^{iDt})^l] v,$$

we reduce the proof of the lemma to the derivation of the inequality

$$\begin{aligned} I = \iint |(1 - e^{i\xi^0 \cdot t})^l v - (1 - e^{iD \cdot t})^l v|^2 |t|^{-n-2s(x)} dx dt \\ \leqslant \varepsilon \iint |\Delta_t^l v(x)|^2 |t|^{-n-2s(x)} dx dt + C_\varepsilon \|v\|^2, \end{aligned} \tag{5.12}$$

where ϵ is arbitrarily small. It is obvious that

$$|t|^{-n-2s(x)} \leqslant |t|^{-n-2\bar{s}} + \chi(t)[|t|^{-n-2\underline{s}} - |t|^{-n-2\bar{s}}], \tag{5.13}$$

where $\chi(t) = 0$ when $|t| < 1$ and $\chi(t) = 1$ when $|t| > 1$. In view of (5.13)

$$I \leqslant \iint |(1 - e^{i\xi^0\cdot t})^l v - \Delta_t^l v|^2 |t|^{-n-2\bar{s}} dx dt + C\|v\|^2$$
$$= \int |\tilde{v}(\xi)|^2 d\xi \int |(1 - e^{i\xi^0\cdot t})^l - (1 - e^{i\xi\cdot t})^l|^2 |t|^{-n-2\bar{s}} dt$$
$$+ C\|v\|^2 \leqslant C_1 \int |\xi^0 - \xi|^{2\bar{s}} |\tilde{v}(\xi)|^2 d\xi + C\|v\|^2.$$

If the number $|\xi^0|$ is large, then by (5.8) we have

$$|\xi^0 - \xi| < C|\xi^0|^\theta < C_2(1 + |\xi|)^\theta, \quad \xi \in \operatorname{supp} \tilde{v}.$$

Hence, in view of Proposition 5.3,

$$\int |\xi^0 - \xi|^{2\bar{s}} |\tilde{v}(\xi)|^2 d\xi \leqslant C_3 \|v\|_{(\bar{s}\theta)} < C_4 |||v|||_{\bar{s}\theta}.$$

If $\theta < \frac{1}{2}$, then by (5.9) we have

$$\theta\bar{s} < \underline{s} - \omega < s(x), \quad \omega > 0.$$

By applying Proposition 5.2 we obtain that for sufficiently large $|\xi^0|$

$$I < C_4 |||v|||_{\bar{s}\theta} + C\|v\| < \varepsilon |||v|||_{s(x)} + C_\varepsilon \|v\|.$$

For finite $|\xi^0|$ the estimate is obvious: if $\lambda(\xi) \in C_0^\infty$ and $\lambda(\xi)\tilde{v}(\xi) = \tilde{v}(\xi)$, then $|||v|||_{s(x)} \leqq \|\lambda(D)v\|_{(s)} \leqq C\|v\|_{(m)}$, where m is arbitrary, whence follows (5.10). The lemma is proved.

LEMMA 5.2. *Let $\{\phi_j(\xi)\}$ be the partition of unity in Theorem* 1.1, *and suppose that $\bar{s} - \underline{s} < \theta$. Then the norm* (5.2) *is equivalent to the norm*

$$\sum \iint |\Delta_t^l \varphi_j(D) u|^2 \frac{dx dt}{|t|^{n+2s(x)}} + \|u\|^2. \tag{5.14}$$

In many respects the proof is reminiscent of the proofs of Theorems 1.1 and 4.1. Because $\sum \phi_j^2 = 1$ we have

$$\iint |\Delta_t^l u|^2 \frac{dx dt}{|t|^{n+2s(x)}} = \int \frac{dt}{|t|^n} \| |t|^{-s(x)} \Delta_t^l u\|^2$$
$$= \sum_j \int \frac{dt}{|t|^n} \|\varphi_j(D) |t|^{-s(x)} \Delta_t^l u\|^2. \tag{5.15}$$

According to Proposition 1.1

$$\varphi_j(D)|t|^{-s(x)} v = |t|^{-s(x)} \varphi_j(D) v$$
$$+ \sum \frac{1}{\alpha!} D^\alpha |t|^{-s(x)} \varphi_j^{(\alpha)}(D) v + R_j^N(t) v.$$

Since

$$|D^\alpha |t|^{-s(x)}| < C|t|^{-\bar{s}-\delta} + C\chi(t)[|t|^{-\underline{s}+\delta} - |t|^{-\bar{s}-\delta}],$$

where $\delta > 0$ is arbitrarily small and $\chi(t)$ is the characteristic function of the exterior of the ball $|t| \leqq 1$, we have

$$\begin{aligned} &\sum_j \int \| D^\alpha |t|^{-s(x)} \Delta_t^l \varphi_j^{(\alpha)}(D) u \|^2 \frac{dt}{|t|^n} \\ &\leqslant \sum_j \int \| \varphi_j^{(\alpha)}(D) \Delta_t^l u \|^2 \frac{dt}{|t|^{n+2(\bar{s}+\delta)}} + C\|u\|^2 \\ &= C_1 \sum_j \int |\xi|^{2(\bar{s}+\delta)} |\varphi_j^{(\alpha)}(\xi)|^2 |\tilde{u}(\xi)|^2 d\xi + C\|u\|^2 \\ &\leqslant C_2 \|u\|^2_{(\bar{s}+\delta-\theta|\alpha|)} + C\|u\|^2. \end{aligned} \tag{5.16}$$

(Here we have used the result that $|\phi_j^{(\alpha)}(\xi)| < \text{const}\,(1+|\xi|)^{-\theta|\alpha|}$.) By the hypothesis of the lemma, for sufficiently small δ

$$\bar{s} + \delta - \theta|\alpha| \leqslant \bar{s} + \delta - \theta < \underline{s} - \delta.$$

Hence in view of Proposition 5.2 the right-hand side of (5.16) does not exceed

$$\varepsilon \||u\||^2_{s(x)} + C_\varepsilon \|u\|^2.$$

We next show that for sufficiently large N

$$\sum \int \| R_j^N(t) \Delta_t^l u \|^2 |t|^{-n} dt \leqslant \text{const} \|u\|^2. \tag{5.17}$$

To do this we take the Fourier transform of the left-hand side of (5.17). By Proposition 1.1 $R_j^N(t)$ is an integral operator whose kernel is

$$r_j^N(\xi, \eta, t) = \Big[\varphi_j(\xi) - \sum_{|\alpha|<N} \frac{(\xi-\eta)^\alpha}{\alpha!} \varphi_j^{(\alpha)}(\eta)\Big] \tilde{L}(\xi-\eta, t)(1-e^{i\eta\cdot t})^l, \tag{5.18}$$

where $\tilde{L}(\xi, t)$ is the Fourier transform (with respect to x) of $L(x,t) = |t|^{-s(x)} - |t|^{-s}$, and s is the value of $s(x)$ when $|x| \to \infty$. Thus, on the left-hand side of (5.17) stands the expression

$$I = \int d\xi \int |t|^{-n} dt \sum \Big| \int r_j^N(\xi, \eta, t) \tilde{u}(\eta) d\eta \Big|^2. \tag{5.19}$$

By applying Minkowski's integral inequality to (5.9) we find that

$$I \leqslant \int d\xi \Big[\int K_N(\xi, \eta) |\tilde{u}(\eta)| d\eta\Big]^2, \tag{5.20}$$

where

$$K_N(\xi, \eta) = \Big(\sum_j \int |r_j^N(\xi, \eta, t)|^2 |t|^{-n} dt\Big)^{1/2}. \tag{5.21}$$

We will show that

$$K_N(\xi, \eta) < \text{const}\,(1 + |\xi - \eta|)^{-n-1}, \tag{5.22}$$

and (5.17) will follow from this result.

To prove (5.22) let us first note that for fixed ξ and η the kernel $r_j^N(\xi,\eta,t)$ is nonzero for only a finite set of values of j (the number of these j does not depend on ξ and η). Hence it is sufficient to prove the estimate (5.22) for each term on the right-hand side of (5.21). We rewrite the right-hand side of (5.18) in the form

$$r_j^N(\xi,\eta,t)=\sum_{|\alpha|=N}(\xi-\eta)^\alpha\varphi_j^{(\alpha)}(\zeta)(1-e^{i\eta\cdot t})^l\widetilde{L}(\xi-\eta,t)/\alpha!,$$

whence we have

$$|r_j^N(\xi,\eta,t)|\leqslant C|1-e^{i\eta\cdot t}|^l(1+|\xi-\eta|)^N(1+|\zeta|)^{-\theta N}|\widetilde{L}(\xi-\eta,t|$$
$$<C_1|1-e^{i\eta\cdot t}|^l(1+|\xi-\eta|)^{N+\theta N}(1+|\eta|)^{-\theta N}|\widetilde{L}(\xi-\eta,t)|. \quad (5.23)$$

Since $L(x,t)\equiv 0$ for large $|x|$, for any $\delta>0$ we have

$$|\widetilde{L}(\xi-\eta,t)|<C_{M,\delta}(1+|\xi-\eta|)^{-M}[|t|^{-\bar{s}-\delta}+\chi(t)(|t|^{-\underline{s}+\delta}-|t|^{-\bar{s}-\delta})].$$

By substituting this estimate into (5.23), squaring, and integrating with respect to t, we find that

$$\left(\int|r_j^N(\xi,\eta,t)|^2|t|^{-n}dt\right)^{1/2}\leqslant\text{const}(1+|\xi-\eta|)^{-M+N+\theta N}(1+|\eta|)^{-\theta N+\bar{s}+\delta}.$$

By putting $M=N+\theta N+n+1$ and choosing N so that $\theta N>\bar{s}+\delta$ we arrive at (5.22); hence our lemma is completely proved.

Thus, if (5.7) holds, then by Lemmas 5.1 and 5.2 we have

$$|||u|||_{s(x)}^2\approx\sum|||\varphi_j(D)u|||_{s(x)}^2+\|u\|^2\approx\sum\||\xi^j|^{s(x)}\varphi_j(D)u\|^2+\|u\|^2, \quad (5.24)$$

where $\{\phi_j(\xi)\}$ is the partition of unity in Theorem 1.1 and the number θ corresponding to it satisfies $\theta<\theta_0$ (θ_0 is given in (5.7)). Since $(1+|\xi|^2)^{s(x)}\in\mathfrak{M}_{1,\delta}$, where δ is arbitrarily small, Theorem 4.1 holds for the norm $\|\ \|_{(s(x))}$ and we can take a partition $\{\phi_j\}$ corresponding to any $\theta<1$; in particular, we can take $\theta<\theta_0$. Hence the norm on the right-hand side of (5.24) is equivalent to the norm

$$\|(1+|D|^2)^{s(x)/2}u\|^2+\|u\|^2,$$

that is, we have proved Theorem 5.1 under the assumption (5.7).

3. Proof of Theorem 5.1 in the general case. We require two more lemmas.

LEMMA 5.3. *The space $W^{s(x)}$ is a module over C_0^∞; that is,*

$$|||au|||_{s(x)}\leqslant C(a)|||u|||_{s(x)},\quad a\in C_0^\infty.$$

PROOF. By Leibniz's rule for differences

$$\Delta_t^l(a(x+(l-k)t)u(x))=\sum\binom{l}{k}\Delta_t^k a(x)\Delta_t^{l-k}u(x),$$

where l is as in (5.2). If $l-k\geqq[\bar{s}]+1$, then by Proposition 5.1

$$\iint |\Delta_t^k a \Delta_t^{l-k} u|^2 |t|^{-n-2s(x)} dxdt$$
$$\leqslant \max |\Delta_t^k a| \iint |\Delta_t^{l-k} u|^2 |t|^{-n-2s(x)} dxdt \leqslant C(a) |||u|||_{s(x)}. \tag{5.25}$$

In view of Proposition 5.2 we can assume in (5.2) that $l > 2[s] + 1$. But under these conditions, if $l - k \leqq [\bar{s}]$, then $k > [\bar{s}] + 1$. Since

$$|\Delta_t^k a(x + (l-k)t)| < \text{const} |t|^k, \quad |t| \leqslant 1,$$

the integral on the left-hand side of (5.25) does not exceed const $\|u\|^2$.

LEMMA 5.4. *Suppose that the functions $s(x)$ and $\sigma(x)$ coincide in the domain Ω_1 and let $\Omega \subset \bar{\Omega} \subset \Omega_1$. If $v \in H^{(\infty)}$ and* supp $v \subset \Omega$, *then*

$$|||v|||_{\sigma(x)} \leqslant |||v|||_{s(x)} + C\|v\|. \tag{5.26}$$

PROOF. We choose l so that

$$l \geqslant 1 + \max \{[s(x)], [\sigma(x)]\},$$

and let ϵ be sufficiently small that the $l\epsilon$-neighborhood of Ω is contained in Ω_1. We have

$$\iint \frac{|\Delta_t^l v(x)|^2}{|t|^{n+2\sigma(x)}} dxdt = \int \int_{|t|<\varepsilon} + \int \int_{|t|>\varepsilon} = I_1 + I_2. \tag{5.27}$$

The second integral on the right-hand side of (5.27) does not exceed const $\|v\|^2$. As regards the first integral, the integration with respect to x is, in fact, over Ω_1. Hence

$$I_1 = \int_{\Omega_1} \int_{|t|<\varepsilon} \frac{|\Delta_t^l v(x)|^2}{|t|^{n+2\sigma(x)}} dxdt = \int_{\Omega_1} \int_{|t|<\varepsilon} \frac{|\Delta_t^l v(x)|^2}{|t|^{n+2s(x)}} dxdt < |||v|||_{s(x)}.$$

The lemma is proved.

Let us return to Theorem 5.1. Since $s(x)$ varies only in a finite domain we can choose in R^n a finite covering $\{U_j\}$, $j = 1, \cdots, J$, where $s(x)$ satisfies the conditions (5.7) in each of these neighborhoods. It is clear that we can pick out those functions $s_j(x) \in C^\infty$ that are stabilized at infinity, satisfy conditions of the type (5.7) and are such that

$$s_j(x) = s(x), \quad x \in U_j.$$

Let $\psi_j(x)$ be the partition of unity connected with the covering $\{U_j\}$. In view of Lemma 5.4 we have

$$|||u|||_{s(x)} \leqslant \sum |||\psi_j u|||_{s(x)} \leqslant \sum |||\psi_j u|||_{s_j(x)} + C\|u\|. \tag{5.28}$$

Since the oscillation of the functions $s_j(x)$ is small, from what has been proved we have

$$|||\psi_j u|||_{s_j(x)} \leqslant C_j \|\psi_j u\|_{(s_j(x))}. \tag{5.29}$$

Because the symbols $(1+|\xi|^2)^{s(x)/2}$ and $(1+|\xi|^2)^{s_j(x)/2}$ coincide in the support of $\psi_j u$, by Theorem 3.2

$$\|\psi_j u\|_{(s_j(x))} \leqslant \text{const}\, \|\psi_j u\|_{(s(x))}. \tag{5.30}$$

Since $\psi_j(x)$ is either finite or unity for large $|x|$, in view of Corollary 5 to Lemma 2.3 we have

$$\|\psi_j u\|_{(s(x))} \leqslant C_j \|u\|_{(s(x))}. \tag{5.31}$$

By combining the inequalities (5.28)—(5.31) we prove that

$$|||u|||_{s(x)} \leqslant \text{const}\, \|u\|_{(s(x))}.$$

Conversely,

$$\|u\|_{(s(x))} \leqslant \sum \|\psi_j u\|_{(s(x))} \leqslant \sum C_j \|\psi_j u\|_{(s_j(x))}$$

$$\leqslant \sum C_j' \,|||\psi_j u|||_{s_j(x)} \leqslant \sum C_j'' \,|||\psi_j u|||_{s(x)} < \text{const}\, |||u|||_{s(x)}.$$

Thus Theorem 5.1 is completely proved.

APPENDIX*

We outline a proof of Proposition 1.1 in the general case. Let $A(x,z)$, $B(x,z)$ and $B_{(\alpha)}(x,z)$ denote the inverse Fourier transforms with respect to ξ (in the sense of the theory of distributions) of the symbols $a(x,\xi)$, $b(x,\xi)$ and $D^\alpha b(x,\xi)$. It is not difficult to show (see [**12**]) that these distributions have finite order and belong to C^∞ outside the point $z=0$, and that as $|z|\to\infty$ they decrease more rapidly than any power of z. Let κ_0 be the least integer such that

$$2\kappa_0 > m_2 + N\delta_2 + n. \tag{1}$$

We put $b^0(x,\xi) = b(x,\xi)(1+|\xi|^2)^{-\kappa_0}$. It is not difficult to verify that

$$T_N u = \iint A(x,x-z)\Big[B^0(z,z-y) - \sum_{|\alpha|\leq N-1} \frac{(z-x)^\alpha}{\alpha!} B^0_{(\alpha)}(x,z-y)\Big] \tag{2}$$
$$\times (1+|D|^2)^{\kappa_0} u(y)\,dz\,dy.$$

By Taylor's formula

$$T_N u = \sum_{|\alpha|=N} \frac{N}{\alpha!}\int_0^1 (1-t)^{N-1} \iint A^{(\alpha)}(x,x-z) \tag{3}$$
$$\times B^0_{(\alpha)}(x+t(z-x), z-y)(1+|D|^2)^{\kappa_0} u(y)\,dz\,dy\,dt,$$

where

$$A^{(\alpha)}(x,z) = (-z)^\alpha A(x,z) = (2\pi)^{-n/2}\int e^{iz\cdot\xi}\left(\frac{1}{i}\frac{\partial}{\partial\xi}\right)^\alpha a(x,\xi)\,d\xi. \tag{4}$$

---

*Added by the authors for the English translation.

We put

$$T_\alpha(x,y) = (1+|D|^2)^{\kappa_0} \int A^{(\alpha)}(x,x-z)\, B^0_{(\alpha)}(x+t(z-x), z-y)\, dz \tag{5}$$

and will show that $T_{\alpha t}(x,y)$ is a smooth function and that the following estimates hold:

$$|D_x^\beta D_y^\gamma T_{\alpha t}(x,y)| < \text{const}\,(1+|x-y|)^{-n-1}, \tag{6}$$

$$|\beta| + |\gamma| \leqq [(\rho_1-\delta_2)N - m_1 - m_2] - 2n - 3 = l. \tag{6$'$}$$

First note that it follows from (3) and from the membership $\alpha(x,\xi) \in S^{m_1}_{\rho_1,\delta_1}$ that for $|\alpha| = N$ and for any k

$$|D_z^\lambda D_x^\mu A^{(\alpha)}(x,x-z)| < \text{const}\,(1+|x-z|)^{-k}, \tag{7}$$

$$|\lambda| + |\mu|/\delta_1 < N\rho_1 - m_1 - n. \tag{7$'$}$$

If we differentiate with respect to y under the integral sign in (5) and integrate by parts we represent (5) as a linear combination of expressions of the form

$$\int D_z^\nu A(x,x-z)\, B^0_{(\mu)}(x+t(z-x), z-y)\, dz, \qquad |\nu| + |\mu| \leqq 2\kappa_0. \tag{8}$$

We represent $b^0_{(\mu)}(x,\xi)$ as $b^1(x,\xi)(1+|\xi|^2)^{\kappa_1}$, where κ_1 is the least nonnegative integer such that

$$2\kappa_1 \geqq 2\delta_2|\mu|. \tag{9}$$

Applying the operator $(1+|D|^2)^{\kappa_1}$ to (8) and integrating by parts, we represent (8) as a linear combination of the integrals

$$\int D_z^{\nu'} A(x,x-z)\, B^1_{(\mu')}(x+t(z-x), z-y)\, dz, \qquad |\nu'| + |\mu'| \leqq 2\kappa_1. \tag{10}$$

Since, to within fractional parts, $\kappa_1 \leqq \delta_2\kappa_0$ and $\kappa_1 < \kappa_0$, on continuing this process we can finally write (5) as a linear combination of the integrals

$$\int D_z^\lambda A^{(\alpha)}(x,x-y)\, C(x+t(z-x), z-y)\, dz, \tag{11}$$

where $|\lambda| \leqq 2\kappa_0 + 1$ and $C(x,z)$ is the inverse Fourier transform of a symbol belonging to $S^{-n-\epsilon}_{\rho_2,\delta_2}$, $\epsilon > 0$. In view of (7$'$) we can differentiate $D_z^\lambda A^{(\alpha)}(x,x-z)$ with respect to x and z not more than $[N\rho_1 - m_1] - n - 2\kappa_0 - 1$ times. By using the inequality (1) for κ_0 we find that we can differentiate $D_z^\lambda A^{(\alpha)}$ l times, where l is equal to the right-hand side of (6$'$). We differentiate (5) l times with respect to x and to y, and, using the above algorithm, we can rewrite it as a linear combination of

$$\int D_z^\lambda D_x^\mu A^{(\alpha)}(x,x-z)\, D(x+t(z-x), z-y)\, dz, \tag{12}$$

where the numbers $|\lambda|$ and $|\mu|$ satisfy (7$'$) and $D(x,z)$ is the inverse Fourier transform of a symbol belonging to $S^{-n-\epsilon}_{\rho_2,\delta_2}$, $\epsilon > 0$. Since for any k_1 and k_2

$$|D_z^\lambda D_x^\mu A^{(\alpha)}| < c_{k_1}(1+|x-z|)^{-k_1}|D(\;)| \leqq c_{k_2}(1+|z-y|)^{-k_2},$$

the estimate (6) holds for the integral (12).

We now turn to the proof of our assertion; that is, of the estimate

$$\| T_N(a,b)u\|_{(s+l)} \leqq \text{const}\ \|u\|_{(s)}, \tag{13}$$

where l is given by (6′). When $s=0$, (13) follows directly from (6). Since we are not interested in the exact value of l, we can restrict ourselves to integral s (if s is a fraction, in the left-hand side of (13) we replace s by $[s]+1$ and l by $l-1$; we then substitute $[s]$ for s in the right-hand side). If $s+l>0$ we replace u by $v=(1+|D|^2)^{s/2}u$ and $b(x,D)$ by $\bar b(x,D)(1+|D|^2)^{-s/2}$; we obtain that

$$\| T_N(a,b)u\|_{(s+l)} = \| T_N(a,\bar b)v\|_{(s+l)} \leqq \text{const}\ \|v\| = \text{const}\ \|u\|_{(s)}.$$

Next suppose that s is large, negative and, for simplicity, even. In the proof of the boundedness of the operator

$$(1+|D|^2)^{(s+l)/2}T_N(1+|D|^2)^{-s/2}$$

by integrating by parts we can represent $T_N(x,y)(1+|D_y|^2)^{-s/2}$ as a linear combination of the kernels $P(D_x)\overline{T}_N(x,y)$, where $\overline{T}_N(x,y)$ satisfies conditions of the type (7), (7′) and the order of the operator P does not exceed $-s$. But then $(1+|D|^2)^{s/2}P(D)$ is a bounded operator and the l-multiple differentiation of $\overline{T}_N$ also leads to a bounded operator.

REMARK. It follows from the above arguments that the operator (1.7) is continuous for $N>N_0$, where N_0 depends on l and is independent of s.

Received 1 NOV 1967

BIBLIOGRAPHY

[1] M. I. Višik and G. I. Èskin, *Singular elliptic equations and systems of variable order,* Dokl. Akad. Nauk SSSR **156** (1964), 243-246 = Soviet Math. Dokl. **5** (1964), 615-619. MR **29** #2515.

[2] ———, *Elliptic convolution equations in a bounded region and their applications,* Uspehi Mat. Nauk **22** (1967), no. 1 (133), 15-76 = Russian Math. Surveys **22** (1967), no. 1, 13-75. MR **35** #5757.

[3] L. R. Volevič, *Local properties of inhomogeneous pseudo-differential operators,* Trudy Moskov. Mat. Obsc. **16** (1967), 51-98 = Trans. Moscow Math. Soc. **1967**, 53-105. MR **37** #1761.

[4] ———, *Hypoelliptic and partially hypoelliptic pseudo-differential operators,* Proc. Internat. Congress Math. (Moscow, 1966), "Mir", Moscow, 1968. MR **38** #960.

[5] L. R. Volevič and B. P. Panejah, *Certain spaces of generalized functions and imbedding theorems,* Uspehi Mat. Nauk **20** (1965), no. 1 (121), 3-74 = Russian Math. Surveys **20** (1965), no. 1, 1-73. MR **30** #5160.

[6] ———, *A new approach to imbedding theorems,* Izv. Akad. Nauk Azerbaĭdžan. SSR. Ser. Fiz.-Tehn. Mat. Nauk **1966**, no. 1, 5-13. (Russian) MR **34** #1841.

[7] Ju. V. Egorov, *Hypoelliptic pseudodifferential operators,* Trudy Moskov. Mat. Obšč. **16** (1967), 99-108 = Trans. Moscow Math. Soc. **1967**, 107-116. MR **36** #6985.

[8] J. J. Kohn and L. Nirenberg, *An algebra of pseudodifferential operators,* Comm. Pure Appl. Math. **18** (1965), 269-305; Russian transl., in *Pseudodifferential operators,* "Mir", Moscow, 1967, pp. 9-62. MR **31** #636.

[9] L. N. Slobodeckiĭ, *Generalized Sobolev spaces and their application to boundary problems for partial differential equations*, Leningrad. Gos. Ped. Inst. Učen. Zap. **197** (1958), 54-112; English transl., Amer. Math. Soc. Transl. (2) **57** (1966), 207-276. MR **34** # 3075.

[10] L. Hörmander, *Hypoelliptic differential operators*, Ann. Inst. Fourier Grenoble **11** (1961), 477-492. MR **23** # A3368.

[11] ———, *Linear partial differential operators*, Die Grundlehren der math. Wissenschaften, Band 116, Academic Press, New York and Springer-Verlag, Berlin, 1963; Russian transl., "Mir", Moscow, 1965. MR **28** # 4221; **37** # 6595.

[12] ———, *Pseudodifferential operators and hypoelliptic equations*, Proc. Sympos. Pure Math., vol. 10, Amer. Math. Soc., Providence, R. I., 1967, pp. 138-183; Russian transl., in *Pseudodifferential operators*, "Mir", Moscow, 1967, pp. 297-367.

[13] N. Aronszajn and K. T. Smith, *Theory of Bessel potentials*. I, Ann. Inst. Fourier Grenoble **11** (1961), 385-475. MR **26** # 1485.

[14] N. Aronszajn, F. Mulla and P. Szeptycki, *On spaces of potentials connected with L^p classes*, Ann. Inst. Fourier Grenoble **13** (1963), 211-306. MR **31** # 5076.

[15] A. Unterberger and J. Bokobza, *Les operateurs pseudo-différentiels d'ordre variable*, C. R. Acad. Sci. Paris **261** (1965), 2271-2273. MR **32** # 4567.

Translated by:
L. W. Longdon